The Politics of Large Numbers

The Politics of Large Numbers

A History of Statistical Reasoning

Alain Desrosières

TRANSLATED BY CAMILLE NAISH

HARVARD UNIVERSITY PRESS

Cambridge, Massachusetts, and London, England

Printed in the United States of America
Originally published as *La politique des grands nombres: Histoire de la raison statistique,* © Éditions La Découverte, Paris, 1993.
Published with the assistance of the French Ministry of Culture.

This book has been digitally reprinted. The content remains identical to that of previous printings

Library of Congress Cataloging-in-Publication Data

Desrosières, Alain.
[Politique des grands nombres. English]
The politics of large numbers : a history of statistical reasoning / Alain Desrosières : translated by Camille Naish.
p. cm.
Includes bibliographical references and index.
ISBN 0-674-68932-1 (alk. paper)
1. Statistics—History. 2. Statistical services—History.
I. Title.
HA19.D4713 1998
519.5′09—dc21 98-3199

To the memory of Michaël Pollak,
whose moral rigor and work
on the politics of the social sciences
have profoundly influenced this book

Contents

Acknowledgments

One of the sources of this book has been a long-standing interest in the social conditions under which information on the social world is produced, an interest kindled by the teachings of Pierre Bourdieu. It owes much to conversations through the years with statisticians, demographers, and economists (Joëlle Affichard, Michel Armatte, Denis Bayart, Annie Cot, Jean-Jacques Droesbeke, François Eymard-Duvernay, Annie Fouquet, Michel Gollac, François Héran, Jacques Magaud, Maryse Marpsat, Pascal Mazodier, Robert Salais, Philippe Tassi, Michel Volle, Elisabeth Zucker); with historians and philosophers (Marie-Noëlle Bourguet, Stéphane Callens, François Ewald, Anne Fagot-Largeault, François Fourquet, Bernard Lécuyer, Jean-Claude and Michelle Perrot); and also with certain British, German, and American specialists, whose advice has been most valuable (Margo Anderson, Martin Bulmer, Lorraine Daston, Gerd Gigerenzer, Ian Hacking, Donald MacKenzie, Mary Morgan, Ted Porter, Stephen Stigler, Simon Szreter). It has also benefited from a seminar on the history of probabilities and statistics, organized at the École des Hautes Études (EHESS) in Paris by Marc Barbut, Bernard Bru, and Ernest Coumet. The sociology-of-science perspective was influenced by work carried out in the Groupe de Sociologie Politique et Morale (EHESS) with Luc Boltanski, Nicolas Dodier, Michaël Pollak (who died in 1992), and Laurent Thévenot, whose ideas on statistical formatting were essential—as was the research undertaken at the Centre de Sociologie de l'Innovation de l'École des Mines (Michel Callon and Bruno Latour). Finally, this book could not have been written without the warm welcome and assistance provided by the research department at the Institut National de la Statistique et des Études Économiques (INSEE), the French statistical office. The pertinent criticisms expressed by members of this department—especially by Francis Kramarz—were extremely helpful. I am also indebted to Elisabeth Garcia for her painstaking work in preparing the manuscript, with the help of Dominique d'Humières.

Stader: . . . Listen: my institute uses recent scientific methods. It studies graphology, pathology, hereditary flaws, probability theory, statistics, psychoanalysis, and experimental psychology. We look for the scientific elements of action, for everything that happens in the world happens according to laws. According to eternal laws! The reputation of my institute rests on them. Countless young scholars and students work in my services. I do not ask myself questions about the inept singularities of a case: I am provided with the nomic elements determining a man and I know what he must have done under any given circumstances! Science and modern methods of detection are ever increasingly reducing the domain of chance, of disorder, of what is supposedly personal. There is no chance! There are no facts! There are only scientific connections . . . In scientific circles, my institute does not yet enjoy the understanding it deserves. Your help in this matter would be irreplaceable: the constitution of methods of detection as a theory explaining the life of the superior man of science. It is only an institute of detectives, but its goal is also to provide a scientific structure for the image of the world. We discover connections, we note facts, we insist on the observation of laws . . . My chief hope is this: the statistical and methodological consideration of human states, which results from our work . . .
Thomas: My dear friend, you definitely came into the world far too soon. And you overestimate me. I am a child of this time. I must be content to sit on the ground between the two chairs of knowledge and of ignorance.

ROBERT MUSIL, *DIE SCHWÄRMER*

Introduction: Arguing from Social Facts

Unemployment, inflation, growth, poverty, fertility: these objective phenomena, and the statistics that measure them, support descriptions of economic situations, denunciations of social injustices, and justifications for political actions. They are inscribed in routinized practices that, by providing a stable and widely accepted language to give voice to the debate, help to establish the reality of the picture described. But this implies a paradox. As references, these objects must be perceived as indisputable, above the fray. How then should we conceive a debate that turns on precisely these objects? How can we dispute the undisputable? These questions are often raised in a critical context. Do statistics lie? How many people are really unemployed? What is the real rate of productivity? These measurements, which are reference points in the debate, are also subject to debate themselves.

The controversies can be placed into two categories, depending on whether they concern only the measurement, or the object itself. In the first case, the reality of the thing being measured is independent of the measuring process. It is not called into question. The discussion hinges on the way in which the measurement is made, on the "reliability" of the statistical process, according to models provided by the physical sciences or by industry. In the second case, however, the existence and definition of the object are seen as conventions, subject to debate. The tension between these two points of view—one viewing the objects to be described as real things, and the other as the result of conventions in the actual work—has long been a feature in the history of human sciences, of the social uses they are put to, and the debates concerning them. This book analyzes the relationship between these two interpretations: it is difficult to think *simultaneously* that the objects being measured really do exist, and that this is only a convention.

"The first and most fundamental rule is: Consider social facts as things." In thus formulating his rule of sociological method in 1894, Durkheim placed the social sciences in a perspective of objectification, characteristic of the natural sciences. But this formula is ambiguous. It can be read in two different ways, as a statement of reality or as a methodological choice: either "social facts *are* things," or "social facts must be treated *as if* they were things." In the second reading, the important words are *treat* and *as if.* They imply an instrumentalist attitude, which subordinates the question of the reality of things. What matters is the method, and the conventions it involves, in order to behave "as if."

These difficulties are analogous to those encountered, in the course of history, by the inventors of the very same statistical languages that enable us to establish social facts as things. Today these languages are based on clearly formalized synthetic concepts: averages, standard deviations, probability, identical categories or "equivalences," correlation, regression, sampling, national income, estimates, tests, residuals, maximum likelihood, simultaneous equations. The student, research worker, or statistical-data user receives compact concepts, encapsulated into concise and economical formulas—even though these tools are the result of a historical gestation punctuated by hesitations, retranslations, and conflicting interpretations. To master them, the apprentice must ask himself—and speedily resolve—questions that have been debated for decades or even centuries. The reopening of these debates is not merely a result of scholarly curiosity, tacked on like some supplement of feeling to the acquisition of formalized techniques; rather, it provides a path for and helps in the process of comprehension and apprenticeship. The obstacles encountered by the innovators of yesteryear in transforming social facts into things are close to those that, even today, are faced by the student or that make it difficult to conceive simultaneously the two interpretations—realist and nonrealist—of Durkheim's rule. History makes us understand how social facts became things and, accordingly, how they become things for everyone who uses statistical techniques.

These techniques are intended to back up scientific and political arguments. The history of their gestation allows us to outline, as we recall old controversies and debates, a connecting space between the technical languages and their uses in the social debate. Statistical reasoning can only be reintegrated into a reflexive scientific culture on condition that we return to these translations and debates, by rediscovering uncertain paths and moments of innovation, which always form new junctions between old schemes.

Statistical tools allow the discovery or creation of entities that support our descriptions of the world and the way we act on it. Of these objects, we may say both that they are real and that they have been constructed, once they have been repeated in other assemblages and circulated as such, cut off from their origins—which is after all the fate of numerous products. The history and sociology of statistics will be evoked as we follow closely the way in which these objects are made and unmade, introduced into realist or nonrealist rhetorics, to further knowledge and action. Depending on the case, the antirealist (or simply nonrealist) perspective is termed nominalist, skeptical, relativist, instrumentalist, or constructivist. There are numerous possible attitudes in regard to scientific constructs. They often differ in theory and in practice. This suggests that, instead of taking a stand on one in order to denounce the others, one may more fruitfully study how each of them is coherently inscribed in a general configuration, in a system of reports. The question of reality is linked to the solidity of this system, and to its ability to resist criticism. The more extensive and dense it is, the more reality it has. Science is an immense system, one that is immensely real. Statistics and probability calculus occupy an essential place among the tools for inventing, constructing, and proving scientific facts, both in the natural and social sciences. That we take realist and nonrealist attitudes toward statistical techniques equally seriously allows us to describe a greater variety of situations, or at least to narrate more unexpected stories than those permitted by a narrative form favoring one or another of these points of view.

An Anthropological View of the Sciences

In choosing to study practices that connect science and action in a most particular way, we shall be showing not what this interaction *should* be, but what it *has been*, historically and socially. To this end we must reconstruct contexts of debate, alternative or concurrent ways of speaking and acting, and follow shifts and reinterpretations of objects in changing contexts. But precisely because this field of study is a field of interaction between the worlds of knowledge and power, of description and decision—"there is" and "we must"—it already enjoys a particular relationship with history, prior to the research. History may be invoked to provide a basis for tradition, to foster the founding story of a community and affirm its identity. But it can also be invoked for political ends, in moments or situations of conflict or crisis, to denounce some hidden aspect. These two ways of appealing to history can be termed unilateral or partial, since they are oriented and fashioned by their intentions—here, to affirm identity or to denounce something. However, it is impossible to claim, contrary to these methods, that the stories are exhaustively detailed, for they are always more numerous and varied than all those we can imagine.

On the other hand, we can reconstruct the contexts of debate, the lines of tension along which diverse points of view are placed and intermix. This involves restoring each of them in a terminology similar to that of the actors, while at the same time objectifying this terminology; in other words, making it visible. For example, in noting how a community anxious to recall its tradition makes use of history, I could have employed words such as "self-celebration," or "apologetic discourse." I preferred to say that it "affirmed its identity," since that is the meaning given by the actors to this use of history. Like polemic usage, this use constitutes material for the desired anthropological reconstruction. The question is no longer whether or not a story is true but, rather, what its place is among numerous other stories.

There is a risk of being overwhelmed by the abundance of these stories. The one that follows is not constructed in a linear fashion, like that of enlightened science triumphing over darkness: in this latter method, the description of the past appears as a way of sorting out what already existed and what did not yet exist, or as a search for precursors. Instead of defining an unequivocal direction of progress ordering and describing successive constructs, I shall suggest, by way of a preamble, a few lines of tension that, one way or another, structure the debates encountered. These oppositions fluctuate across time. They are often retranslations or metamorphoses of each other: description and decision, objective and subjective probabilities, frequentism and epistemism, realism and nominalism, errors in measurement and natural dispersion. But a *complete* comprehension of the themes mentioned in this introduction is not absolutely necessary in order to read the chapters that follow. Here I am simply trying to establish links among seemingly disparate narrative elements, while addressing readers whose cultural backgrounds are equally diverse. This diversity, which makes the exercise so difficult, is linked to the place of statistical culture in the scientific culture, and of scientific culture in the general culture. It is part of the object under study.

The history and sociology of the sciences have long been studied according to two extremely different, if not opposite, perspectives, termed "internalist" and "externalist." According to the first perspective, this history is that of knowledge itself, of instruments and results, of theorems and their demonstrations. On the whole it is created by specialists in the actual disciplines (physicists, mathematicians). The second, in contrast, is the history of the social conditions that have made possible or impeded the course of the first: of laboratories, institutions, fundings, individual scientific careers, and connections with industry or public powers. It is usually created by historians or sociologists. The relationship between internal and external history has been the subject of numerous debates and has itself a complex history (Pollak, 1985). During the 1950s and 1960s, the separation of the tasks was clearly preached by, for example, Merton, who studied the normal rules of functioning of an efficient scientific community: professionalization; institutionalization and autonomy of research; rivalry among research workers; clarity of results; judgments offered by one's peers.

From the 1970s on, this division of labor was questioned both in Great Britain (Bloor, 1982) and in France (Callon, 1989; Latour, 1989). Their "strong program" focused on science "in the making," through the sum of its practical operations, within the actual laboratory. These operations are described in terms of the recording and stabilizing of objects, and in terms of establishing connections and ever more extensive and solid networks of alliances between objects and human beings. In this perspective, the distinction between technical and social objects—underlying the separation between internal and external history—disappears, and sociology studies both the sum of these objects and these networks. This line of research has unsettled some observers, especially in the scientific community, for one of its characteristics is that it relegates the question of truth into parentheses. In science in the making (or "hot" science), truth is still a wager, a subject of debate; only gradually, when science cools down again, are certain results encapsulated, becoming "recognized facts," while others disappear altogether.

This program has given rise to misunderstandings. That is because, in situating the question of truth as such outside the field—thus favoring the analysis of social mechanisms of strife in order to transform certain results into recognized facts—it seems to negate the very possibility of truth, favoring instead a form of relativism in which everything becomes a matter of opinion or balance of power. But its direction is more subtle, and just as Durkheim's motto "Consider social facts as things" must not be taken solely as a statement of reality, but also as a *methodological choice,* we can also follow this line of research to point out other things. In the field of statistics and probabilities, which has always mixed the problems of the state and of decision making, on the one hand, with those of knowledge and explanation, on the other, the wager of a program that transcends the separation between "internal" and "external" history compels recognition even more than for theoretical physics or mathematics.[1]

Description and Decision

The tension between these two points of view, descriptive and prescriptive, can be kept as a narrative mainspring in a history of probability calculus and of statistical techniques. The rationality of a decision, be it individual or collective, is linked to its ability to derive support from things that have a stable meaning, allowing comparisons to be made and equivalences to be established. This requirement is as valid for someone trying to guarantee the continuity of his identity in time (for example, taking risks, lending money with interest, insurance, betting) as it is for persons seeking to construct, from common sense and objectivity, everything that enables us to guarantee a social existence transcending individual contingencies. A description can thus be assimilated into a story told by one person or by a group of people, a story sufficiently stable and objectified to be used again in different circumstances, and especially to support decisions, for oneself or others.

This was already true of forms of description more general than those offered by probability and statistical techniques since the late seventeenth century. For example, it was true of descriptions that had a theological basis. But the type of objectivity that originated in the seventeenth century with the advent of scientific academies, professional scholars, and experiments that could be reproduced—and therefore separated from the person conducting them—was linked to the social and argumentative autonomy of a new space of description, that of science. Basing its originality on its autonomy in relation to other languages—religious, juridical, philosophical or political—scientific language has a contradictory relationship with them. On the one hand it preaches an objectivity and, thereby, a universality which, if this claim is successful, provide points of support and *common reference* for debates occurring elsewhere; this is the "incontrovertible" aspect of science. On the other hand this authority, which finds its justification in the actual process of objectification and in its strict demands for universality, can only be exercised to the extent that it is party to the world of action, decision making, and change. This motivates the process, if only through the questions that must be solved, the mental structures linked to these questions, and the material means for registering new things in transmissible forms.

The question is therefore not to know whether a pure science, freed of its impure uses, can simply be imagined, even as some impossible ideal. Rather, it is to study how the tension between the claim to objectivity and universality, on the one hand, and the powerful conjunction with the world of action, on the other, is the source of the very dynamics of science, and of the transformations and retranslations of its cognitive schemes and technical instruments. Linked successively to the domestication of risks, the management of states, the mastery of the biological or economic reproduction of societies, or the governance of military and administrative operations, the history of probability calculus and statistics teems with examples of such transformations. In the case of probability, there is the switch from the idea of "reason to believe" to that of "frequency limit as more lots are drawn" (Chapter 2); in the case of statistical techniques, the retranslation of the interpretation of averages and the method of least squares, ranging from the theory of errors in astronomy, to Quetelet's average man (Chapter 3), to Pearson's analysis of heredity or Yule's analysis of poverty (Chapter 4).

The complex connection between prescriptive and descriptive points of view is particularly marked in the history of probability calculus, with the recurrent opposition between *subjective* and *objective* probability; or, according to a different terminology, between *epistemic* and *frequentist* probability (Hacking, 1975). In the epistemic perspective, probability is a degree of belief. The uncertainty the future holds, or the incompleteness of our knowledge of the universe, leads to wagers on the future and the universe, and probabilities provide a reasonable person with rules of behavior when information is lacking. But in the frequentist view, diversity and risk are part of nature itself, and not simply the result of incomplete knowledge. They are external to mankind and part of the essence of things. It falls to science to describe the frequencies observed.

Numerous constructs have connected these two concepts of probability. First are the diverse formulations of the "law of large numbers," beginning with Jacques Bernoulli (1713). This law forms a keystone holding together the two views, with the important reservation that random events are supposed to be indefinitely reproducible in identical conditions (tossing a coin heads or tails, throwing dice), which represents only a limited part of the situations involving uncertainty. For other situations, Bayes's theorem (1765)—which links the partial information provided by the occurrence of a few events with the hypothesis of "a priori probability"—leads to an "a posteriori probability" that is more certain, improving the rationality of a decision based on incomplete knowledge. This reasoning, plausible from the standpoint of rationalizing behavior (epistemic), is no longer plausible from a descriptive (frequentist) point of view in which "a priori probability" has no basis. This tension runs through the entire history of statistics. It is crucial to the opposition between two points of view: in one case, one "makes do" because one has to make a decision; in the other, one is not content with an unjustified hypothesis, designed only to guide actions.

The discussion on the status of knowledge amassed by the bureaus of official statistics instituted after the first half of the nineteenth century is also linked to the tension between these two points of view, prescriptive and descriptive. From its very origins, because of its demands, its rules of functioning, and its manifest goals, the administrative production of statistical information found itself in an unusual position, combining the norms of the scientific world with those of the modern, rational state, which are centered on the general interest and efficiency. The value systems of these two worlds are not antinomic, but are nonetheless different. Public statistics services subtly combine these two types of authority, conferred by the science of the state (Chapters 1, 5, and 6).

As the etymology of the word shows, statistics is connected with the construction of the state, with its unification and administration. All this involves the establishment of general forms, of categories of equivalence, and terminologies that transcend the singularities of individual situations, either through the categories of *law* (the judicial point of view) or through *norms* and *standards* (the standpoint of economy of management and efficiency). The process of encoding, which appropriates singular cases into whole classes, is performed by the state through its administrations. These two processes—defining classes of equivalence and encoding—constitute the essential stages of statistical work (Chapter 8). This work is not only a subproduct of administrative activity, designed to further knowledge; it is also directly conditioned by this activity, as shown by the history of censuses, surveys conducted by means of samplings (Chapter 7), indexes, national accounting—all inseparable tools of knowledge and decision making.

The link betwen description and management becomes clearly apparent when several states undertake—as is the case today with the European Union—to harmonize their fiscal, social, and economic legislation in order to permit the free circulation of people, goods, and capital. A comparison of statistical systems reveals numerous differences: to harmonize them involves an enormous amount of work, similar to that necessitated by the unification of rules of law, norms, and standards. Creating a political space involves and makes possible the creation of a space of common measurement, within which things may be compared, because the categories and encoding procedures are identical. The task of standardizing the territory was one of the essential labors of the French Revolution of 1789, with its unified system of weights and measures, division of territory into departments, creation of the secular civil state, and Civil Code.

Making Things That Hold

Modern statistics derives from the recombining of scientific and administrative practices that were initially far apart. This book seeks to link narratives that are habitually separated: technical history and cognitive schemes, and the social history of institutions and statistical sources. The thread that binds them is the development—through a costly *investment*—of technical and social *forms*. This enables us to make disparate things hold together, thus generating things of another order (Thévenot, 1986). This line of research can be summarized schematically, before being developed in the nine chapters of this book.

In the two seemingly different fields of the history of probabilistic thinking and of administrative statistics, scholars have emphasized the ambiguity of work that is simultaneously oriented toward knowledge and action, and toward description and prescription. These two dimensions are distinct but indispensable to each other, and the distinction between them is itself indispensable. It is because the moment of objectification can be made autonomous that the moment of action can be based on firmly established objects. The bond linking the two worlds of science and practice is thus the task of objectifying, of making *things that hold*, either because they are predictable or because, if unpredictable, their unpredictability can be mastered to some extent, thanks to the calculation of probability. This trail makes clear the relationship between probabilities and the way they reflect the play of chance and wagers, and the macrosocial descriptions of state-controlled statistics. These two domains are constantly intersecting, meeting, and diverging, depending on the period. Intersection had already occurred in the eighteenth century, with the use of mortality tables to support insurance systems, or with Laplace's estimates of the population of France, based on a "sampling" conducted in a few parishes (Chapter 1).

But it was Quetelet who, in the 1830s and 1840s, largely disseminated the argument connecting the theory of probability and statistical observations. This construct held together both the random, unpredictable aspect of individual behavior and the contrasting regularity (and consequent predictability) of the statistical summation of these individual acts, through the notion of the *average man*. It was based both on the generality of Gaussian probability distribution (the future "normal law") and on sets of "moral statistics" (marriages, crimes, suicides) developed by the bureaus of statistics. This form of argument would long send probability theory swinging from its subjective, epistemic side, expressed in terms of "reason to believe," toward its objective, frequentist side: the regularity of averages, opposed to the chaos and unpredictability of individual acts, provided an extremely powerful tool of objectification. This crucial moment is analyzed in Chapter 3.

Things engendered by the calculations of averages are endowed with a stability that introduces rigor and natural science methods into human sciences. We can understand the enthusiasm aroused by this possibility among those who, between 1830 and 1860, established the bureaus of statistics and the international congresses designed to propagate this new, universal language and to unify recording methods. The process of objectification, providing solid things on which the managing of the social world is based, results from the reuniting of two distinct realms. On the one hand, probabilistic thinking aims at mastering uncertainty; on the other, the creation of administrative and political spaces of equivalence allows a large number of events to be recorded and summarized according to standard norms. The possibility of drawing representative samples from urns, in order to describe socioeconomic phenomena at less expense, thanks to sampling surveys, results from this reunion. It was because the spaces of equivalence, which were practical before being cognitive, were constructed politically that probability systems involving urns could be conceived and used. Before drawing lots, one must first compose the urn and the actual balls, and define the terminology and procedures that allow them to be classified.

The fact of centering attention on the process of objectification allows us to exit the debate—which is a standard one in the sociology of sciences—between objectivists and relativists. For the objectivists, objects do exist, and it is up to science to clarify their structures. For the relativists, objects result from formats created by scientists: other formats would result in different objects. Now if construction does indeed occur, it is part of the social and historical process that science must describe. The amplitude of *the investment in forms* realized in the past is what conditions the solidity, durability, and space of validity of objects thus constructed: this idea is interesting precisely in that it connects the two dimensions, economic and cognitive, of the construction of a system of equivalences. The stability and permanence of the cognitive forms are in relation to the breadth of investment (in a general sense) that produced them. This relationship is of prime importance in following the creation of a statistical system (Héran, 1984).

The consistency of objects is tested by means of statistical techniques born of probability models. The status of such models is a subject of debate, and the choice is still open between epistemic or frequentist interpretations of these models, in terms of rationalizing a decision or of description. The choice of an interpretation results not from a philosophical debate on the essence of things, but from the general construction in which the model finds a place. It is logical for actors in daily life to reason *as if* things existed because first, in the historical, envisaged space of action, the process of prior construction causes them to exist and second, a different way of thinking would prohibit any form of action on the world.

Similarly, the practice of statistical adjustments, intended to calculate the parameters of a probability model in such a way that the model adopted is the one that confers the greatest possible probability on the observed results, is a way of leaving the two interpretations open. The calculation of an arithmetical mean, which allows the likelihood of an object to be maximized, can be seen either as truly justifying the existence of this object, with deviations being treated as errors (the frequentist view, adopted by Quetelet), or as the best means of using the observations to optimize a decision (the epistemic view), the deviations then being seen in terms of dispersion.

The existence of an object results simultaneously from a social procedure of recording and encoding and from a cognitive procedure of formatting that reduces a large quantity to a small number of characteristics, described as *attributes of the object* in a frequentist perspective, and *parameters of a model* in an epistemic view. Despite the precautions that a good teacher of statistics must take in order to make his or her students aware of the various possible statuses of a probability model, current language and the social uses of these methods often slip inadvertently from one interpretation to another. This choice depends on the consistency of a general argument, in which the statistical resource used is one element conjoined with other rhetorical resources. Depending on the case, the existence of an object is normal and necessary or, on the contrary, its fabricated status can and must be remembered. Such ambiguity is inevitable: one cannot separate the object from its use.

The question of the consistency and objectivity of statistical measurements is often raised. The perspective I propose is intended to avoid the recurrent dilemmas encountered by the people preparing the figures, if they wish to answer it fully. On the one hand, they will specify that the measurement *depends on conventions* concerning the definition of the object and the encoding procedures. But on the other hand, they will add that their measurement *reflects a reality.* The paradox is that although these two statements are incompatible, it is nonetheless impossible to give a different answer. By replacing the question of *objectivity* with that of *objectification,* one creates the possibility of viewing this contradiction in another way. Reality appears as the product of a series of material recordings: the more general the recordings—in other words, the more firmly established the conventions of equivalence on which they are founded, as a result of broader investments—the greater the reality of the product. Now these investments only acquire meaning in a logic of action that encompasses the seemingly cognitive logic of measurement. If the thing measured is seen as relative to such a logic, it is simultaneously real—since this action can be based on it (which is a good criterion of reality)—and constructed, in the context of this logic.

Two Types of Historical Research

The diverse uses of the words "statistics" and "statistician" reflect the tension between the viewpoints of reality and method. For some people, statistics is an administrative activity involving the recording of various data, leading to incontestable figures that are adopted in the social debate and determine a course of action. For others, it is a branch of mathematics, taught at a university and used by other scientists: biologists, doctors, economists, psychologists. The autonomy of these two meanings dates from the beginning of the twentieth century, when the techniques of regression and correlation were routinized and diffused, beginning with Karl Pearson's center of biometrics (estimates, tests, variance analysis) developed in Ronald Fisher's experimental agricultural laboratory. From that time on, statistics has appeared as a branch of applied mathematics.

But even before then, another profession had begun to assume a life of its own within the administration: that of state statisticians, responsible for the bureaus of official statistics, whose spokesman and organizer Quetelet had been for forty years. Until the 1940s, the mathematical techniques used by these bureaus were rudimentary, and the two professions were quite different. This situation subsequently changed, with the use of sampling methods, econometrics, then other increasingly varied techniques. But the autonomy of the different skills remained, and helped maintain the open tension between the administrative and scientific sides of these professions. The aim of statistics is to reduce the abundance of situations and to provide a summarized description of them that can be remembered and used as a basis for action. That involves both the construction of a political space of equivalence and encoding and a mathematical processing, often based on probability calculus. But these two dimensions of statistics are in general seen as two separate activities, and research into their history is distinct.

I have decided in this book to follow these two guiding threads simultaneously, precisely in order to study their points of intersection and exchange. They did not come together until after the 1940s and 1950s. To this end, I use two categories of historical research. The first concerns statistical institutions and systems. In the case of France the most important works—apart from the two volumes of *Pour une histoire de la statistique* published by INSEE in 1987—are those of J. C. Perrot (1992) and Bourguet (1988) on the eighteenth century and the beginning of the nineteenth century; also, those of Armatte (1991), Brian (1989), and Lécuyer (1982) on the nineteenth century; or of Fourquet (1980) and Volle (1982) on the twentieth century. In regard to Great Britain, the research of Szreter (1984, 1991) deals with the General Register Office and the public health movement. As for the United States, Anderson (1988) and Duncan and Shelton (1978) describe the slow growth of administrative statistics, then its transformation during the 1930s. This transformation led to the present organizations based on four major innovations: coordination by terminology; sampling surveys; national accounts; machine tabulation and, later, computer science.

The second category of works has to do with mathematical statistics and probability. This field of historical research was active in the 1980s: first in France, with the original but solitary book by Benzécri (1982); then in England, following a collective piece of work carried out during 1982–83 in Bielefeld (Germany), which brought together scholars from several countries. The book, entitled *The Probabilistic Revolution* (vol. 1, edited by Krüger, Daston, Heidelberger, 1987, and vol. 2, edited by Krüger, Gigerenzer, Morgan, 1987), was followed by several others: Stigler (1986), Porter (1986), Daston (1988), Gigerenzer et al. (1989) and Hacking (1990). During this same period the history of econometrics has been studied by Epstein (1987), Morgan (1990), and in a collective issue of the *Oxford Economic Papers* (1989).

This blossoming of research on the history of statistics, probability, and econometrics makes it possible to undertake a general interpretation within the perspective of the sociology of sciences. This reading is both comparative and historical. Four countries have been selected as examples: France, Great Britain, Germany, and the United States. They were chosen because of the available documentation concerning them, and because the most significant episodes occurred in those countries. The historical narrative covers events as far as the 1940s. At that point institutions and technologies appeared, the physionomy of which is close to those existing today. An interpretation of their developments since this period requires statistical research of an entirely different nature. Statistical methods are now used in very different fields, and are introduced into the most varied scientific, social, and political constructs. The recent history of statistical bureaus has been little studied as yet, but materials for a study of the relevant French agencies have been assembled in *Pour une histoire de la statistique* (INSEE, 1987). Mathematical statistics, probability, and econometrics have developed in such numerous and different directions that it has become difficult to imagine a historical overview comparable to Stigler's account of the eighteenth and nineteenth centuries.

The nine chapters of the present book follow the developments of the two faces of statistics, scientific and administrative. They examine several branches of the genealogical tree of modern statistics and econometrics: a diagram of this tree, with a summary of the paths I retrace, is provided at the beginning of Chapter 9. The first chapter describes the birth of administrative statistics in Germany, England, and France. The second describes the appearance, in the seventeenth and eighteenth centuries, of probability calculus, its applications to measurement problems in astronomy, and the formulation of the normal law and method of least squares. The third and fourth chapters focus on averages and correlation, through the works of Quetelet, Galton, and Pearson. In the fifth and sixth chapters, I analyze the relationships between statistics and the French, British, German, and American states. The seventh chapter describes the social conditions in which sampling techniques originated.[2] In the eighth chapter I discuss problems of terminology and encoding, especially in the context of research previously carried out with Laurent Thévenot. The ninth chapter explores the difficulties involved in uniting the four traditions leading to modern econometrics: economic theory, descriptive historicist statistics, mathematical statistics resulting from biometrics, and probability calculus. My conclusion briefly evokes the development of and subsequent crisis in statistical language since the 1950s.[3]

1

Prefects and Geometers

What do *statistics*—a set of administrative routines needed to describe a state and its population; *the calculus of probabilities*—a subtle manner of guiding choices in case of uncertainty, conceived circa 1660 by Huyghens and Pascal; and the *estimates* of physical and astronomical constants based on disparate empirical observations, carried out around 1750, all have in common? Only during the nineteenth century, after a series of retranslations of the tools and questions involved, did these various traditions intersect and then combine, through mutual exchanges between the techniques of administrative management, the human sciences (then known as "moral sciences"), and the natural sciences.

The need to know a nation in order to govern it led to the organization of official bureaus of statistics, developed from the very different languages of English *political arithmetic* and German *Statistik*. Furthermore, thinking on the justice and rationality of human behavior developed through the ideas of expectation and probability. Then again, an endeavor to formulate laws of nature that accounted for fluctuating empirical records resulted in increasingly precise work on ideas of the "middle road," *le milieu qu'il faut prendre,* the mean (or central value), and on the method of least squares. The first two chapters of this book will describe these three traditions which, despite their seeming heterogeneity, are all concerned with creating forms that everyone can agree on: *objects,* accessible to common knowledge. For a long time, however, bureaus of official statistics ignored research on probability or the theory of errors. Probability is the subject of this chapter, whereas the theory of errors will be discussed in Chapter 2.

In emphasizing, in my introduction, the idea that the social world is a constructed phenomenon, I did not mean to imply that the descriptions of it given by statistics were merely artifacts. Quite the reverse: these descriptions are only valid to the extent that the objects they exhibit are consistent. But this consistency is not given in advance. It has been created. The aim of surveys is to analyze what makes things hold in such a way that they constitute shared representations, which can be affected by actions with a common meaning. Such a language is necessary to describe and create societies, and modern statistics is an important element of it, being especially renowned for its factualness, its objectivity, and its ability to provide references and factual support.

How did this distinctive reputation—enjoyed in particular by statistics among the modes of knowledge—come about? This credit is the result of an unusual interaction, brought about by history, between two forms of authority that are clearly separate: that of science and that of the state. In the seventeenth and eighteenth centuries a conceptual framework took shape, in which to think out both the *reasons for believing* something (supporting decisions that involved the future) and the *degrees of certainty* of scientific knowledge, through the theory of errors. The authority of "natural philosophy" (the science of the time) gradually became separate from the authority of religion and of princes. The division between the constitution of things and the constitution of mankind became more and more pronounced, with the former loudly stating its autonomy (Latour, 1991).

At the same time, however, the modes of exercising princely authority were evolving. They developed differently from one country to the next, depending on how the relationship between state and society was changing. Thus specific branches of knowledge were formed, useful both to a prince and to his administration, and produced by their activities. Moreover, just as a civil society distinct from the state was acquiring its autonomy (with forms and rhythms that differed according to the states) and public spaces were forming, other specific branches of this society's knowledge of itself were also taking shape. All these constructions resulted (essentially but not exclusively) from work done by the state. They would subsequently constitute the second source of the unusual credit enjoyed by modern statistics—at least in the more or less unified meaning the word had in the nineteenth century, when it signified a cognitive space of equivalence constructed for practical purposes, to describe, manage, or transform human societies.

But these branches of knowledge themselves had forms and origins that varied with each state, depending on how those states were constructed and conjoined with society. I shall mention the case of Germany, which gave us the word *statistics* and a tradition of general descriptions of the states; and also that of England, which through its *political arithmetic* bequeathed the counting of religious and administrative records together with techniques of calculation allowing them to be analyzed and extrapolated. Lastly, in France, the centralization and unification that took place first under the absolute monarchy, then during the Revolution and the Empire, provided a political framework for conceiving and establishing a model for the office of "General Statistics" in 1800 (although certain other countries, such as Sweden in 1756, had already done as much). More generally, it provided an original form of "state-related sciences," with corps of engineers issuing from the major academies *(grandes écoles)*, rather than from the university.

The opposition between German *descriptive statistics* and English *political arithmetic* is a standard theme in works dealing with the history of statistics or with demography. Some authors particularly emphasize the failure and collapse of the former in the early nineteenth century; they also point out that the latter—although at that time inheriting only the *name* of its rival ("statistics")—was really the true ancestor of today's methods (Westergaard, 1932; Hecht, 1977; Dupaquier, 1985). Others, however, see in the methods of German statistics an interesting foretaste of certain problems in modern sociology (Lazarsfeld, 1970), or a significant effort to think out and describe the territorial diversity of a national state (J. C. Perrot, 1977; Bourguet, 1988). Here I shall try, rather, to reconstruct the circumstances in which these methods of description (in which languages and subjects are completely different) were developed, having only been compared with each other since 1800.

From the standpoint of the history of the accumulation of statistical techniques, the English style of political arithmetic has certainly provided the tools. These are analyzing parochial records containing baptisms, marriages, and deaths (Graunt, in 1662); creating mortality tables and calculating life expectancy (Huyghens, in 1669); and estimating a population based on a sample with a calculated *probable margin of error* (Laplace, in 1785). German statistics, in contrast—a formal framework for a general description of the power of the states—failed to emphasize quantitative methods, and thus transmitted nothing comparable. It is therefore logical that a history seen as chronicling the birth of techniques should emphasize political arithmetic, and treat the German tradition as an outmoded literary creation of scant interest.

German Statistics: Identifying the States

Nonetheless, in a perspective intended to clarify the relative position and cultural significance of the statistical mode of thought among the various ways of representing the social world, the pole constituted by this German "statistics" (which has little to do with the statistics of today) is significant. It expresses a synthetic, comprehensive will to understand some human community (a state, a region, and later on a town or profession) seen as a *whole,* endowed with a *particular power,* and describable only through the combined expression of numerous features: climate, natural resources, economic organization, population, laws, customs, political system. For an analytical mind concerned with directly linking the tool it uses to a clearly identified question, this holistic view of the community being described offers a major drawback: the pertinent descriptive traits are present in a potentially unlimited number, and there seems to be no particular reason for keeping one rather than another. But political arithmetic, which focuses attention on a small number of estimates put to direct uses, can easily lay claim to legitimacy and social recognition. Mortality rates, for example, serve as a basis for annuities or life insurance premiums. Estimates of population in the various provinces were indispensable for levying taxes or conscripting soldiers.

But German statistics met other needs. It offered the prince or official a framework for organizing the multiform branches of knowledge available for a particular state; in other words, a *nomenclature* or terminology inspired by Aristotelian logic. This form was codified around 1660 by Conring (1606–1681). It was then transmitted, throughout the eighteenth century, by the University of Göttingen and its "school of statistics"—notably by Achenwall (1719–1772), thought to be the inventor of the word "statistics," and then by Schlözer (1735–1809), who succeeded him as chairman of the department of statistics. Schlözer was the author of a *Treatise on Statistics* translated into French in 1804 by Donnant (thereby introducing this mode of German thought into early nineteenth-century France); he was the first of this school to recommend the use of precise figures rather than information expressed in literary terms, although he did not do so himself (Hecht, 1977). One of Schlözer's maxims indicates the rather structuralist and synchronic style of German statistics: "Statistics is history without motion; history is statistics in motion."

Conring conceived his statistics as a way of classifying heteroclitic knowledge. As Lazarsfeld (1970) remarks, "He sought a system that would make facts easier to remember, easier to teach, easier for men in government to use." Memorizing, teaching, implementing things in order to govern: we are not far from objectification, the effort to exteriorize things, to consign them to books, in order to reuse them ourselves or hand them on to others. This organizational and taxonomic aspect is just as characteristic of modern statistics as is its calculative aspect, which political arithmetic opened up. But the classificatory framework, organized from the viewpoint of the active state, was very general. It followed the order of the four causes in Aristotelian logic, causes which are themselves subdivided systematically (Hoock, 1977). *Material* cause describes territory and population. *Formal* cause brings together law, the constitution, legislation, and customs. *Final* cause has to do with the goals of the state: increasing the population, guaranteeing the national defense, modernizing agriculture, developing trade. And last, *efficient* cause gives account of the means available to the state: the administrative and political personnel, the judicial system, the general staff, and various élites (Bourguet, 1988). This Aristotelian distinction among material forces, mode of union, and effective organization is summarized in a Latin motto of Schlözer's: *vires unitae agunt,* "united forces act." This formula recalls the link between, on the one hand, the construction of equivalence needed for addition as an arithmetical operation and, on the other hand, coalition—the uniting of disparate forces into a superior force. In both cases processes of representation play a part: some typical or representative element of the class of equivalence, or the existence of spokesmen or representatives in the case of united forces (Latour, 1984).

Lazarsfeld (1970) relates this descriptive system to the situation of Germany during the second half of the seventeenth century, after the Thirty Years' War. The Empire was at that time divided up into nearly three hundred micro-states, all poverty-stricken and at odds with one another. The questions of the definition or redefinition of the rights and duties of all were essential. For all legal disputes over problems of territory, marriage, or succession, decisions had to be made by referring to case laws and examining the archives. Such situations conferred authority and prestige on minds more inclined toward systematic cataloguing, rather than the construction of new things, and this helped to prolong scholarly traditions already less influential in other domains. The weakness and need for self-definition of these micro-states led to this framework of thought—a kind of cognitive patchwork that later unraveled when, in the nineteenth century, powerful states (notably Prussia) emerged, erecting bureaucracies sufficiently complex to manage "bureaus of statistics" comparable to the French bureau created in 1800 (Hoock, 1977).

Before dying out, this tradition gave rise to significant controversy in the early nineteenth century. Certain officials proposed using the formal, detailed framework of descriptive statistics to present comparisons between the states, by constructing *cross tables* on which the countries appeared in rows, and the different (literary) elements of the description appeared in columns. An observer could thus take in at a glance the diversity of these states, presented according to the different points of view. The possibility of using the two dimensions of a page in a book to classify objects—thus enabling the observer to view them simultaneously—established a radical distinction between written and oral material, between graphic reasoning and spoken reasoning (Goody, 1979). But this conquest of the two-dimensional space of the table was not without its difficulties, for it required the construction of spaces of comparison, common referents, and *criteria*. It also exposed itself to the general criticism of reducing the objects described, and making them lose their singularity. Now this is precisely the kind of objection that the method of cross tables gave rise to, especially since this form of presentation encouraged the inclusion of numbers in the rows of the table, capable of direct comparison—whereas initially the classifiable information had been literary. It was thus the tabular form itself that prompted the quest for and comparison of numbers. This form literally created the equivalence space that led to quantitative statistics.

The fact of having to select certain characteristics to effect comparisons between countries or people can always give rise to a holistic type of criticism, because a particular country or individual person cannot be reduced to characteristics selected precisely in order to permit comparison. This kind of criticism of the formulation of equivalence possesses a high degree of generality, and one of the main concerns of the present work is to track down the recurrent modalities of this type of debate, and the common ground between protagonists upholding either position. The example of the controversy surrounding the "table makers" of this statistical school is significant. The partisans of tables adopted an overhanging position, allowing them to see different countries at the same time and through the same *grid*. Their opponents established a distinction between "subtle and distinguished" statistics, and "vulgar" statistics. According to them, vulgar statistics

> "degraded this great art, reducing it to a stupid piece of work . . ." "These poor fools are spreading the crazy idea that one can understand the power of a state simply by a superficial knowledge of its population, its national income, and the number of animals nibbling in its fields." "The machinations in which these criminal statistician-politicians indulge in their efforts to express everything through figures . . . are ridiculous and contemptible beyond all words." (Quoted by Lazarsfeld, 1970, from *Göttingen gelehrte Anzeiger,* c. 1807, itself taken from John, 1884)

Later, the same controversy can be found in the positions adopted by the "historical school" of German statisticians in the nineteenth century, in opposition to the diverse forms of abstract universalism, whether economic (English) or political (French). It was also characteristic of the debates prompted by the use of the "numerical method" in medicine (around 1835), of the use of statistics in psychology, or in regard to "docimology" (the science of examinations). In every case, a form of singularity (historical, national, individual) is appealed to, referring back to ways of describing things—in other words, of constructing totalities, different from those of statistics. Thus the tables created and criticized within the Göttingen school can be read *in a column*—that is, by comparing a *variable* (as the idea appears) for different countries; but also *in a row,* by describing a country in all its aspects and by seeking the source of its unity and its specificity. These two methods of interpretation each possess their own coherence. The second method is no more "singular" than the first, but involves another manner of adding up the elemental records.

But reading the tables in columns, and thus comparing countries, implies that a certain exteriority and distance can be adopted in relation to the state—qualities scarcely inherent in the position of the German statisticians, who argue from the viewpoint of the power and activity of the state itself. Identifying with that state, they are not prepared to conceive of a civil society distinct from the state, nor to adopt the oversight position implied by the creation and interpretation of tables. That is precisely what distinguishes them from the English *political arithmeticians.* In late seventeenth-century England a new relationship between the monarchic state and the various classes of society was taking shape, allowing these classes to go about their activities in a relatively autonomous way in relation to the monarch, with the two Houses of Parliament ensuring the representation of the aristocracy and middle class. In Germany, however, these distinctions only came about much later, and in other forms.

English Political Arithmetic: The Origins of Expertise

It was in this context of England in the 1660s—where the state became a part of society, and not its totality as in Germany—that a set of techniques originated, techniques of recording and calculation designated by the term political arithmetic. Inspired by the work done by Graunt (1620–1674) on *bills of mortality,* these methods were systematized and theorized first by Petty (1623–1687), then by Davenant (1656–1714). From the standpoint of our inquiry into the birth of the material procedures of objectification, they involve three important stages: keeping written records; scrutinizing and assembling them according to a predetermined grid; and interpreting them in terms of "numbers, weights, and measures."

Keeping written records of baptisms, marriages, and burials is linked to a concern with determining a person's identity, to judicial or administrative ends. This was the basic act of all statistical work (in the modern sense of the term), implying definite, identified, and stable unities. Thus the function of writing things down was to stabilize and prove (like a notarized act) the existence and permanence of a person and his or her links of kinship with a mother, father, spouse, and children. Just as evaluations of probability are linked to the concern with fixing and certifying (in other words, objectifying) "reasons to believe" and degrees of certainty, the entries in parochial records were intended to fix and attest to the existence of individuals and their family ties:

> It is entirely probable that the appearance and generalization of records is situated in the period when—and were caused by the fact that—in late Medieval law, written proof tended to prevail over oral proof. The old legal maxim that "witnesses surpass letters" was now replaced by a new one, "letters surpass witnesses." (Mols, 1954, quoted by Dupaquier, 1985)

These records were made obligatory by royal decrees, at approximately the same time in both England (1538) and France (edict of Villers-Cotterêts, 1539). Later on, other lists were made public: for example, during epidemics, announcements of burials were posted. It was on surveys of this kind that Graunt and Petty constructed their political arithmetic, in which they calculated the total population, or the numbers of "deceased" in various towns, by means of successive hypotheses as to the structures of families and households. They tried to introduce methods that had been proved elsewhere. Thus Petty explains:

> The method I employ to this end is not yet very common, for instead of simply using terms in the comparative and superlative and purely rational arguments, I've adopted the method (as a specimen of the political arithmetic I have long had in mind) which consists of expressing oneself in terms of numbers, weights, and measures. (Petty, 1690, quoted by Hecht, 1977)

These calculations are presented as practical methods for resolving concrete problems. Graunt speaks of "shopkeeper's arithmetic." Davenant mentions "the art of reasoning by means of figures on objects relative to government." Their difference from the German statisticians is clear: these were not academic theorists constructing an overall, logical description of the state in general, but men of diverse origins who had forged a certain practical knowledge in the course of their activities and were offering it to the "government." Graunt had been a tradesman; Petty, a doctor, a mathematician, a member of Parliament, an official, and a business man—in that order. Davenant had been an official and a Tory member of Parliament (Schumpeter, 1983). Thus a new social role took shape: the expert with a precise field of competence who suggests techniques to those in power while trying to convince them that, in order to realize their intentions, they must first go through him. These men offered a precisely articulated language, whereas German statisticians, who identified with the state, offered a general, all-encompassing language.

One of the reasons why the English political arithmeticians had to resort to indirect methods and roundabout calculations to reach their goals was linked to the liberal concept of the state and the limits of its prerogatives, which forbade it from organizing vast direct surveys—as certain continental countries, notably France, had already done. Thus in 1753 a plan to take a census of the population was violently denounced by the Whig party as "utterly ruining the last freedoms of the English people." For this reason, too, the systematization of a quantified description (which was not yet termed "statistics") stagnated in England during the second half of the eighteenth century, whereas Sweden conducted a census in 1749. In Holland, the calculation of probabilities was applied to the human life span (Huyghens, in 1669), was used to estimate the purchase price of an annuity, with the help of mortality tables (De Witt, in 1671), and was used to evaluate the size of the population, based on the annual numbers of births and life-expectancy at birth (Kersseboom, in 1738). A census was taken in Amsterdam in 1672 (Dupaquier, 1985).

Among the techniques handed down by eighteenth-century political arithmetic, the most famous (and the most controversial, in the following century) was that of the *population multiplier.* The problem was to evaluate the total population of a country, taking into account the fact that one could not conduct a census, but that the numbers of annual *births* were provided by parish registers throughout the land. The method consisted of taking a census in a few places, calculating the relationship between that population and the annual number of births in these same places, assuming that this relationship was more or less the same everywhere else, and estimating the total population by multiplying the general sum of births by this number, which was usually between twenty-five and thirty. This calculation, widely used in eighteenth-century Europe, was perfected by Laplace in 1785. From hypotheses concerning the probability distribution of the multiplier, he deduced a *probable error* for the estimated population (Bru, 1988).

This technique, the ancestor of random samplings, was vigorously attacked in the nineteenth century, and until the beginning of the twentieth century *exhaustive* censuses were preferred. The main criticism leveled at it concerned the hypothetical uniformity of the multiplier in regard to the entire territory. The idea that the kingdom could constitute a single probabilistic urn, endowed with a constant relationship between population and birth, proved problematical. How to construe the national territory as a single space of equivalence was to become an overriding concern in France, particularly after 1789; it was one of the principal questions in the great "survey of prefects" of 1800, the aim of which was to assess the disparities between departments, to try to reduce them, and to approach that One and Indivisible Republic dreamed of by the Revolution.

French Statistics of the Ancien Régime: Scholars and Intendants

In the field of statistics, France under the absolute monarchy did not leave a stereotyped intellectual tradition recorded in specific treatises and then taken up later by academic culture, as was the case in Germany with Conring, Achenwall, and Schlözer, and in England with Graunt and Petty.

But it did bequeath the periods that followed, particularly the Revolution and the Empire, both an *administrative* tradition alive with memoirs and surveys—almost leading, during the 1780s, to the establishment of a specific statistical institution (as actually happened in 1800)—and an effervescence of *learning and scholarship,* external to the actual state, involving empirical descriptions and systems for organizing them. Applying various requirements inherent to the two traditions in both England and Germany (general, logical taxonomic description in the one case, and quantification and mathematicization in the other), it prepared the path for the syntheses that would come later.

To describe this fermentation, we shall follow the construction of a strong, centralized state and the various ways of describing both the state and the society associated with it, before 1789 on the one hand, and between 1789 and 1815 on the other (Bourguet, 1988). In the case of royal power, descriptions of the country were intended to educate the prince, and administrative surveys, linked to management, already involved quantitative analyses. Outside the state, travelers, doctors, local scholars, learned men, and philosophers all produced research that had not yet been codified in accordance with precise disciplines. After the revolutionary period, however, a comparison of statistical experiments conducted in France during the Consulate and the Empire show how the word "statistical" lost its eighteenth-century German meaning and acquired its modern sense, of a system of quantified description.

What particularly distinguished France from Germany and England was that, from roughly 1660, the power of the monarchy was very strong; it possessed a fairly centralized administration, even though provincial disparities still subsisted in the form of laws and customs that were denounced and abolished in 1789. Tocqueville (1856) has shown how the unifying tradition of the Jacobins was deeply rooted in the absolute monarchy, and how the Revolution and the Empire pursued and amplified characteristics already present in the Ancien Régime. Thus the role and behavior of the intendants anticipated those of departmental prefects in the nineteenth and twentieth centuries. Ever since the time of Richelieu in 1630, Colbert in 1663, and at regular intervals thereafter, these intendants were charged with sending the king descriptions of their provinces in accordance with increasingly codified modalities. This system of surveying dated back to the medieval tradition of the "prince's mirror," intended to instruct him and show him the reflection of his grandeur in the form of his kingdom—the metaphorical extension of his own body. Gradually, this system split into two branches, comprising on the one hand a general descriptive table reserved for the king, and on the other an assemblage of particular pieces of information, quantified and periodic, intended for the administrators.

For the king, this table consisted of a methodical presentation, the spirit and contents of which rather resembled those of German descriptive statistics; it showed what constituted his power, as measured by the sum of taxes and the functioning of institutions, within a static and judicial perspective. Thus were defined the framework and limits of his action. An immutable order was described in this table. The diversity of mores was recorded in it, but there was no question of modifying them. The analysis was conducted from the point of view of the king and his power, and thus had little to do with the state of the society, its economy, or a precise enumeration of its inhabitants. An archetype of this kind of description is provided by the series of memoirs, written by the intendants between 1697 and 1700 to educate and inform the duke of Burgundy, the heir to the throne. This program was inspired by Fénelon.

An entirely different form of information was collected after the end of the seventeenth century, by and for the bureaus of administration, to ends more immediate and practical than educational. Linked to the development of the administrative monarchy and its services, these inquiries were less localized, and more specialized and quantitative; they had to do with counting the populations, the inventory of materials, and prices. Often they had fiscal objectives. In 1686 Vauban composed, in regard to tax reform, a "General and easy method for census taking," repeated later in his *royal tithe*. In 1694 a census of the entire population was proposed, as a basis for the first *poll tax*. Emergency situations created by famines, epidemics, or wars prompted partial surveys of the population and supplies, first in 1693 and then in 1720 (a plague in Marseilles). Then, gradually, regular and precise statistics were produced that had nothing to do with emergencies or fiscal reforms. The most important ones were the annual readings of *births, marriages,* and *deaths* (this measure, adopted by the abbé Terray in 1772, was the origin of statistics concerning trends in the population, produced by the registry office); the recording of the *prices of agricultural and industrial products* (dispatched every week to Paris, these records enabled a "general table of the kingdom" to be composed); and finally, between 1775 and 1786, Montyon's compilation of *criminal condemnations,* the ancestor of Quetelet's moral statistics.

Thus systems of accounting and regular statistics were set up, connected with precise fields that were national in character and did not pass through the detour of local descriptions; they were intended particularly to describe historical changes, and were established on the basis of records associated with the permanent management of state services. All these characteristics formed a type of construct different from the literary descriptions of Conring or Fénelon, and anticipated the practices of bureaus of statistics in the nineteenth century. But an essential difference remained: these descriptions, whether intended for the king or his administration, were secret and linked to royal prerogative. They were not intended to inform a civil society distinct from the state, nor an autonomous public opinion. These latter entities received greater expression after the 1750s, and of their own accord produced forms of knowledge separate from those of the government.

Outside the government, a *private* tradition of social description was developing. Accounts of journeys, geographical analyses of specific places, compilations of documents about the soil, mores, and the economy were produced by local scholars, learned men, doctors, and legal minds. These men were driven by the new philosophy of the Enlightenment and were grouped into societies and reform clubs, which debated and formulated the dominant themes of 1789. Among them the group formed by doctors was very significant, for its influence lasted late into the nineteenth century in the hygienics movement (Lécuyer, 1982), which had comparable ideas. These doctors developed theories of air and climate inspired by Galen and Hippocrates, for whom illnesses could be interpreted according to geographical surroundings. This encouraged them to organize detailed local surveys that related pathologies to various natural, economic, and social characteristics of such places. Thus, in 1776, Vicq d'Azyr, general secretary of the Royal Society of Medicine, conducted a survey of all French doctors, to establish:

> A geographical and medical map of France in which the temperament, constitution, and illnesses of the inhabitants of each province or canton would be considered in relation to the nature and the use of the soil. (Quoted by Bourguet, 1988, p. 39)

The secrecy that shrouded the results of these administrative surveys had the effect of stimulating these scholars to produce estimates based on partial information, on samples and circuitous means of calculation, like that of the multiplier, in accordance with methods that resembled those of English arithmetic. But the causes of these "algebraist's" artifices, resorted to in the absence of empirical data, were not the same in the two countries. In England, this lack was the sign of liberal trends in power, whereas in France it resulted from the secretive nature of royal absolutism that kept information for itself: we thus find two opposing methods of establishing the state.

At the same time as the power of the state, the optimistic notion was developing that a rationalism based both on mathematics and on empirical observations could make possible an objectivity—and therefore a clarity—that were both a matter of descriptions and decisions. The former, descriptive aspect was represented by Laplace's work on the theory of errors of observation in physics, or on the population multiplier. The second, decisional branch, appears in the research of Condorcet. Its aim was to create an algebra of mankind in society, a social mathematics that expressed in probabilistic terms the decisions of trial juries or of representative assemblies.

These formalizations can affect particular problems of estimation or decision by offering them precise solutions. But they can also have a more general and systematic aim, which in that respect resembles German statistics, but which uses different tools. This was the case with physiocrats who denounced the "all too easy temptation of calculation." Unlike the traditional German statisticians mentioned above, however, they criticized not so much the recourse to calculation as the choice of the magnitudes calculated, and the fact that these dimensions were not connected to a pertinent global construct (according to them). Thus in his "Letter on the need to make the calculation of supplies tally with that of the population," Dupont de Nemours (1766) ironically observed:

> All those writers who toil laboriously in their studies, adding up records of births and deaths, and making arbitrary multiplications to count men . . ., who imagine that by their calculations, which have nothing to do with wealth, they can judge the power and prosperity of the nation; and who, persisting in this idea, neglect to apply their zeal and toil to learning about the advances and work achieved in culture, the state of its products and especially the net product. (Dupont de Nemours, quoted by Bourguet, 1988, p. 42)

With Quesnay, the idea appeared of a general construct that was not simply a formal logical system, such as used by the Germans of Göttingen, but a descriptive framework that combined various evaluations in a "tableau économique" (Schumpeter, 1983). This goal—which was very comparable to that declared by national accountants beginning in the 1940s—combined the demand for *totality*, for at least potential completeness, of systems resulting from German scholasticism with the arithmeticians' demand for *measurement*. Now, to measure something is also to test its consistency, by endowing it with a certain exteriority and independence in relation to its inventor or observer (depending on the point of view, relativist or realist). Thus made consistent (or objective) by measurement, a thing can be included in a machine, a system of things that holds together independently of its constructor. In this case the machine was a *model*, which simulated society not only by means of terminology but also through measurements. The debate over the realism of the object takes on a new dimension here: that of the realism of the machine, in other words, of the model. The idea of a model has diverse connotations: descriptive (simplified schema); causal (a chain of explanations); normative (a form to be imitated); probabilistic (a hypothetical system of distributions of random variables). Several of these were already included in Quesnay's construct, which tried to be simultaneously descriptive (analysis of economic agents and measurement of their exchanges), explanatory (the role of agriculture), and prescriptive (liberation from the shackles of commerce and industry). In this way the idea of the empirical model was born, but the tools for testing its solidity would not be constructed for more than a century.

In their state-controlled or private modes, the various forms of description and calculation practiced in France during the Ancien Régime occupied many positions in the space—which their contemporaries were beginning to perceive—between the two opposing poles of the German and English traditions. Thus, in the works of the physiocrats, we find the system-oriented goals of the German tradition, and the concern with quantified objectification of the English school. But the dominant fact of this late monarchic period is that there was still a disjunction between the surveys conducted by the royal administration and reserved for its use and the investigations carried out independently of the state. These latter were marked by the new spirit of Enlightenment, which held that the circulation and promulgation of knowledge were essential conditions for social progress. The requirements of these independent surveys were assimilated fairly easily into the new type of state established after 1789 and, after numerous stumbles, proved a decisive factor in redefining the word statistics and in giving it fresh content, even though this definition remained subject to debate throughout the nineteenth century (Armatte, 1991).

Revolution and Empire: The "Adunation" of France

The period between 1789 and 1815 was decisive in the formation of political, cognitive, and administrative tools that gave the statistical description of the social world its originality, as compared with other modes of description, and that gave French statistics its originality as compared with that of other countries. This period witnessed confrontations—sometimes harsh ones—between the opposing concepts mentioned above. These confrontations occurred during times of contrast, in which emergencies, grand descriptive goals, and last, almost routinized sets followed in succession (Woolf, 1981). Between 1789 and 1795 censuses and special surveys were conceived. Nothing came of them, however, because they were launched in emergency situations of penury and war, and because an adequate administrative infrastructure was lacking. Then from 1795 to 1806 general surveys of the new departments were organized, in a spirit comparable to that of German statistics. Finally, between 1806 and 1815, regular quantitative statistics were instituted, especially agricultural and administrative ones.

The process of creating equivalence was, during the quarter century that spanned the Revolution and the Empire, particularly spectacular. At few points in the history of the world has this process been willed, thought out, and initiated so systematically in so short a time and in so many areas: the metric system was instituted, as well as the unification of weights and measures (the same everywhere, and logically expressed in relation to the meter); the French language was generalized, and the number of provincial dialects reduced (by means of the army and the schools); the universality of the rights of man was proclaimed ("All men are born and remain free and equal"); nobiliary privileges and professional guilds were abolished; the Civil Code was instituted (inspired by a natural law of man in general, rather than being linked to a particular society); and the national territory (homogenized by the abolition of rights particular to certain provinces) was divided for administrative purposes into departments, all organized identically and of approximately the same size. But some of these attempts to transform the referential framework of the natural and social world failed: for example the revolutionary calendar, perhaps because unlike the other reforms this one did not introduce a more universal, rational, and economical way of coding time than before. In this instance, the cost of the investment needed to change the Christian calendar—a form already firmly unified for centuries by the papacy—was not compensated by a subsequent, significant savings, as was the case with the other successful reforms: the double dimension, cognitive and economic, of investments in form is clearly visible here.

All these metrological, judicial, and taxonomical constructs had the effect of making physical measurements, judgments, and encodings theoretically independent of singular and local circumstances, by making them transportable, generalizable, and repeatable to an identical degree. They were intended as much to guarantee the fairness of the relationships among men as to guarantee the accuracy of standardizations. Thus the universality and clarity of the system of weights and measures enabled people to avoid deceit in commerce and trade, whereas the administrative and judicial forms of encoding were indispensable in lending an objective consistency to things that could not otherwise be counted: marriages, crimes, suicides, and later on businesses, work-related accidents, and the number of people unemployed.

The most visible manifestation of this process of homogenizing and codifying many aspects of human existence was the unification of the national territory, since many things and rules that were then redefined and generalized had previously been specified at a local or provincial level. This complex, expensive, and often tedious work was termed *adunation* by Siéyès, one of the principal actors involved: by this he meant the deliberate unification of systems of reference. A major instance of this was the division into departments, carried out within the space of a few months in late 1789 by the Constituent Assembly (Ozouf-Marignier, 1986). Its principle was to share *(départir)* an already unified whole (the nation), and not to bring together entities (the provinces) that had previously existed with singular characteristics. That is why this departmentalization was carried out in accordance with general criteria defined by the Assembly, and not according to local contingencies. (One extreme plan was to establish a network of squares defined by lines of latitude and longitude.)

Among the criteria, the surface areas of the departments had to be of much the same order of size, and their prefectures situated in such a way that people could reach them in a single day from anywhere in the department, and the subprefectures in such a way that people could travel there and back within the same day. The names of the departments were created from the names of rivers or mountains, while names of former provinces were avoided. At the urging of envoys sent by their regions, the deputies occasionally tried to influence certain choices; but this ran contrary to the basic principle by which they were, collectively, the elected body of the entire nation, and not just the delegates of their own province. This national rule accordingly constrained them to resist these demands, and that was what made it possible to carry out the work in so little time. The general principle was to make a *tabula rasa* of a society that had previously been characterized by its privileges, by fiscal arrangements that differed according to the province, and by local superstitions. The departments (in 1789), and subsequently their prefects (instituted in 1800), had to be the instruments of this *adunation*—the politico-cognitive construction of a space of common measurement to the scale of the One and Indivisible Nation. This process was launched by a monumental statistical survey, the responsibility for which fell to these new prefects.

Between 1789 and 1800, France went through a period in which hopes of refounding society upon fresh bases went hand in hand with crises in the political, economical, and military spheres. These lofty goals led to a strong demand for descriptions of society in all its aspects, the better to transform it. This explains the numerous plans for censuses and thorough surveys, especially those that would provide content for the new framework of departments. But the urgency of the crises prompted an inconsistent succession of demands for information by the central government, ill-sustained and ill-regulated, and generally unfruitful (Gille, 1964).

The 18th of Brumaire (November 9, 1799) resulted in the establishment of the strong, authoritarian power of Napoléon Bonaparte, one that would realize the potential of and previous ambitious projects for efficient institutions: the Civil Code, universities, lycées, prefectorial administration, bureaus of statistics, and censuses. In statistics, however, two quite different periods followed, significant in their contrast: surveying in the German manner; and later, limited statistics that could be put to direct use. The various means of describing and formalizing the social world, so recently debated in philosophical circles and outside the royal administration, could now be summoned up by those who (especially in the Ministry of the Interior) had both to prepare for the most urgent needs and lay down the bases for a general mode of describing French society. Such was the case of François de Neufchâteau, minister between 1797 and 1799, who regularly sent circulars to the municipalities and departments asking for information of every kind.

At the request of the new administration, therefore, all this work by local scholars, learned societies, doctors, and philanthropists was gathered together, having previously flourished in an uncoordinated manner in every corner of the kingdom. The appetite for knowledge of the social groups that carried the Revolution between 1789 and 1795 was thus requisitioned in the service of the state, and these were the groups addressed first by François de Neufchâteau, then by Chaptal, after 1800. An important aspect of this new way of doing statistics was that, unlike the administrative work of the Ancien Régime, it was intended for publication. The first to do so was Sébastien Bottin, who in 1799 brought out a *Political and Economical Almanach of the Department of Bas-Rhin,* before launching an almanac business that was subsequently bought by Didot, the editor of "Bottins" (directories) (Marietti, 1947). Neufchâteau hailed it as the "first really statistical work of this kind that we have in France," adding this prediction: "I do not despair of seeing him append his name to this kind of work, and one day people will say the 'Bottin' of a department when referring to an informative, complete almanac, just as one says a 'Barème' when speaking of accounts."[1]

The unification of the nation was accompanied by a broad diffusion of information on the soils that composed it, on the new techniques of agricultural and industrial production, and on potential markets. At this point, statistics passed from manuscripts cloistered in administrative archives to printed matter theoretically intended for a large public. This shift was linked to the fact that the *republican* state, having literally become the *public thing,* represented society in its entirety, by means of electoral representation, and also through statistics, which now became the "mirror of the nation" and no longer simply the "mirror of the prince." The aim of offering society a mirror image of itself through a network of prefectorial surveys was the first goal of the new "bureau of statistics of the Republic," created in 1800 by the then minister of the interior, Lucien Bonaparte, who was swiftly replaced by Chaptal.

De Ferrière and Peuchet, the two men mainly responsible for this bureau until 1805, were rather literary in culture, drawn to the German style of statistics (Schlözer's treatise had been translated by Donnant), and reticent in regard to English political "algebra." But, within the actual bureau, they were challenged by Duvillard, a mathematician specializing in mortality tables and their use in calculating life annuities. Two cultures, modes of knowledge, and lists of requirements clashed in mutual incomprehension at a moment when the "human sciences" had not yet been structured into clearly distinct academic disciplines and when the emerging languages were in direct competition.

Peuchet and Duvillard: Written Descriptions, or Calculations?

Whereas Peuchet promoted written description as a form that allowed narrative and memorization, and criticized the reductive nature of tables, which he compared to skeletons without a body, Duvillard preached the precision of numbers, which could be cross-checked and which had laws that could be represented by equations. It is interesting to read the discourses thus typified, not with a view to asking which was right but from the standpoint of their internal consistency. It is also interesting to see what social and political forces they sought to join, as if saying to them, "See how much you need me," and to see what arguments they put forward to that end.

In 1805 Peuchet published a work whose full title gives an idea of the subject: *An Elementary Statistics of France, Containing the Principles of This Science and Their Application to the Analysis of the Wealth, Forces, and Power of the French Empire, for the Use of Those Who Intend to Pursue the Study of Administration.* After his name he appended a list of agricultural and commercial societies and political and administrative authorities to which he belonged. He used the word "administration" in a general sense, meaning the management of public or commercial business. Directly addressing these notables, of which he himself was one, he offered them a general descriptive discourse, easy to read and memorize, on the "wealth, forces, and power of the Empire." In a preface "on the manner of writing statistics," he emphasized the stylistic qualities that were appropriate:

> To French minds ever impatient to know the purpose of a piece of work, and unable to tolerate the dryness of tables, no matter how accurate they may be . . . General considerations, useful applications, clear definitions, everything that sustains reflection through the attraction of discourse and elocution, all this necessarily enters into the French manner of instruction. (Peuchet, 1805)

He seems to lump together German statistics—guilty of "amassing, while smothering them, a host of positive pieces of information or reasoning, in a context that does not belong to them . . . in terms that have almost no application"—and the calculations of "algebraists" and "geometricians." But he devotes the main part of his attack to the latter:

> If we have criticized the method that perverts statistics by confusing or mixing together pieces of information that are foreign or useless in teaching it, we think ourselves even more justified in rejecting the method which, by using enigmatic formulas, algebraic calculations, or figures of geometry, tries to present or analyze something that is far simpler to say naturally and without obscurity . . . These remarks are all the more pertinent in that relatively enlightened people have in good faith thought they were contributing to the progress of political economics and lending solidity to its maxims by making it bristle with algebraic calculations. It is impossible to grasp how these calculations are applied to the object of this self-complicating science, a science one must avoid making even more obscure by an excess of difficulties and metaphysical abstractions. (Peuchet, 1805)

It would seem that Peuchet was himself rather unfamiliar with the methods of arithmeticians, and ill at ease with them. But what matters is that he offered his public, whom he knew well, a legible and memorizable discourse, the parts of which held together by virtue of a narrative thread, sustained by a unifying project: to analyze the "power of the Empire" while successively describing its territory, population, agriculture, industry, commerce, navigation, state budget, and its army. Furthermore, he did not forbid himself to make detailed use of the works of the "algebraists" whom he criticizes elsewhere, but whom he studied closely; for example, he mentions an "estimate of the total consumption, according to the estimated consumption of each individual," and he compares three methods of calculation typical of these algebraists. His vehement attacks on the latter can be read as a way of anticipating the reaction of his readers and their presumed reticence in regard to the "dryness of tables." He thus acted rather as an intermediary—or translator (Callon, 1989)—between the formalizations of the arithmeticians and the questions that the "administrators" were asking themselves. But his vehement criticism of the arithmeticians was undoubtedly clumsy, and prevented him from forming an alliance with them. In the end, his party turned out to be the losing one, when De Ferrière had to leave the bureau of statistics in January 1806.

Duvillard, who then replaced him for a while, had an entirely different strategy. A mathematician by training, he had been employed at the general auditing office and in the Treasury prior to 1789. He had established *mortality tables* (used by insurance companies until 1880) and had become adept in applying them to the problems of liquidating annuities, calculating retirement pensions, and redeeming public debts. In 1791 he was appointed director of the Bureau of Political Arithmetic, created by the Constituent Assembly at the instigation of Condorcet and Lavoisier. Throughout the Revolution and the Consulate, he found numerous occasions to demonstrate that his techniques were indispensable for solving many problems of the public Treasury. In 1805 De Gérando, secretary general of the Ministry of the Interior, appointed him subchief of the bureau of statistics, charged with evaluating the work done by De Ferrière and his subordinates. Duvillard was shocked by what appeared to him as a complete absence of rigor in the compilation of tables, especially in view of the incomplete and inconsistent replies of the prefects' survey carried out in 1800. He expressed his indignation on January 13, 1806, in a *Memoir on the Work of the Bureau of Statistics.* De Ferrière left, but Duvillard did not succeed in replacing him. A prudent and realistic administrator, Coquebert de Montbret, was appointed in April 1806. In November, Duvillard wrote a *Mémoire pour le rétablissement de la place de géomètre calculateur* (Memoir in favor of reestablishing the position of calculating geometrician), in which he described his career and the services he had rendered, expressing the wish that his competence be institutionalized through the creation of a special bureau, directed by himself. He concluded both memoirs by presenting himself as the "penniless father of a family," and asked for his talents to be recognized (Reinhart, 1965; J. C. Perrot, 1977).

In his January memoir Duvillard accurately explains what in his view a bureau of statistics should do. First, he observes that no one had thought to test the consistency of objects by comparing them with one another:

> No one, it seems, in this office has suspected that facts might be used to verify one another. However, facts all have essential and necessary relationships with one another. The same causes that modify some, also introduce differences into the others. After having attentively considered their relationships, one can often represent their relationships and laws by means of equations. (Duvillard, 1806)

He then gives a concrete description of the considerable investment involved—for an administration that was still scarcely routinized—in constructing equivalences that did not exist a priori: the endless correspondence with prefects, the care needed in the mechanical work of the bureau:

> The main function of the director of this bureau should have been to examine attentively the bulletins sent by the prefects; to discuss, compare, and verify the facts; to communicate to the prefects the remarks they should have made; to invite them to make fresh observations and search for the causes that could have given rise to seemingly absurd or extraordinary results. Now, not only has this function not been fulfilled, but the form of the bulletins soliciting the facts has also been deficient. The numerous omissions, or mistakes in addition made in the incomplete printed tables showing the state of manufacturing, of the population and changes in the population make these tables useless, and indicate that the mechanical work of the Bureau was not carried out with sufficient care. (Duvillard, 1806)

Then Duvillard notes that the prefects could only give rigorous replies if the administration "kept records"; that is, if there was a preexisting form of records and of encoding, of which the general registry constituted the prototype. Failing this, the statistician had to proceed indirectly, by means of *reasoning and calculation* (this being the type of algebra that Peuchet had denounced, but used anyway):

> One can only expect prefects to have accurate knowledge of facts of which public and particular administrations keep a record. There are a host of other important facts that it will always be difficult to know completely through observation. Examples would be: the duration of marriages, or widowhood; the inventory of stocks and shares, of industrial products, of raw and worked materials, and knowledge of their intended purpose. But often, if one has the necessary data, reason and calculation can discover that which cannot be counted or measured immediately, through the methodical combination of facts. The physico-mathematical sciences offer many examples of this. (Duvillard, 1806)

Finally, Duvillard answers Peuchet's criticisms of "dry tables" point by point, observing that this form "facilitates comparisons and mental views," and waxing ironic about men who shine by virtue of the "seductive polish of an elegant style":

> Isolated facts, those that can only be obtained by rough estimate and that require development, can only be presented in memoirs; but those that can be presented in a body, with details, and on whose accuracy one can rely, may be expounded in tables. This form, which

> displays the facts, facilitates comparisons, knowledge of relationships, and mental views; to this end, however, one should keep records, as I did for the population, and this has not yet been done . . .
>
> In a country where one lives by rough estimates and people are more concerned with the form rather than the substance of things (because knowledge rarely leads to fortune) there is no dearth of men who have the seductive varnish of an elegant style. But experience proves that it is not enough to know how to make plans, summaries, and rough statistical estimates to be a good statistician . . . However intelligent a person may be, it is impossible for him to improvise a science that requires preliminary studies and almost an entire lifetime of experience: when one considers the extent of knowledge in economics, in political arithmetic, in transcendental mathematics, and in statistics, together with the shrewdness, talent, and genius combined with the spirit of order and perseverance needed to fill this position, it seems that in respect of usefulness and difficulty, the position would not be too far above men most distinguished by their writings. (Duvillard, 1806)

These two men were thus far more complex than is suggested by the stereotyped images they give of themselves. Peuchet used the results obtained by the algebraists when it served his purpose. Duvillard knew how to write well, and his style was not lacking in bite or in humor, as is shown by his emphasis of "by their writings" in a phrase that obviously alludes to Peuchet. When one of them reproaches the other for his "dry tables" and "hermetic calculations," and then is mocked in return for the "seductive polish of his elegant style," one can perceive two recurrent ways in which, beyond the classical opposition between literary and scientific culture, statisticians try to demonstrate that they are necessary. In one case, the aim is to put across a simple and easily remembered message, to produce things that can be used readily, on which constructions of another rhetorical nature—for example, political or administrative—can be based: Peuchet's remark about the "wealth, forces, and power of the Empire" is of this order. But in the other case, emphasis is placed on the technique and professionalism involved in producing and interpreting results that are neither gratuitous nor transparent. In the course of time, the expression of these two modes of discourse became more refined, and the opposition between them become less brutal than that between Peuchet and Duvillard. However, this basic tension is inherent in the very nature of bureaus of administrative statistics, the credibility of which depends both on their visibility and on their technical aspects. The way in which this dual requirement is approached and retransformed according to the era or the country involved is a major topic in the history of these bureaus.

In the case of the Napoleonic bureau of statistics of 1806, the two protagonists defended their points of view in too radical a fashion, and neither of them prevailed. A senior official close to the administration and its needs, Coquebert de Montbret, became the bureau's director. The economic consequences of the continental blockade against England led to a state of emergency, and all efforts were devoted to instituting statistics of production, in both industry and agriculture. Then, perhaps because it had not managed to respond with the requisite speed to a demand by Napoleon for detailed information on the machinery of production, the bureau of statistics was closed down in 1812 (Woolf, 1981). Traces of this period survived in the "memoirs of the prefects"—responses to Chaptal's survey of 1800, publication of which was stopped in 1806—and in an attempt to create series of economic statistics, which were also suspended (Gille, 1964).

Prefectorial Statistics: Conceiving Diversity

The departmental memoirs written by the prefects, in response to Chaptal's questionnaire, were collected and published by the bureau of statistics until 1806. Others were printed later by private publishers, until 1830. Historians long considered them to be heteroclitic, incomplete documents, unserviceable as a source of numerical data. That is true from the standpoint of economic and social quantitative history that developed between 1930 and 1960, following the works of Simiand and Labrousse. For these historians, the construction of statistical series consisting, for example, of market prices or of agricultural production implied that preexistent, rigorous conditions would be met: that methods of recording would remain constant in time and space, and that the recorded objects would be identical. The task of criticizing sources consisted precisely in checking these conditions—or rather, in supposing that the objects and the circumstances in which they were recorded were sufficiently equivalent for it to be pertinent to reduce them to a single class, by means of a debate on the links between sufficient equivalence and pertinence. This question is of prime importance in the creation of long series dealing with professions or economic sectors. It is also important if data concerning the different regions of a state are being collected, and if the conditions of recording have not been clearly codified: that is precisely the criticism that Duvillard leveled against his predecessors, although he did concede that the prefects could only "have accurate knowledge of facts of which the administrations kept a record."

But the interest offered by prefectorial memoirs changes if one chooses the actual *process of adunation* as a topic of historical research, observing that it was one of the most important aspects of the French Revolution, one whose consequences would prove the most durable, regardless of the judgment applied to such a project. In this perspective, Chaptal's inquiry appears as an enormous effort to describe the diversity of France in 1800, and to measure the extent of the task required by this "adunation." Not only does the prefects' view of their departments offer precise information on the actual departments, it also and above all shows how the protagonists in this venture portrayed it to themselves, how they perceived the diversity of France, and the possible obstacles to this political and cognitive undertaking. In this respect, these documents, analyzed by Marie-Noëlle Bourguet (1988), offer unique material to the historian.

The survey can be interpreted in several ways. First, one might ask: what was the situation of France in 1801? Like the description of a journey, it offers a great number of observations, of an interest more ethnographical than statistical in the modern sense. Second, how was that situation *seen*? How were the supposedly pertinent characteristics selected? Third, what obstacles were perceived to the political plan of transforming and unifying the territory? The resistances this plan encountered reveal aspects of society which previously had no reason to be made explicit. It was precisely because there was a will to act on things that it was necessary to name and describe them. More precisely, the passage from prerevolutionary to postrevolutionary France involved changing not only the territory but also the words and the tools used to describe it: a striking aspect of the prefectorial memoirs was the telescoping of rival analytical grids, expressed in rather muddled fashion by the prefects' pens. I shall mention two exemplary cases of this taxonomic confusion. How was the division and order between the social groups to be conceived? How were the homogeneity or inner heterogeneity of each of these groups to be assessed?

To describe social groups, three quite different grids were available. The first one had been handed down by the France of the Ancien Régime, and was supposed to have been completely eliminated in 1789: the nobility, the clergy, the third estate. The society of three orders then disappeared, replaced by an egalitarian society in which men were "born free and equal under the law." The new official grid was based on property ownership and source of revenue. The sale of national possessions and the division of the land into numerous new properties gave this group of landowners great importance, and the distinction between "real estate owners" and all the others formed the essential criterion of the grid proposed by the circular of 19th Germinal, year IX (April 9, 1801), in which Chaptal sent the prefects the questionnaire they would have to answer. They had to indicate the number of

1. real estate owners
2. persons living solely from the proceeds of their real estate
3. persons living entirely from monetary income
4. persons employed or paid by the state
5. persons living off their work, either mechanical or industrial
6. unskilled or casual laborers
7. beggars

This second grid, published thus in an administrative circular, gave groups consistency according to a clearly objectified criterion: that of the source of income. It placed landowners at the head of the list, then investors and officials. On the other hand, wage earners in the modern sense were not yet seen as a distinct group, since Category 5 groups together *compagnons* (journeymen) and *maîtres* (masters) (in the terminology of the corporations). The future working class was even less in evidence, since craftsmen are included in Category 5, and unskilled laborers in Category 6.[2]

But from the comments made by the prefects in regard to social differences among the various populations of their departments, it becomes clear that this grid offered a major drawback for them: it did not distinguish *enlightened persons*—that is, rather urbane and cultivated people, with habits and interests that separated them fairly clearly from the *common people*. The third kind of grid is therefore apparent in the descriptions of lifestyles, but it is hard to objectify and its boundaries are always presented as vague. The contradiction between the two grids is in fact mentioned. Certain landowners (notably in the country) were not very "civilized" (and were sometimes quite poor). On the other hand, "talented persons" (doctors, teachers) often were not landowners.

This distinction between enlightened persons and the common people corresponded to a significant vacillation in the analysis of the internal heterogeneity of the two groups: which of these large masses was the more homogeneous? Or rather: in what manner should this homogeneity be determined? The ambiguity of the answers given to this question reflects the numerous ways of establishing equivalence. In certain cases, the educated élite were presented as being the same everywhere: it was useless to describe them in detail, since their refined mores had been standardized by the same requirements, by the same process of *civilizing mores* (Elias, 1973). In contrast to the élite, the ways in which ordinary people lived were splintered into numerous local customs, characterized by dialects, festivals, and rituals that differed greatly, not only from one region to another, but even from one parish to another. Yet in other cases, the prefects interpreted their realities in opposite ways: only cultivated people could have a distinct individuality and personal modes of living, whereas the common people were defined as a group, in a large mass, and were all the same.

However, these readings appear less contradictory if we observe—to take up Dumont's terminology again (1983)—that in both cases the common people were described according to a *holistic* grid, based on the community they belonged to. In contrast, the élite were described according to an *individualist* grid that abstracted individuals from their group, making them theoretically equal. This, then, is the individual of the declaration of the Rights of Man, and of modern urban society. In this individualistic vision, all men were different because they were *free,* and all men the same because they were *equal under the law.* This opposition between holistic and individualistic readings is a classical schema in sociology, found for example in Tönnies's distinction between *community* and *society.* It is interesting from the historical standpoint of the *objectification of statistics,* for it corresponds to two tendencies in the use and interpretation of social statistics. The first extends from Quetelet and Durkheim to part of modern macrosociology. It argues about groups conceived as totalities endowed with collective traits, which statistics describes by means of averages. The second, intent on describing the distributions of individual traits, extends from Galton and Pearson to other contemporary trends, and refuses to grant a group a status distinct from the conjunction of individuals that compose it.

In their answers to Chaptal's survey, the prefects were constantly wavering between the different methods of gathering knowledge (archival research; written questionnaires; and direct observations). Sometimes the circular required quantitative replies (population, professions, prices, equipment, production), and sometimes literary descriptions (religions, customs, ways of living). They themselves hesitated between different analytical grids. In all these aspects, this survey proves dismaying to the historian or statistician concerned with reliable data. But we must be aware that in order for reliable data to be produced, the country being described must already be "adunated," equipped with codes for recording and circulating elementary, standardized facts. The a posteriori interest of such a survey is precisely that it shows things in the process of taking shape, before they solidify—even though they never grow entirely solid. An index of subsequent evolution would be that, in the nineteenth and twentieth centuries, the territorial aspect gradually lost its importance in national statistics, which relied on summations other than those of the departments. No longer would the prefect be a man who explored his department on behalf of a central, still somewhat virtual authority: he became a man who instituted administrative *measures,* measures formulated by an authority that was henceforth solidly based, and informed by statistical *measurements* made possible by this unification of the land.

2

Judges and Astronomers

The calculation of probabilities—a procedure intended to ground the rationality of choices made in situations of uncertainty—was born during a precise period, between 1650 and 1660.[1] Describing this "emergence" of probability, Ian Hacking (1975) emphasizes the initial duality of this tool, simultaneously prescriptive and descriptive, epistemic and frequentist, and contrasts the idea of *reason to believe* with that of *chance*. The same reflection on this duality is found in Condorcet, who distinguished "reason to believe" from "facility"; in Cournot, who speaks of "chance" and "probability"; and in Carnap, who contrasted "inductive" probabilities and "statistical" ones.

The predominance of one or the other of these two interpretations is often presented in historical terms. Thus the decisional aspect seems to have been very important in the eighteenth century ("the classical era of probabilities"), notably with the procedures resulting from Bayes's theorem. Bayes proposed a way of taking note of incomplete information regarding previous events in order to estimate a "probability of causes" governing a decision. These procedures were subsequently contested in the nineteenth century, from a frequentist point of view. This perspective made a complete distinction between decisions based on nonquantifiable assessments (such as those of a trial jury), and decisions based on repeated observations, such as those provided by the emerging institutions of official statistics. For these "frequentists" the Bayesian procedure—which combined a small number of observations with a purely conjectural "a priori probability" so as to infer a more certain "a posteriori probability"—appeared as a fantasy. In the twentieth century, in contrast, reflection on the manner in which decisions are made in situations of uncertainty experienced a renewal with the works of Keynes, De Finetti, and Savage. Discussions on Bayesianism and its significance became a matter of prime importance once again. Research in experimental psychology was undertaken to observe whether the human brain functions effectively in accordance with such procedures (Gigerenzer and Murray, 1987).

This alternation between decisional and descriptive points of view characterized the entire history of probability formalizations, which began around 1660. But the questions that this new language was attempting to resolve had themselves resulted from much older arguments on the possibility of speculating on chance, either by allowing it to decide difficult cases, or by integrating the assessment of an uncertain future into the present time. This archeology of probabilities was reconstructed by Ernest Coumet (1970). The value of Coumet's article is that he presents this mode of calculation and the *equivalences* it implies as responses to a problem of *equity*.

Aleatory Contracts and Fair-Minded Conventions

This article by Coumet, entitled "Is the Theory of Chance the Result of Chance?" shows how the question of fairness was the source of the conventions of equivalence between *expectations,* an idea which thus appears to precede that of probability. The difficulty arose from the fact that chance—that is, the uncertain occurrence of a future event—was at that time viewed as a sacred attribute, which mankind would never be able to master, except by consulting God. Thomistic theology thus acknowledged, in very precise cases of necessity, the possibility of drawing lots, and distinguished between advisory lots, divinatory lots, and divisor lots. Divisor lots could be used "to decide what should fall to someone, or be attributed to them, possessions, honors, or dignities." In every other case, resorting to "games of chance" was a serious sin.

Having recourse to divinity—expressed here by the drawing of lots—to decide particularly problematic cases of litigation thus seems to have been a fair-minded solution, since it was based on a convention that transcended the individual litigants, and was thus acceptable to them. It had already been mentioned by St. Augustine:

> Suppose, for example, that you have a superfluous possession. You ought really to give it to someone who doesn't have one. But you can't give it to two people at once. So if two people appear, neither of whom has a more pressing claim by virtue of need or the bonds of friendship, isn't the most equitable solution to draw lots, to see which

of the two shall receive what you cannot give to both of them? (St. Augustine, *Of Christian Doctrine,* quoted by Coumet, 1970, p. 578)

Drawing lots is a way for judges to carry out what, even today, is their imperative duty: the obligation to judge. A judge can decide in accordance with an article of law or according to his intimate convictions; but the one thing he cannot do is fail to decide. One can readily imagine that in the more dramatic cases chance, as the expression of a superhuman will, could long have been a way of alleviating this heavy burden. This obligation to pronounce judgement, even in cases where doubt remains, may be compared to the obligation to *encode* which is the statistician's duty, as he divides the answers (or nonanswers) of an inquiry into classes of equivalence. Questionable cases are burdensome for statisticians too, and there may be a strong temptation to resort to chance, thus sparing themselves this duty. This recourse to chance is formalized and systematized in so-called hot-deck procedures, which assign random answers when no answer is given, according to laws of conditional probability construed on the basis of cases in which answers have been obtained. Thus divisor lots, the untamed random chance of weary encoders, or *hot-deck* procedures may be seen as economical ways of extricating oneself from the predicament of problematic cases.

Apart from these specific instances in which judges can rely on chance decisions as a last resort, chance appears in various other situations, the common feature of which is that they involve uncertain gains in the future: the investment of capital in shipping; insurances; annuities; games of chance. In all these cases, something one is sure of in the present is exchanged for some uncertain profit in the future. Is such an "unjustified enrichment" permissible? As in the case of usury or loans made at interest, theologians engaged in bitter debate about the fairness of conventions binding people by means of future events. Thus St. Francis de Sales questioned gains that did not result from some effort on the part of those contracting them:

We agreed on it, you'll tell me. That is valid for showing that the person who wins is not harming a third party. But it does not thence follow that the convention itself is not unreasonable, and gambling too: for gain which should be the reward of industry becomes the reward of chance, which deserves no reward since it in no way depends on us. (St. Francis de Sales, *Introduction to the Religious Life,* 1628, quoted by Coumet, 1970, p. 577)

Whereas this condemnation mainly concerned gambling, the risky investment of capital necessary to the shipping industry could, in contrast, involve some reward, by virtue of this very risk. For medieval theology, this even constituted a possible justification for loans made at interest, which were generally prohibited and associated with usury (Le Goff, 1962). Moreover, insurance contracts were deemed permissible.

Thus a whole body of work came into being, one that dealt with different fields (divisor lots, loans with risk attached, insurance contracts, games of chance), but that tended to formalize the idea of a *fair convention*. Divisor lots, for example, could be moved from the domain of the sacred and justified as a basis for such conventions. Similarly, in order to provide a solid judicial basis for contracts involving risky events in the future or to justify a game of chance, it was necessary for the contracting parties or gamblers to enjoy "equal conditions." This demand for equality, required by justice, opened the way to the construction of a conceptual framework common to activities that were completely separate in other respects: playing with dice, life insurance, and profits expected from uncertain trade. Gambling games were only one example among *aleatory contracts:* such contracts "relied on voluntary conventions according to which the acquisition of wealth depended on the uncertain outcome of fortune, and to be legitimate, they had to meet certain conditions of fairness" (Coumet, 1970, p. 579). The problem of equity arises almost at same time as the *apportionment,* when a business matter or game is interrupted: how should the profits or the stakes be divided up? A similar question, addressed to Pascal by the chevalier de Méré, formed the basis of Pascal's calculation of probabilities. But this seemingly anodyne question concerning the trivial activity of gambling was in fact linked to old debates on the problem of the fairness of the conventions on which aleatory contracts were based.

In introducing his new formalization, Pascal naturally borrowed from the language of jurists, since that was the language of the debates of that time. But he also created a new way of keeping the role of arbiter above particular interests, a role previously filled by theologians:

> The moralists who were trying to determine what conditions a game must fulfill in order to be equitable placed themselves above the greed and antagonism of the gamblers. A mathematician who tries to calculate a "fair distribution" simply adopts a more rigorous version of the same attitude: he is the judge. (Coumet, 1970, p. 580)

But on what could this objectification be founded, that enabled Pascal to define this position of arbiter—as in his reply to the chevalier de Méré's question (which concerned the equitable distribution of the stakes, in case the game was interrupted)? The *expectations* (or *utilities*) calculated had to be such that the various alternatives (stopping or continuing the game) were unimportant to the players:

> The regulation of what should be theirs must be so proportionate to what they could justifiably expect from fortune, that it's a matter of indifference to them whether they take what is assigned or continue with the luck of the game . . . (Pascal, quoted by Coumet, 1970, p. 584)

It was thus, by first establishing equivalent expectations—conceivably interchangeable among the players, being equal in value—that the idea of probability would subsequently be created, by dividing this expectation by the sum total of the wager. These expectations were common knowledge among the players, providing a pause amid controversy, a rule that allowed the game to be interrupted without anyone's feeling injured. This affinity of concern between judges and geometers involved a certain distance on the part of both in relation to the litigation or decision; they had to put themselves in the position of spectators, of impartial umpires or counselors to a prince.

This distance could in fact be adopted to different ends: either to arbitrate between two players in case the game was suspended, or else to advise the man of action faced with choosing among several decisions, the consequences of which depended on uncertain events in the future. They were the same calculations of expectation or utility that determined the sentences of judges or the advice of counselors. In the first case, fairness was intended, allowing harmonious relationships among men. In the second, it was the internal consistency of decisions made by men of action, the quest for a rationalism encompassing the different moments of their life, that encouraged the adoption of this new mode of calculation, which could be objectified and transposed from one case to another.

We should, however, realize how unpalatable and unacceptable the creation of such expectations of equivalence could be, concerning as they did events in the future, which had not yet occurred, events that were incompatible and heterogeneous. One even feels this when rereading the argument of Pascal's wager, which was built on this type of comparison. One also senses it in seeming, now well-worn paradoxes: for example, having to choose between situation A (winning one million dollars, *guaranteed*) and situation B (*one chance in ten* of winning twenty million dollars). A calculation of one's chances leads one to choose B, but few "reasonable" men would make such a choice, and many would prefer A: that is, the guaranteed, if lesser gain. We now know that making this calculation in terms of expectation (which in this case would lead one to choose B) is not really plausible except in a frequentist perspective: if this (fictitious!) choice were repeated many times (ten times? twenty times?), one would obviously choose B. This might also be the case if small sums were involved (winning ten dollars for certain, or one chance in ten of winning two hundred dollars).

It becomes clear from these various cases (arbitration, help in decision making) that the probabilistic rationale, often not very intuitive, justified itself in terms of an overlooking viewpoint, of generality: that of a judge poised above the litigants, a banker or insurer dealing with numerous risks, or even of an isolated individual faced with micro-decisions that do not require much commitment (choosing between small sums of money). There was thus a close connection between the faculty of conceiving and constructing these spaces of equivalence and the possibility of occupying such overlooking positions. The tension between objective and subjective probabilities can be retranslated in terms of point of view: one based on a unique, contingent choice, or on an all-encompassing and generalizable position.

Thus, in proceeding from the viewpoint of a jurist or counselor helping with decisions, the first probabilists had to overcome great difficulties in establishing and getting people to accept a geometry that dealt with *disparate dimensions*. Leibniz evokes these hesitant attempts, and the difficulty of simultaneously calculating possible gain, probability, and of compounding them into a single utility, the way the surface of a rectangle results from length multiplied by breadth:

> As the size of the consequence and that of the consequent are two heterogeneous considerations (or considerations that cannot be compared together), moralists who have tried to compare them have become rather confused, as is apparent in the case of those who have dealt with probability. The truth is that in this matter—as in other disparate and heterogeneous estimates involving, as it were, more than one dimension—the dimension in question is the composite product of one or the other estimate, as in a rectangle, in which there are two considerations, that of length and that of breadth; and as for the dimension of the consequence and the degrees of probability, we are still lacking this part of logic which must bring about their esti-

mate. (Leibniz, *New Essays on Human Understanding,* quoted by Coumet, 1970, p. 595)

This effort to create a space of comparison for heterogeneous sizes thus resulted from the debates of seventeenth-century jurists intent on justifying the equity of aleatory contracts. It would be taken up and broadened in the eighteenth century, in order to establish a homogenous space of degrees of certainty, itself linked to the necessities of action and decision making.

Constructive Skepticism and Degrees of Conviction

The history of this unfolding of probabilistic thinking in a series of practical problems involving uncertainty has been synthesized by Lorraine Daston in a book published in 1988. The book continues Coumet's work, presenting another root of this tool that enabled the various aspects of uncertainty to be thought out within the same framework: the debates on certainty and knowledge resulting from the Reformation and the Counter-Reformation.

These debates, which emphasized the foundations of belief (revelation, in the case of Protestants; tradition, for the Catholics), gave rise to reciprocal denunciations that gradually undermined the various components of this belief and led to skepticism. Pyrrhonism—an extreme form of skepticism expressed by erudite libertines—denied even the evidence of sensation and of mathematical demonstration. At the same time, midway between the dogmatic fideists adhering to the certainties of true faith and these most corrosive skeptics, several authors tried to define an idea of what was "simply probable," and of the "degree of conviction required to prompt a prudent business man to take action . . . , a degree of conviction dependent on an intuitive judgement of the plan and possible risk involved" (Daston, 1989).

These "constructive skeptics" (to use an expression coined by Popkin, 1964) thus considered action the foundation of knowledge, rather than the reverse. They were less interested in *equity* (as the jurists who inspired Pascal had been) than in the *rational belief* that shaped a decision. However, they too made use of the doctrine of aleatory contracts, using examples showing that it was sometimes reasonable to exchange a guaranteed possession in the present for an uncertain possession in the future. This philosophy of knowledge gave probability a clearly "epistemic" role, since it was directed at an aspect of insufficient knowledge rather than a risk. But by incorporating games of chance, risky activities (trade, vaccination), and a jury's decisions as to the guilt of a defendant into the same model, they prepared the pathway from one aspect to another.

It is interesting to compare the philosophical attitude of these *constructive skeptics,* standing between the fideists and the extreme skeptics, with the position I suggested in the introduction. I suggested a modern sociology of knowledge, distinguished from both scientific objectivism—for which "facts are facts"—and relativism, for which objects and accounts of things depend entirely on contingent situations. The two historical configurations of the seventeenth and twentieth centuries differ radically, however, if only because the pole of certainty is embodied by religion in one case and by science in the other. From this point of view the probabilistic way of proceeding, which objectifies uncertainty by quantifying it, belongs to a process of secularization. That is why nowadays both believers and scientists (whether or not they believe) feel uncomfortable on rereading the argument of Pascal's wager, for both groups feel that two henceforth separate orders of reasoning overlap in this famous text.

The distance that the "constructive skeptics" adopted in relation to the nihilistic skepticism of the Pyrrhonians (today they would be known as radical relativists) was intended to produce objects capable of providing a basis for action. These objects were "degrees of conviction," that is, of *probabilized beliefs.* As Lorraine Daston remarks: "This emphasis on action as the foundation of belief, rather than the other way round, is the key to the defense against skepticism. Writers like Wilkins often observed that the most convinced skeptic has eaten his dinner *just as if* the outer world really existed." (Wilkins quotes the example of a merchant accepting the risks of a long journey in the hope of increased profit, and recommends following such rules of action in scientific and religious questions.) The important words in the story of the skeptical diner are *just as if,* which refer not to a problem of essential reality (as a fideist would, or a realist of today), but to practical behavior, to a logic of action.

These authors therefore created a framework allowing the joint conceptualization of forms of certainty that were previously distinct: the *mathematical* certainty of demonstration, the *physical* certainty of sensory evidence, and the *moral* certainty of testimony and conjecture. They arranged these different forms of certainty according to an ordered scale, and inferred that most things are only certain at the lowest level, that of testimony. Thus the word *probability,* which in the Middle Ages meant "an opinion certified by authority," came to mean a "degree of consent proportionate to the evidence of things and witnesses." Then Leibniz and Nicolas Bernoulli, both of whom were mathematicians and jurists, combined these three levels of certainty into a continuum, including every degree of consent from incredulity to complete conviction.

These three levels of certainty corresponded to three very different ways of evaluating probabilities: (1) *equal possibilities* based on physical symmetry (appropriate only for games of chance); (2) the *observed frequency* of events (which presupposed a collection of statistics having adequate temporal stability, such as Graunt's mortality rates in 1662); and finally, (3) degrees of *subjective certainty* or belief (for example, judicial practices assessing the weight of indexes and presumptions). This construct is striking in that it resembles a distinction that has subsequently again been made between so-called *objective* probabilities, connected with *states of the world* (the first two levels), and *subjective* probabilities, corresponding to *states of mind* (the third level). But this distinction is only valid if we ignore the problem of how conviction is born: how do we pass from belief, in level three, to the objective certainties of levels two and one? This question—somewhat reminiscent of the one posed by partisans of the "strong program" in the sociology of science—haunted the philosophers of the Enlightenment.

Thus Locke, Hartley, and Hume taught an associationist theory, considering the mind as an adding machine, counting up the numbers of past events and computing the possibility that they would be repeated. Hartley imagined that repeated sensations hollowed out and stabilized "channels in the brain." Hume spoke of the "added vividness" that a repeated experience gave to a mental image, and emphasized the notion of *habit*, comparable to Bourdieu's *habitus* (Héran, 1987). The aim of these constructs was to link the certainties of level 3 (beliefs) with those of level 2 (frequency, repeated *sensations*). These questions recur in the subject chosen by a sociology of science aiming to describe science in the making and the way in which a conviction is created. The scientist must record experiments within standard protocols, which are explicit and thus repeatable, otherwise he will not be recognized by his peers. Similarly, in legal proceedings a single witness is considered worthless. Historians and journalists must "cross-check their sources." Repetitions are indications of objectivity and may be invoked as proof.

The "classical probabilities" of the eighteenth century in some way precede the great division between, on the one hand, objective scientific knowledge describing things existing independently of mankind and, on the other hand, beliefs characteristic of prescientific thinking. These scholars did in fact integrate into a single framework the probabilities of casting dice, the reliability of demographic statistics, and the private convictions of judges. But through this same effort at integration, they prepared the great division by seeking to extend the territory of what could be objectified—including problems such as the decisions of trial juries, which Condorcet and Poisson would discuss.

The division was anticipated by several debates, during which doubts were expressed as to the rational nature of behavior dictated solely by the calculation of expectations. The "St. Petersburg paradox" was a famous example (Jorland, 1987). Peter throws "heads or tails" until heads appears once. He gives one dollar to Jack if the first throw is heads; two dollars if heads appears only on the second throw, four dollars if it appears on the third throw . . . and 2^{n-1} dollars if heads first appears only at the *n*th throwing. Jack's expectation of winning is:

$$\frac{1}{2} + 2\left(\frac{1}{2}\right)^2 + 2^2\left(\frac{1}{2}\right)^3 + \ldots + 2^{n-1}\left(\frac{1}{2}\right)^n + \ldots$$

It is therefore infinite. Thus, according to the theory of expectations, it would be in Jack's interest to wager *any sum at all* in this game, since the "probable expected gain" is always larger than this sum. But common sense, in showing that nobody would bet more than a few dollars in such a game, contradicted the theory, and this was extremely troubling to the scholars of the time. The problem was set by Nicolas Bernoulli in 1713. A solution was proposed by his cousin, Daniel Bernoulli, in 1738 at the academy of St. Petersburg (whence the problem's name).

This paradox gave rise to a most animated debate, especially between the two cousins, Nicolas and Daniel Bernoulli. The debate, which has been analyzed by Jorland and Daston, had the merit of showing the various possible meanings of the calculation of probabilities. Without entering into the details of this discussion, the contrast between the points of view is significant. For Daniel, a classic calculation of expectations would be suggested by a disinterested judge ignorant of the gambler's individual characteristics, whereas for the gambler it was less a case of fairness than of prudence. Daniel thus opposed a "moral" expectation—produced by the probability of the event through its "utility" (in the sense of economic theory)—to "mathematical" expectation.

Daniel drew his argument from the world of business, whereas Nicolas, a jurist, objected that this "moral expectation" did not conform to "equality and justice." Daniel retorted that his reasoning "was perfectly in accordance with experience." In fact, Nicolas was basing his remarks on the sense of equality produced by aleatory contracts, whereas Daniel was upholding a kind of commercial prudence. The cunning merchant stood opposed to the disinterested judge. As for the first, simpler paradox mentioned above, we find on the one hand a judge in the position of being able to look down from on high (or even a Utopian gambler of unlimited wealth, who could play the game an infinite number of times, while wagering enormous sums), and on the other hand the "normal" player, likened to the prudent merchant of limited fortune, who could not allow himself to bet a large sum against an enormous but very unlikely profit.

The drawback of these paradoxes was that they had to do with fictional problems, and seemed like intellectual games. This was not the case with the debate over inoculation against smallpox, which pitted scholars against one another at around the same time (Benzécri, 1982). This preventive measure significantly reduced incidences of the illness, but unfortunately caused the death of one in every three hundred people in the year following the inoculation. On balance, however, the results were still positive, and Daniel Bernoulli calculated that, despite these tedious accidents, the life expectancy of inoculated persons was three years more than it was for everyone else. From the point of view of public health this measure could thus be made obligatory, or at least be strongly recommended. Understandably, however, some individuals (often described as "fathers of a family") were more than reticent, either on their own account or in regard to their children. It can be seen from this example that the frequentist viewpoint went hand in hand with a macrosocial position (the state, or some public authority) whereas the epistemic position was that of someone having to decide for himself. The problem recurs in the nineteenth century in other debates dealing with the use of statistical method in medicine.

This type of discussion cast doubt on the use of "expectations" to rationalize human decisions, and prepared the way for the division that later arose between the two ways of envisaging probability. In the nineteenth century, a provisional boundary was erected, rejecting the "state of mind" side of probability and confining itself to the "state of the world" aspect, and to the frequentist current in particular. When Auguste Comte, among others, attacked probabilities in general, accusing them of being unable to describe the complexity of human behavior, he was in fact attacking the epistemic interpretation of this calculation, which had sustained the deliberations of classical probabilists.

The examples that could be used to support this type of criticism were of several kinds. Some, such as the St. Petersburg paradox or the problem of smallpox, questioned the utility function associated with objective probability (geometrical, in the case of "heads or tails," frequentist in the case of inoculation), itself seldom discussed. Other cases, however—such as the verdicts of criminal juries based on votes, which involved an assessment of guilt, given indications and presumption but not complete proof—cast doubt even on the possibility of estimating the (subjective) probability of a "cause" (the guilt of the accused), knowing that certain "effects" had been observed (indications, or dubious testimony). This question of the *probability of causes* (or *inverse probability*) played an important role in the eighteenth century with Bayes and Laplace. It was then largely ignored in the nineteenth century, before being revived in the twentieth in the context of game theory, decision making, and the cognitive sciences.

The Bayesian Moment

The problem of justice, defined by the equality of expectations, dominated the works of the founders of probability theory; the *Logique* of Port Royal, for example, made extensive use of it. The frequentist point of view, implicit in the case of games of chance, led Jacques Bernoulli (1654–1705) to the first demonstration of the law of large numbers, published after his death in 1713: the frequency of appearance of a phenomenon with a given probability tends toward this probability when the number of attempts is multiplied (Meusnier, 1987). Then, in 1738, Abraham De Moivre (1667–1754) completed this demonstration by calculating the probability that this frequency would occur in as small an interval as was desired, after a sufficient number of drawings. In the process, he gave the first explanation of the future "normal law," basing his remarks on Stirling's asymptotic formula for $n!$ (factorial $n = 1$ times 2 times 3 times . . . times n).

The preceding results allowed expected effects to be deduced from a known cause (the composition of an urn); that is, the probabilities of observing different frequencies. But the contrary problem was often posed in all those cases when it was desirable to say something about (unknown) causes on the basis of observed events. First Bayes (1702–1761), then Laplace (1749–1827) set about answering this question, in terms that would continue to arouse debate.

The problem of "inverse probability," thus posed by Bayes and Laplace, would play (even today) a crucial role in the history of the use of probabilities and of statistics in the various sciences. In fact it found itself at the turning point between an objectivist point of view (science *unveils* things, or draws closer to a hidden reality) and a constructivist point of view (science *constructs* objects, models, and so forth using criteria and tools that lend it relative stability). The process of inferring things on the basis of recorded events could admit both these interpretations: revealed or constructed causes.

Bayesian formalizations aimed at estimating *reasons to believe,* given previous experiences, with a view to evaluating a concrete situation and making a decision. Bayes's text was published in 1764 under the title *An Essay toward Solving a Problem in the Doctrine of Chances.* Its opening sentence, which posed the opposite problem from that of Bernoulli and De Moivre, used words in a way that seems strange today (the "chance of a probability"). Such ambiguity, however, can help us to understand what it was that scholars were looking for:

> Given the number of times in which an unknown event has happened or failed: required the chance that the probability of its happening in a single trial lies somewhere between any two degrees of probability that can be named. (Bayes, 1764, reprinted in Pearson and Kendall, 1970)

The words "probability" and "chance" are then defined in terms of the theory of expectations; in other words, by a relationship of values to be estimated in order for a wager to be accurate.

> The probability of any event is the ratio between the value at which an expectation depending on the happening of the event ought to be computed, and the value of the thing expected upon its happening. By chance, I mean the same as probability. (Bayes, 1764)

Thus, the words *chance* and *probability* are given the same meaning—which scarcely clarifies the initial sentence. However, the reasoning that follows shows that the word "chance" is used like "reason to believe" with a view to a decision, whereas the word "probability" has an objective sense, like that of "the composition of the urn." It is thus a matter of "probability" that the composition (the relationship between the number of black and white balls) of this unknown urn may be within a given interval—that is, of a "probability of cause" which, just like the "probability" of a defendant's guilt, can only be interpreted in terms of the "degree of certainty" necessary for making a choice. The only way of giving a formalized meaning to this "probability of cause" would be to suppose that the urn itself was drawn from among a large number of urns of different composition. But that returns us to the question of the distribution of these compositions—in other words, to an "a priori probability." It was precisely on this point that Bayesian procedure would be most criticized.

In creating the idea of "conditional probability," this procedure introduces the irreversibility of time (A *then* B) into the formulation, and that is the cause of its ambivalence (Clero, 1986). Indeed, the reasoning can be constructed on the basis of this double equality:

$$P(A \text{ and } B) = P(A \text{ if } B) \times P(B) = P(B \text{ if } A) \times P(A)$$

whence

$$P(A \text{ if } B) = P(B \text{ if } A) \times \frac{P(A)}{P(B)}$$

Transposed to the problem of the probability of a cause H_i (among a series of n possible, mutually exclusive causes) for an event E, this can be written, using more modern notation:

$$P(H_i|E) = \frac{P(E \cap H_i)}{P(E)} = \frac{P(E|H_i).P(H_i)}{\sum_{i=1}^{n} P(E|H_i).P(H_i)}$$

This formula was clarified by Laplace in 1774 in a long sentence that is now rather difficult to understand: its difficulty can be compared to that, of another order, of the foregoing mathematical formula.

> If an event can be produced by a number n of different causes, then the probabilities of these causes given the event are to each other as the probabilities of the event given the causes, and the probability of the existence of each of these is equal to the probability of the event given that cause, divided by the sum of all the probabilities of the event given each of these causes. (Laplace, 1774, *Mémoire sur la probabilité des causes par les événements,* no. 2, in *Oeuvres complètes,* vol. 8, p. 28)

But the important point in Bayes's demonstration is that the symmetry of the double equality defining the conditional probabilities $P(A \text{ if } B)$ and $P(B \text{ if } A)$ did not exist for him, and that the two equalities are demonstrated separately and independently of each other (Stigler, 1986), through two distinct "propositions." These reasonings are based on the *increases* in expectations of profit, introduced by the occurrence of an initial event A. Each demonstration is the narration of a series of hypothetical events, and of their consequences in terms of profits. But these narrations can only reach a conclusion if one gives the a priori probabilities of the causes; that is, in this case, the $P(H_i)$ even before any partial knowledge. The hypotheses of equiprobability of such a priori causes have often been debated, with the help of examples showing that these "equal" probabilities can be chosen in different ways and are thus pure conventions.

The tension and fecundity of the Bayesian procedures comes from the ambiguity of the initial expression: the original double equality is formally symmetrical but logically asymmetrical, since time intervenes, and since the events are *known* whereas the causes are *inferred* from them. This ambiguity was moreover inherent in the equipment Bayes had imagined in support of his argument. In fact, the example of the urn was inadequate, for it was difficult to construct a series of events that involved *first* drawing urns, *then* drawing balls: the asymmetry was too strong. Bayes therefore proposed successively shooting two balls onto a square billiard table, in such a way that the probable densities of points where they came to rest on the green carpet would be uniform. Ball A was shot first, and then after it stopped a vertical straight line was drawn through its stopping point, dividing the square into two rectangles P and Q. Then ball B was shot, and the probability that it might stop in rectangle Q was studied. There was therefore a sequence of two events, and one could calculate linked and conditional probabilities, while maintaining both their symmetrical (geometrically) and asymmetrical (temporally) character.

The process of Bayesian inference can be read in the perspective—adopted in the present work—of *construction of classes of equivalence*, of *taxonomy*, and of *encoding*. Indeed, the idea of causality it postulated implied that similar causes can lead to similar consequences, and that future events can be connected to past observations in such a way as to circumscribe their uncertainty. It was thus opposed to an attitude of nominalistic skepticism very much in evidence in eighteenth-century philosophy (for example, Berkeley), which held that nothing could be assimilated into anything else, and no general representation was possible (Clero, 1986). In this respect it adhered to the philosophy of Hume, who described human knowledge as the product of an accumulation of "pencil strokes" creating "grooves in the brain," to which each additional experience lent "heightened vividness." This idea of addition linked to fresh knowledge is very present in Bayes's text.

In this respect it was also typical of *the sciences of clues* that Ginzburg

(1980) contrasts with *Galilean sciences.* Whereas these latter simultaneously treat a large mass of information by mathematical and statistical methods so as to infer general laws from them, the former proceed from singular characteristics and create histories, or else attribute cases to families of causes. That is the way of the historian, the policeman, and the doctor who suggests a diagnosis on the basis of symptoms. Thus the encoding of "causes of death," by processing the medical statements in death certificates according to the International Classification of Diseases (ICD), can be described (Fagot-Largeault, 1989) as a Bayesian procedure, in that it involves both singular symptoms (events) and the degree of diffusion, already known, of a disease (a priori probability). In this instance, which offers a refined analysis of the genesis and evolution of the ICD (its terminology), the encoding of causes of death appears as a convention. Indeed, a cause deemed *interesting to describe in statistical terms* must, among the succession of events preceding death, be neither too anterior (for then its effect would be indirect, weak, and diluted), nor too closely linked to death (with which it would then be synonymous: the heart no longer beating). It must be a category in between the two preceding ones, whose effect is to significantly increase the probability of death without, however, making it certain. It is assumed that one can *act on this cause,* either by prevention, or by the appropriate therapeutic measures, so as to reduce this probability. This analysis of the medical diagnosis and encoding of a death certificate (taken from the dissertation of Anne Fagot-Largeault) is doubly typical of Bayesian procedure. Typical on the one hand are the conventions defining causality by a categorizing of events—knowledge of which implies variations of probabilities—and, on the other hand, the treatment of the *moment of encoding,* seen as a *decision* integrating both singular events and a priori probabilities.

The *Bayesian* moment was thus an essential episode in the history of probabilities (Bayes and Laplace), in that it was the origin of an idea of *inverse probability,* or of *probability of causes.* This meant nothing in a purely axiomatic theory of probability (like that of Kolmogorov), but assumed its full significance in the analysis of many forms of decision making and creating classes of equivalence, which constitutes the key stage in statistical activity. This moment of encoding and its particular constraints are forgotten when, subsequently, the statistical tables are treated and interpreted—just as Bayesianism has long been rejected from statistical thinking.

The Middle Road: Averages and Least Squares

Among the techniques which, nowadays, help build and stabilize the social world, statistics play a double role. On the one hand, they stabilize *objects*, by determining their equivalences through standardized definitions. This enables them to be measured, while specifying through probabilistic language the degree of confidence that can be attached to these measurements. On the other hand, they provide forms for describing the *relationship* between objects thus constructed, and for testing the consistency of these links. These two aspects—the construction of objects and the analysis of their relationships—seem closely connected. They nonetheless derive from two quite distinct traditions, which only converged at the beginning of the twentieth century. Sometimes, the same formalism could be used in regard to quite different questions, but it took a full century for this mental equipment to be transferred from one field to another, and a strong retranslation was required.

An example is provided by the method of adjustment known as *least squares* (Armatte, 1991). It had been formulated by Legendre in 1805, in answer to a question asked throughout the eighteenth century by astronomers and geodesists: how could one combine observations made in different conditions to obtain the best possible estimates of *several* astronomical or terrestrial quantities, themselves connected by a linear relationship? These distances had been measured, with imperfect instruments, in different circumstances, for example, in different historical periods or at several points on the earth's surface. How could one best use these copious measurements, given that they never perfectly confirmed the theoretical geometric relationship, but allowed a slight *deviation* (or *error*, or *residue*) to remain just where a value of zero should have occurred? In other words, the two or three unknown quantities appeared as solutions to a system that had too many equations (as many equations as it had observation points). One would thus have to find the optimal combination of these equations in order to reach an estimate of the desired quantities. This was the problem that Legendre solved in 1805, by a method that consisted in *minimizing the sum of the squares* of these deviations (Stigler, 1986).

The aim was therefore to provide the best and most precise foundation possible for the measurement of objects, through the best combination of the various observations made of the same quantity. In contrast, the problem posed and resolved during the 1890s by the English eugenicists Galton and Pearson—the inventors of *regression* and *correlation*—was completely different: how was it possible to describe the relationships and partial links between objects that were neither independent nor totally dependent on each other, as in the case of heredity? Adjusting one variable against another by a model of linear regression nonetheless led to a system of equations and mode of resolution formally analogous to those of Legendre. But the meanings of this mathematical construction were so different in each of the two cases that the transfer of Legendre's formalism, completed in 1810 by means of the probabilistic interpretation due to Gauss and Laplace, did not really take place before the 1930s.

The synthesis effected in around 1810 by Laplace and Gauss itself resulted from the union of two quite separate traditions. On the one hand, astronomers and physicists were used to making empirical combinations of imperfect observations—for example, through calculating averages (the "middle road")—in order to estimate sizes in nature as accurately as possible. On the other hand, probabilistic mathematicians and philosophers had worked on the problem of the *degree of certainty* attributable to some knowledge or belief. The mathematicians and philosophers were thus led to question the astronomers' and physicists' use of the expression "as accurately as possible": how could one estimate the degree of confidence that an estimate deserved? Before Gauss and Laplace, no one had provided an answer to this question.

The first of these traditions, the one concerned with measuring astronomical or terrestrial quantities, already had a long history (Stigler, 1986). This problem could have significant economic or military implications. Thus, throughout the eighteenth century, the concern with perfecting techniques that enabled ships to calculate their position (latitude and longitude) gave rise to numerous forms of research. Since 1700, the calculation of latitude (based on the height of fixed stars) had been relatively easy. That of longitude, however, caused considerable difficulty. In 1714 in England, a commission was created to study this question and to subsidize research contributing to its solution (more than £100,000 were spent to this end between 1714 and 1815). Two techniques were then developed: accurate time-keeping, enabling Greenwich time to be observed on board ships; and the establishment of tables giving detailed descriptions of the positions of the moon.

The problem in this second case came from the fact that the moon does not always present the same face to the earth, and so slight oscillations in its rotation (known as "librations") greatly complicated the task of calculating its position. In 1750 the German astronomer Tobias Mayer published a solution containing an ingenious method of combining observations. Calculations had led him to observe accurately the position of a certain crater on the moon at different moments, and this observation led to the measurement of three distinct astronomical quantities, connected by an equation born of spherical trigonometry. Since he had made these observations twenty-seven times, it was necessary to solve an overdetermined system of twenty-seven equations for three unknowns.

Having no rule for minimizing the deviations between the calculated values and estimates by a possible adjustment, Mayer rearranged his twenty-seven equations into three groups of nine equations. Then, by adding each of the three groups separately, he finally obtained a system of three equations for three unknowns, which provided him with the estimates he was seeking. The pertinence of Mayer's method came from the judicious choice of the three subclusters of points thus replaced by their centers of gravity, in such a way as to preserve the greatest possible amount of the information initially provided by the twenty-seven observations. The fact that Mayer himself had taken the measurements and was very familiar with them gave him the audacity needed to rearrange the equations, and the intuition necessary to do so cleverly. But the characteristic of such an empirical solution was that it was not based on any general criterion. This made it hard to transpose: it was an ad hoc solution, typical of an artisan.

A general criterion for optimizing an adjustment was suggested later, in 1755, by Roger Boscovich, in regard to another problem that troubled many eighteenth-century scholars: that of the "figure of the earth." It was known that the earth was not a perfect sphere, but was slightly flattened at the poles and bulged at the equator (some scholars held the opposite theory). The resolution of this problem involved measuring the terrestrial arcs at widely separated latitudes. This was done in Paris, Rome, Quito, Lapland, and at the cape of Good Hope. In this case, it was necessary to solve a system of five equations for two unknowns.

Perhaps because the number of measurements involved was smaller than for Mayer, Boscovich reasoned in quite a different way. He invented a geometrical technique for minimizing the *sum of absolute values* of the residues—that is, of the deviations between observed values and adjusted ones. But if this criterion was general, it proved very hard to manipulate, and the "geometrical" solution was only possible by virtue of the small number of observations and of unknown sizes (Stigler, 1986, pp. 47–48).

Laplace tried to treat this sum of absolute values mathematically, but was forced to desist in view of the complexity of the calculations involved.

The solution of minimizing the *sum of the deviations' squares* seems to have been used first by Gauss as early as 1795 (so, at least, he claims), although he did not formulate it explicitly. It was conceived, formulated, and published by Legendre in 1805, independently of Gauss. This led to a lively quarrel over which of them was first (Plackett, 1972).[2] Gauss declared he had used this criterion in 1795 but—as he would later say, during the controversy—had found it banal, and had not deemed it useful either to publish it or give it a name: the *method of least squares* (as it would be known in posterity). For him, this was only a means of calculation, the essential thing being the results of his research. But Legendre in 1805, Gauss himself in 1809, and Laplace in 1810 all made use of the very special properties of this method, especially, for Gauss and Laplace, those concerning the links that could finally be established with the laws of probability, and with "Gaussian" law (later known as "normal law") in particular.

We must now retrace our steps in order briefly to follow the other tradition leading to the Gauss-Laplace synthesis: that of the philosophers working on the degree of certainty of knowledge, based on probabilistic schemas. To formulate the law of probability for a statistical estimate, one must first agree on certain laws for the mathematical errors of each observation, and then combine them mathematically, from this deducing a law for the statistics being calculated. Various forms for the distribution of elementary errors were proposed. Simpson (1757) tried a linear form leading to an isosceles triangle: $-a|x|+b$. Laplace proposed first, in 1774, an exponential $[(m/2)(e^{-m|x|})]$, and then, in 1777, a logarithm $[1/2a \log(a/|x|)]$. While working on this question of the theoretical distribution of errors in view of an empirical distribution, Laplace was led to pose the problem of *inverse probability*, or *probability of causes*, from a point of view quite close to that of Bayes.

In 1810 Gauss and Laplace effected a synthesis between the two perspectives, empirical and probabilistic. Henceforth the Gaussian formula e^{-x^2} became almost completely predominant, by virtue of its mathematical properties and the fact that it tallied with the observations. The question of the distribution of elementary errors had, moreover, lost some of its importance with Laplace's demonstration in 1810 of the *central limit theorem*. This showed that, even if the distribution of the probability of errors did not follow a normal law, the distribution of their *mean* generally inclined toward such a law, when the number of observations increased indefinitely. This gave the Gaussian form a decided advantage that was to support the whole of nineteenth-century statistics, beginning with Quetelet and his *average man*.

Thus the results of Gauss and Laplace established an extremely solid synthesis that sustained the experimental sciences of the nineteenth century. It united, on the one hand, the empirical works leading to the method of least squares and, on the other, the probabilistic formulations culminating in the normal law and its numerous mathematical properties. However, for reasons I shall presently explain, a full century elapsed before these techniques were taken up and formalized in the social sciences, especially in economics. Among the possible reasons for explaining this delay is the theory that as yet there were no procedures for recording data, which were connected with the creation of the modern states and of the spaces of equivalence—institutional, legal, statutory, or customary—that these states involved.

Adjusted Measurements Permitting Agreement

In the seventeenth century, if a judge was asked to settle a conflict involving the occurrence of subsequent, and therefore unknown events, his decision would require the litigating parties to agree that their expectations were equivalent. In this case probability—the ratio between expectation and wager—appeared as a measurement on which agreement could be based. In the eighteenth century, geometricians, astronomers, and physicists endeavored to summarize their multitudinous observations of nature into measurements that could be taken up by other scholars. To this end, they created procedures (calculation of averages, adjustment by least squares) that had properties of optimality allowing agreement among scholars. Then, in the early nineteenth century, it was the prefects who, traveling cane in hand along the byways of their departments while noting their particular features, contributed, through these still clumsy observations, to the establishment of a new administration in the future. The task of this administration was to provide the entire nation with measurements allowing enough agreement to provide a firm basis for public debate and decision making.

In these four cases, involving judges, astronomers, geometricians, and prefects, the word *measure* may appear to be used in different senses. Judges made measured decisions; astronomers and geometricians optimized their measurements; and prefects enforced the measures proposed by their ministers. But the connection among these various meanings is not fortuitous, once these "measures" are seen as being intended to bring about *agreement,* among litigants, scholars, or citizens. Thus these seemingly heteroclitic characters and situations—presented here as ancestors in this genealogical tree of the modern procedures of statistical objectification—all have in common that they link the construction of objectivity to that of "technologies of intersubjectivity" (Hacking, 1991); in other words, *formulas of agreement.*

Around 1820, however, there were as yet no unified statistical procedures to give these formulas of agreement the solidity that makes them omnipresent today. Their later history can be narrated as a series of marriages between traditions that were a priori distinct. The first such example has been provided by the Laplace-Gauss synthesis. This combined the probabilistic formulation resulting from the binomial laws of Bernoulli and De Moivre with Legendre's adjustment by means of least squares. Other marriages would follow. During the 1830s, Quetelet compared the regularities observed by the bureaus of administrative statistics with those presented by astronomical measurements, and from them he deduced statistical laws that were autonomous in relation to individuals. Between 1880 and 1900 Galton and Pearson brought together the questions on heredity born of Darwinian theory; the normal distributions of human characteristics, born of Quetelet; and the techniques of adjustment resulting from the theory of errors in measurement.[3]

3

Averages and the Realism of Aggregates

How can single units be made out of a multiple? How can this unity be decomposed to make a new diversity? And to what end? These three different but inseparable questions recur throughout the slow development of the statistical tools used in objectifying the social world. Use of the verb "to make," in their formulation, is not intended to imply that this process of producing reality is artificial (and therefore false); rather, it is meant to recall the continuity between the two aspects—cognitive and active—of the analysis. This close overlap, characteristic of the probabilistic and statistical mode of knowledge, may explain why these techniques are seldom described with any subtlety by the philosophy of sciences. The apparent complexity of the field—the technical nature of which alone could justify this relative silence—is itself the result of this particular situation, in which the worlds of action and knowledge are conjoined.

The history of these techniques comprises a long series of intellectual distillations, purifications intended to produce key tools, freed of the various contingencies that presided over their birth. The animated debates that took place throughout the nineteenth century on the idea of the *average,* its status and interpretation, are instructive from this point of view. Beyond the seeming triviality of the mode of calculation of this basic tool of statistics, far more was at stake: these debates concerned the nature of the new object resulting from such calculation, and also the possibility of endowing this object with an autonomous existence in relation to the individual elements. The debate surrounding the idea of the *average man* was especially animated. It introduced an entirely new apparatus, including the "binomial law" of the distribution of errors in measurement, to deal with a very old philosophical question: that of the realism of aggregates of things or people.

In this chapter I shall describe some of the moments in these debates concerning the real or nominal nature of such combinations, and also the tools—especially the probabilistic ones—called on for this purpose. Quetelet in effect combined three ways of conceiving the unity of diverse phenomena, resulting from very different perspectives. In medieval philosophy, with William of Occam and the opposition between nominalist and realist philosophies, the question of the extent to which a multiple group could be designated by a single name had already arisen. This question is essential to the perspective adopted in this book, of studying the genesis of the conventions of equivalence and of statistical encoding. Seventeenth- and eighteenth-century engineers and economists, such as Vauban, had already calculated averages, with a view both to estimating an existing object and to creating new entities. Then through their questions on errors in measurement and the probability of causes inferred from their effects, the eighteenth-century probabilists mentioned in Chapter 2 had provided powerful tools for creating equivalence, such as Bernoulli's law of large numbers and the Gauss-Laplace synthesis leading to the central limit theorem.

In basing his principles on these diverse traditions and on the ever more copious statistical records produced by institutions, the genesis of which I described in Chapter 1, Quetelet created a new language that enabled new objects to be presented. These objects had to do with *society* and its stability, and no longer concerned *individuals* and the rationality of their decisions, as probabilism did until Laplace and Poisson. The conventions of homogeneity implied by this use of the law of large numbers were discussed by Poisson, Bienaymé, Cournot, and Lexis, and this debate resulted in other instruments for testing the realism of macrosocial objects, and for conducting increasingly refined analyses. Last, Durkheimian sociology adopted these tools to establish the idea of the social group exterior to individuals, although it later rejected them in the name of a concept of totality no longer indebted to calculations and averages. Thus the recurrent opposition between individualistic and holistic perspectives in modern social sciences can be seen in the debates—very lively during the second half of the nineteenth century—over statistics and averages, these latter having been used and criticized in turn by both of the opposing camps.

Nominalism, Realism, and Statistical Magic

Why return to medieval controversies between realists and nominalists, at the risk of amalgamating entirely different intellectual contexts, before discussing the nineteenth-century debates over averages? As it happens, the medieval period is enlightening, both because of the mental systems it employed (the logical anteriority of the whole, or of the individuals composing it) and because these same systems were directly invoked in a quarrel between the papacy and the Franciscan order concerning the ownership of the Franciscans' possessions (Villey, 1975).

The debates on the relationships among universal ideas, words with a general meaning, and individualized things are, of course, as old as classical philosophy. Classical philosophy distinguishes three separate references in the word *homo* (man): the two syllables that make up the actual word; a particular man; and mankind in general (in other words, the actual signifier, and two levels of things signified, singular or general). The fourteenth century witnessed a period of controversy between realists and nominalists (represented metaphorically in Umberto Eco's novel *The Name of the Rose*), with the realists maintaining that only ideas and general concepts had real existence (a view that may seem quite contrary to our present idea of realism [Schumpeter, 1983]). The nominalists, whose principal theoretician was William of Occam, held that there were only singular individuals: words designating a group of individuals or a concept were useful conventions, but did not designate a reality, and were therefore to be mistrusted. Occam thus maintained that "one should not needlessly increase the number of abstract entities," a principle of economy often known as "Occam's razor" (Largeault, 1971).

The nominalist position was important in that it anticipated the decline of the old scholasticism, and opened the way to "subjective law" and to the individualistic philosophers and empiricists of the subsequent centuries. Of course, the meaning of this nominalism would evolve, by emphasizing the abstractive power inherent in the naming of things. Thus, in the view of Paul Vignaux (1991), "The nominalism of modern empiricists has laid emphasis on the active role of the word which, in not mentioning certain characteristics of the thing, is a factor in an abstracting mental attitude." This formulation describes precisely those processes of criterialization and statistical encoding which, in omitting "certain characteristics of the thing," allow greater diversity in abstraction, thus permitting a multiplicity of points of view.

For its part, the realist position also developed from its ontological and idealistic medieval version (realism of ideas) into more materialistic and empirical versions. This development required tools to connect things and make them into realities of a superior level: the terminologies of eighteenth-century naturalists (Linné) and nineteenth-century statistical averages (Quetelet). This new kind of well-tooled realism thus maintained with individualistic nominalism—which was indispensable for recording and encoding singular elements—a tension that was typical of the statistical mode of knowledge. This latter mode, the aim of which was to constitute and support realities of a superior level thus capable of circulating as synthetic substitutes for multiple things (*the* price index, for increases in products; *the* unemployment rate, for unemployed persons) had necessarily to be anchored in nominalist and individualist conventions. This tension is inherent in the magical transmutation of statistical work: this transfer from one level of reality to another also involves a transfer from one language to another (from unemployed persons to unemployment). The status of reality now granted these two levels, which can exist in partly autonomous ways, shows the ground that has been covered since Occam.

This multiplicity of possible registers of reality is today justified by the fact that each of them is integrated into a construction, an ensemble of things. The various mechanisms have autonomous internal consistency (at least in part). "Statistics" (in the sense of the *summary* of a large number of records) often play an important part in establishing this consistency. These complex constructions are cognitive and active at the same time: *a* national unemployment rate was only calculated and published after a national policy to fight unemployment in general was conceived and put into effect. Prior to this, unemployment relief was administered locally (Salais, Baverez, and Reynaud, 1986).

It is tempting, at this stage, to mention the political and philosophical controversy in regard to which Occam stated his nominalist position so decisively (Villey, 1975). What is surprising is the apparently paradoxical nature of the situation, and therefore of the argument involved. The problem originated in the vow of poverty taken in the thirteenth century by St. Francis of Assisi, which was laid down as a rule for the Franciscan order. The Franciscans, however, were so successful that they soon found themselves in charge of numerous monasteries and rich agricultural lands. To allow them to respect their vow of poverty—at least to the letter—the pope had agreed to assume ownership of these possessions, while allowing the monks to make use of them. But in the fourteenth century, this subtle distinction came under harsh criticism. Weary of the administrative burden involved, the pope decided to return the properties to the Franciscans. This would of course have enriched them, but would also have fueled the criticisms uttered even within the order by opponents demanding a return to the poverty implied in St. Francis's initial vow. It was in this complicated context that Occam intervened, upholding the Franciscans' position in regard to the pope. He argued that it was impossible to return these possessions to the order as a whole, since the Franciscan Order was only a *name* designating *individual* Franciscans.

Occam thus denied the existence of collectivities as distinct from individual persons: a question with a fertile future. To the logical individualism of nominalism a moral individualism was therefore added, itself linked to a concept of the freedom of the individual, alone before his Creator (Dumont, 1983). I shall proceed no further in this analysis, which refers to subtle theological and judicial constructs, the architecture and language of which are now largely foreign to us: for this reason, it is rather unwise to make too direct a comparison between seemingly close themes (the realist-nominalist approach), which are five centuries apart. The preceding short narrative is intended simply to show that conceptual schemas, introduced at a given moment into a far larger edifice, can be transmitted by transforming themselves—sometimes even into their opposites—and be integrated anew into other, radically different constructions.

The Whole and Its Semblances

The possibility of manipulating macrosocial objects based on statistical calculations without distorting those objects today enables us to circulate effortlessly among several levels of reality, whose modes of construction are nonetheless very different: this is the statistical magic described above. From this point of view, the connection between two types of calculations of averages, formally identical but logically quite distinct, was a key moment. The connection involved, on the one hand, the best possible approximation of a quantity in nature, based on different measurements resulting from imperfect observations of a single object, with, on the other, the making of a new reality affecting different objects joined for that occasion. This connection, made famous by Quetelet, between the permanence of an object that has been observed several times, in the former case, and the existence of some common element between different objects, lent an entirely new solidity to the link then established between these objects. But this tool was not immediately adopted, and fueled bitter controversies throughout the nineteenth century.

The importance of the step that the probabilistic formulation of the law of large numbers enabled Quetelet to take becomes apparent when we examine, as did Jean-Claude Perrot (1992), the terminology Vauban used around 1700 in regard to various calculations he made with a view to justifying a new national tax, the royal tithe. To this end the minister needed various estimates of the surface area of the kingdom, agricultural yields, and fiscal burdens. In some cases, he had several estimates of an unknown size (the total surface area of France), from which he drew a *proportional average.* But in other cases he used information—concerning agricultural produce, for example—relating to different parishes or to different years. He then performed a calculation analogous to the preceding one. He did not, however, term the result an "average," but a *common value:* a common square league, a common year.

This way of proceeding had long existed in intuitive practices, which typically compensate for deviant values with opposite tendencies. Traces of this appear in several expressions in common speech, some of which were used in the eighteenth century and have now disappeared, while others have survived: "one carrying the other," "the strong carrying the weak" (used by Vauban), "year in, year out," "one inside another," and "*grosso modo.*" The process of addition caused local singularities to disappear and caused a new object of a more general order to appear, eliminating the nonessential contingencies. This leap from one level to another is reflected in expressions such as "when all is said and done" or "all things considered."

Analyzing (thanks mainly to the *Dictionnaire* of the Académie) the eighteenth-century connotations of the terms used by Vauban, Perrot has shown different forms of logic at work in each of these two calculations. An *average* indicated "what was between two extremes." The word *proportional* applied to a quantity "in relation to other quantities of the *same kind.*" For its part the word *kind* was associated precisely with identity, with what caused permanence. The calculation of an average thus implied that the quantities were of the same kind, in this narrow sense. On the other hand, the term "common" referred to something ordinary, well-worn, and universal, or to "something amended during a continual series of occurrences (the common year)." Commenting on the expression "one carrying the other," the Académie introduced a formula scarcely imaginable in Occamian oppositions: a result was obtained by "using one to compensate for the other, and forming a *kind of whole*"—an expression taken up by Vauban to study the yields of seeds. This "kind of whole" designated a vague area between the firmly established object and the diversity of incommensurate objects. Commensurability, and the possibility of calculating a "common value," allowed the concept of a "kind of whole." But it was still impossible to demonstrate the stability of this "common value," and also the similitude of the distribution of errors in both of its instances, and of the "proportional average," to produce a consistent whole: the average man.

These questions completely renewed the old problem of the existence of a general entity logically superior to the elements that constituted it. This embryonic whole—still, for a while longer, the semblance of a whole—appeared when Vauban tried to combine previously disparate elements in a single calculation, the better to conceive and describe the kingdom's productive capacities, and to estimate what a new tax based on them could provide. The new means of creating equivalences, linked to properties characteristic of addition, was proposed at the same time as a new kind of state was being created, involving a radical rethinking of the origin and circuitry of its fiscal resources. But the conceptual tool for accomplishing this had not yet been formalized. It could not therefore circulate and be reused in different contexts without some further effort. One of the reasons why this formalization was still rather unlikely was that there was no centralized system for collecting records and making elementary calculations; in other words, there was no national statistical apparatus, even in an embryonic form.

Quetelet and the "Average Man"

The existence of such a system for accumulating measurements and individual encodings was in fact needed if the connection between the two calculations offered by Vauban was to be made, since this connection involved having recourse, in two different ways, to a large number of observations. On the one hand, the *regularity* of the annual rates of births, deaths, marriages, crimes, or suicides in a given country, opposed to the contingent and random nature of each of these occurrences, suggested that these additions were endowed with properties of consistency quite different from those of the occurrences themselves. On the other hand, the striking resemblance between the *forms of distribution* of large numbers of measurements—whether these were repeated measurements of the same object, or one measurement of several different objects (for example, a group of conscripts)—confirmed the idea that these two processes were of the same nature, if one assumed the existence, beyond these individual contingent cases, of an *average man,* of which these cases were imperfect copies. Thus each of these two ways of relying on statistical records and their totals in order to create a new being involved the centralized collection of a large number of cases. The task of reconciling and orchestrating these two different ideas—and also of organizing the censuses and the national and international statistical systems needed to produce these figures—was all accomplished by one man, Adolphe Quetelet (1796–1874), a Belgian astronomer who was the one-man band of nineteenth-century statistics.

The first of these ideas, that of the relative regularity of the rates of births, marriages, and deaths, had already been expressed in the eighteenth century. At that time it was interpreted as a manifestation of providence, or of a divine order directing society as a whole from on high, far above volatile and unpredictable individuals. This was how Arbuthnot (1667–1735)—doctor to the queen of England and translator of the first treatise on probability, written by Huygens in 1657—had interpreted the fact that the birth rate of baby boys was always slightly higher than that of girls. He was thus the first to set social statistics against a probabilistic model, that of "heads or tails." If the sex of every child born was the result of some such chance drawing, there would be one chance in two that for any given year more boys than girls would be born. But since the higher birth rate of boys had been observed for eighty-two years in succession, this could only have occurred with a probability of $(1/2)^{82}$, or 1/4, 8 times 10^{24}—as good as nothing. However, it also appeared that during childhood and youth, the death rate of boys was higher than that of girls. It was therefore tempting to suppose that divine providence, anticipating this higher male mortality, caused more boys than girls to be born, so that a proportion of girls did not later find themselves doomed to celibacy.

The Prussian pastor Sussmilch (1707–1767) had for his part assembled (in a perspective closer to that of the English political arithmeticians than of his German "statistician" compatriots: see Chapter 1) a large body of demographic information. His major work, *The Divine Order*, which had a great impact throughout Europe, interpreted these regularities as the manifestation of an order that was exterior and superior to mankind. For Arbuthnot or Sussmilch, therefore, these statistical results, though based on individual records, were still read in terms of a universal order—that is, of "realism" in the medieval sense previously described. Similarly, first Quetelet then Durkheim would give a "holistic" reading of these regularities: Sussmilch's divine order, Quetelet's average man, and Durkheim's society were realities *sui generis,* different from individuals, and calling for specific methods of analysis.

But even if it was interpreted as a sign of divine order, a construct centered on macrosocial realities left open another question: how was the diversity of individual physical traits or moral behavior to be reconciled with the regularities noted in numerous populations, especially in regard to average heights, weddings, crimes, and suicides? And at a deeper level, wasn't the freedom that people claimed negated by a kind of statistical determinism reflected in these results? This was the question for which Quetelet provided a fresh answer.

First he studied the diversity of physical characteristics, which could easily be measured, seeking unity beneath this diversity. By making a graph of the frequency distribution of these heights divided according to sections (the histogram), he was able to show a form which, despite discontinuities resulting from the fact that the sections of heights were necessarily discrete, nonetheless resembled the distribution of a series of measurements tainted by errors, for a given quantity. Habitually known later by the term "bell curve" or "Gaussian curve," this form then presented itself as a "law of errors" or "law of possibility." Its analytical expression, ke^{-x^2}, was formulated by De Moivre and Gauss as the limit of a binomial law for drawing balls from an urn of given composition, when the number of drawings tended toward infinity. This is why Quetelet used it under the name of a "binomial law," or "possibility curve" (the expression "normal law" only appeared in 1894, penned by Karl Pearson [Porter, 1986]). It provided a good approximation of the distribution of imperfect measurements of a distance existing independently of these measurements (for example, the vertical ascent of a star), or of a distribution of the points of impact of shots fired at a bull's-eye. In these various cases, one can show that this form results from the composition of a large number of small errors that are independent of one another.

From this first "fitting" to a "model" in the contemporary sense, deduced from the resemblance between the distribution of the heights of conscripts and that of the imperfect measurements of a single size, Quetelet inferred that these deviations in regard to the central tendency were of the same kind: imperfections in the effective realization of a "model" in the primary sense of the word. To compare the two forms, he used the metaphorical story of the Prussian king who, filled with admiration for a statue of a gladiator, ordered a thousand copies to be made by a thousand sculptors in his kingdom. Their sculptures were not perfect, and their works were distributed with deviations from the model, in much the same way that particular individuals were distributed around the average man. The Creator was then equivalent to the king of Prussia, and individuals were imperfect realizations of a perfect model:

> Things happen as if the creative cause of man, having shaped the model of the human species, had then like a jealous artist broken his model, leaving inferior artists the task of reproducing it. (Quetelet, quoted by Adolphe Bertillon, 1876)

Thus shifting, thanks to the normal form of distribution, the distinction between what Vauban termed "proportional average" and "common value," Quetelet distinguished *three* kinds of mean values, the presentation of which would provide the core of numerous statistical debates until the beginning of the twentieth century. When Adolphe Bertillon (1876) presented these distinctions thirty years after Quetelet, he designated them quite clearly. The *objective mean* corresponded to a real object, subjected to a certain number of measurements. The *subjective mean* resulted from the calculation of a central tendency, in the case in which the distribution presented a form adjustable to that of the "binomial law" (the case of heights). Only these two cases really deserved the term "mean value." The third case presented itself as if the distribution did not have this "normal" form at all. Bertillon termed it an *arithmetical mean* to emphasize the fact that it was pure fiction (the example he gave was that of the height of the houses in a street, without any such distribution being offered in support), whereas in the second case, the normal form of a frequency histogram attests to the existence of an underlying perfection in the multiplicity of cases, justifying the calculation of a "true" mean value.

After his study of heights, Quetelet continued his measurements of other physical attributes: arms and legs, skulls, and weights, for which he still observed distributions in accordance with binomial law. From this he inferred the existence of an ideal average man, in whom all average characteristics were combined and who constituted the Creator's goal—perfection. This perfection therefore resulted from a first, decisive comparison between several measurements of a single object, and the measurement of several objects.

The second crucial connection in Quetelet's construct allowed him to associate *moral* behavior with the *physical* attributes previously studied. Indeed, both moral and physical attributes present on average, as has been seen, an important regularity, if one only considers the masses. The average heights and forms of the human body show little variety, and that is explained by the law of large numbers, if the diversity of individual cases is interpreted in terms of realizations deviating at random from a model, according to a sum of numerous, small, independent causes. If therefore the heights of individual men are rather far apart, the average heights of two or more groups of men are actually quite close, if only these groups have been composed at random. Now the numbers of marriages, crimes, or suicides present the same kind of stability, despite the highly individual and free nature of each of these three acts. The connection between these two types of regularities, having to do with masses rather than singular cases, some with physical attributes and others with moral characteristics, allows us to clinch the argument: decisions of the *moral* type are manifestations of *tendencies* distributed at random around average types. Combined, they constitute the moral attributes of the average man—an ideal intended by the Creator, and a symbol of perfection.

Constant Cause and Free Will

Quetelet shifted the boundary between the two formally identical—though quite different in interpretation—calculations that Vauban termed *average* and *common value,* respectively. Henceforth there were both real "wholes," for which the calculation of a mean value was fully justified and, in contrast, ensembles of objects that did not follow a Gaussian distribution and did not even form "kinds of wholes," as scholars had put it a century before. The assimilation between the two mean values, known as *objective* and *subjective,* was assured thanks to the idea of *constant cause,* which allowed a continual passage from one to the other. Thus the successive measurements of a real object were a series of operations comprising an element of unpredictable chance *(accidental causes);* but this series was guided by an attempt to fit the object, which constituted the constant cause. In deviating from this model involving a "real object," the points of impact of a series of shots directed at a target involved a share of accidental causes disturbing a constant cause, resulting from the marksman's attempt to hit the bull's-eye. Finally, in the distribution of the attributes of a human population, this constant cause could result from the Creator's intention of reproducing a perfect model (Quetelet's case); from the effect of a material, climatic, or geographical environment (Hippocratic model of the eighteenth century); from natural selection (Darwin); or from a social and cultural milieu (twentieth-century sociology). A normal form distribution of measurements spread around a modal value thus allowed the existence of a constant cause to be inferred, or of an ensemble of causes, the combination of which produced constant effects during the series of these observations. Thanks to this form a class of equivalence could be constructed joining these events, interconnected by the fact that they were partly determined by a cause common to them all, although part of these determinations derived from accidental causes different for each of them. With this construction, the difference between "objective" and "subjective" means disappeared completely. In both cases, the idea of a "common *cause*" for diverse events provided an exteriority in relation to these events: a "real object" independent of its contingent manifestations. In probabilistic terms, this object was the composition of the urn involved in random drawings.

But Quetelet and his successors were so fascinated by the novelty of the macrosocial construction induced by this model that they did not think to reason—as Bayes, Laplace, and Poisson had done—in terms of the probability of causes; that is, of tracing effects back to an assessment of the degree of certainty of their causes. The "frequentist" perspective relies on an *objective* concept of probabilities, linked to things and to the variable combinations between constant causes and accidental causes, whereas the "epistemic" viewpoint of the eighteenth-century probabilists was based on *subjective* probabilities, linked to the mind and to the degree of belief it can confer upon a cause or an event. The speculations of the philosophers of the Enlightenment aimed at making explicit the criteria of rationality for the choices and decisions of an informed person—the embodiment of a universal human nature, based on reason. In the nineteenth century, in contrast, the French Revolution and its unpredictable convulsions substituted questions about society and its opaqueness in place of questions concerning rational people and their judicious choices. Not only was this society henceforth seen as a mysterious whole, but it was seemingly viewed as if from the outside. The objectifications and macrosocial realities shown by Quetelet responded to this kind of anxiety, which was typical of the nascent social sciences of the nineteenth century. The reasonable, prudent human nature embodied by the enlightened eighteenth-century scholar was succeeded by the normal man, the average of a large number of different men, all of whom shared in a sum that exceeded them. Two different ideas of probability were connected with these two perspectives.

A clear line separated Condorcet, Laplace, and Poisson from Quetelet, the Bertillons, and the nineteenth-century "moral statisticians." They did not set themselves the same questions and did not have much to say to each other—as is shown by the tepid communications between Poisson, Cournot, and Quetelet, who were nonetheless partly contemporaries.

The apparition of this new entity, *society*, objectified and seen from the outside, endowed with autonomous laws in relation to individuals, characterizes the thought of all the founding fathers of sociology, a science taking shape precisely at this time. Comte, Marx, Le Play, Tocqueville, Durkheim: despite their differences (Nisbet, 1984), all were confronted with the disorders and the breakdown of the old social fabric brought about by the political upheavals in France and the industrial revolution in England. How were they to rethink the social bonds destroyed by individualism, be it the individualism of the market economy or of democracy? In his work on the "sociological tradition" Nisbet develops this seemingly paradoxical idea by uniting all these very different authors behind a "constant cause": the concern with responding to the worries and crises in society resulting from these two revolutions, the French political revolution and the English economic revolution.

In sketching this gallery of portraits, Nisbet does not mention Quetelet, whose properly so-called sociological thinking can seem rather simplistic compared with that of the others. However, his thinking is situated in a very comparable context of political anxiety. Endowed with every virtue, Quetelet's "average man" was presented as a kind of prudent centrist, who avoided every conceivable form of excess—for perfection lies in moderation. But beyond this naiveté, already sensed by some of his contemporaries, Quetelet's mode of reasoning—which allowed the birth and instrumentation of new entities then capable of circulating autonomously in relation to their elements of origin—would enjoy a posterity at least as important as that of more famous social thinkers. Although his numerous activities were much celebrated at the time (he was often dubbed the "famous Quetelet"), his name has partly disappeared, in accordance with a process not unlike that of statistics, lumping together individuals and the conditions of its birth. The key tool once constituted by the average has now become so trivial that the fact, if not of inventing it, but of having greatly changed its use no longer seems significant; in particular, the intellectual feat that led to the fusion of objective and subjective averages behind the idea of "constant cause" no longer has anything surprising about it. But the reflection on the "consistency" of statistical objects would remain important in other forms, in relation to the question of the identification and description of these objects.

The notoriety Quetelet enjoyed in the nineteenth century contrasts with his relative oblivion during the twentieth century. This, too is partly explained by the fact that, from the standpoint of the history of the mathematical techniques of statistics, his contribution seems far weaker than those made by Gauss and Laplace before him, or by Pearson and Fisher after him. But the celebrity he enjoyed in his own time came from the fact that he had managed to create a vast international sociopolitical network, by connecting in a new way worlds that were previously quite separate. At first he took the works of previous French probabilistic mathematicians as a basis, while retaining only the portion of their work that concerned the law of large numbers and the convergence of binomial distribution, leaving aside their reflection on the probability of causes. In another connection, he created or gave rise to the creation of statistical services, helping, through the reverberation of his multiform activities, to establish the legitimacy and the audiences of these institutions. He organized population censuses. He gathered the statisticians of various countries at "international congresses of statistics," which met regularly between 1853 and 1878; these were direct ancestors of the International Statistical Institute founded in 1885 and which still exists today (Brian, 1989). Finally, by his writings on the average man and "social physics," Quetelet participated indirectly in the then animated debates on political philosophy while introducing an entirely new rhetoric into them.

This debate concerned the apparent contradiction between the determinism and fatalism that seemed to result from macrosocial realities—for example, crime or suicide—and, on the other hand, free will and the moral idea that man, who is responsible for his acts, is not driven by forces that exceed him. This question greatly troubled the commentators of the time and gave rise to lively controversy. In this regard, Quetelet's work found a place in the old debate on realism and nominalism mentioned earlier, on the logical anteriority of the universal and the individual. In the nineteenth century, the universal assumed a new form, society, and ever since Rousseau the question that was raised was: how did individuals (citizens) adhere to this totality? It was discussed at length again in 1912 by Lottin, a Belgian Catholic philosopher, in a synthesis entitled *Quetelet: Statistician and Sociologist.* Lottin's aim was to show that there is no contradiction between individual responsibility and the ineluctability of social laws. He based his argument on two quotations from Quetelet that offer a kind of motion of synthesis inspired by Rousseau and his social contract:

> As a member of the social body, he (man in general) is subject at every instant to the necessity of causes and pays regular tribute to them; but as a man (an individual) mustering all the energy of his intellectual faculties, he can in some way subdue causes, modify their effects, and try to attain a better state. (Quetelet, 1832)
>
> Man can be considered in different aspects; above all, he has his individuality, but is further distinguished by another privilege. He is eminently sociable; he willingly renounces part of his individuality to become a fraction of a large body (the state), which also has a life of its own and different phases . . . The part of individuality that is thus engaged becomes a regulator of the main social events . . . This is what determines the customs, needs, and national spirit of peoples, and regulates the budget of their moral statistics. (Quetelet, 1846)

This tension between collective fatalism and individual liberty recurs in almost the same forms in Durkheim and the twentieth-century sociologists he inspired. What changed, however, was the administrative use of statistical addition, henceforth linked to diagnostics, to standardized procedures and assessments of them. Every kind of macrosocial politics developed since the beginning of this century involved mutually dependent modes of management and knowledge. The alternation between measures in the sense of decisions applicable to a large number, and measurements of the effects of these measures, illustrates this rediscovered equivalence. The establishment of systems of social protection, ensuring a statistical coverage of individual risks, is a good example of this (Ewald, 1986).

Two Controversial Cases: Medical Statistics

In the philosophical debate over the average man, the difficulty of simultaneously conceiving singular situations and statistical regularities is treated in terms of a contradiction between fate and freedom. But at about the same time during the 1830s the use of averages was a subject of controversy that turned on very concrete questions, and not only on philosophical discussions. The field of medicine offers two examples, one having to do with clinics and therapeutic methods, the other with policies of prevention and the fight against epidemics. In both cases, the various relevant positions can be analyzed as specific combinations of cognitive schemes and particular forms of action and insertion into larger networks. More exactly, the identification of the object having the most reality (society, poor people, typhoid, the Paris arrondissements, patients, doctors, miasma, the cholera vibrio, and so on) did not derive from the opposition between determinism and free will. Rather, it derived from the most complete and explicit formulation of the situations into which the protagonists were inserted, and also of the pertinent objects that the actors could rely on and agree among themselves about. Real objects thus functioned as common reference points in mobile constellations that had to be reconstituted.

Illness and how a doctor treats it constitute a unique event, and this singularity was long monopolized by the medical community, reluctant to accept any form of categorization and summation that might violate the "privileged communication" between doctor and patient. But this resistance soon gave way when disease became a collective problem calling for general solutions: epidemics, and how to prevent them. In this case, the sometimes urgent need for measures of public hygiene to prepare for or halt an epidemic involved viewing the epidemic as a whole, the constant causes of which—that is, the *factors* that encouraged it to spread—were being urgently sought. To this end, calculations of average death rates of segments of the populace classified according to different criteria (region of Paris, type of dwelling, level of income, age) provided data for possible preventive policies (groups termed "at risk" in debates on the AIDS epidemic derive from the same logic). This form of "medical statistics" is easily accepted by the medical community, for it allows it to intervene officially in public debates and in the organization of society. A review founded in 1829, *Les Annales d'hygiène publique et de médecine légale* (Annals of public hygiene and legal medicine) was the mainstay of statistical and nonstatistical surveys of the social groups most vulnerable to poverty, illnesses, alcoholism, prostitution, or delinquency (Lécuyer, 1982). The aim of these studies was both to raise the moral standard of these groups and improve their living conditions and level of hygiene, especially through legislation. The best known of these militant scholarly reformers were Villermé (1782–1863, a close friend of Quetelet), and Parent-Duchatelet. The cholera epidemic of 1832 was a time of intense activity for these social researchers.

In contrast, the acceptance of quantitative methods was far less evident in the case of medical clinics, when a proposal was made to subject to statistical analysis the comparative effectiveness of various ways of treating a disease. Thus the proponents of the "numerical method," such as Dr.

Louis, relied on percentages of people healed to prove the superiority of purges over bleedings in the treatment of typhoid fever. But they encountered harsh, if varied, criticisms, leveled at the means of creating equivalence classes. Among these critics we find doctors normally divided on every other issue, doctors as diverse as Risueno d'Amador, who was inspired by eighteenth-century vitalism, and Claude Bernard, who later became the founder of modern experimental medicine (Murphy, 1981). For the more traditional physicians, such as Amador, medicine was an art, based on the intuition and instinct of the practitioner manifest during the specific, privileged communication between doctor and patient, and leading to a prescription resulting from the individuality of every case. In Louis's view, however, one had to classify diseases, evaluate treatments according to standards, and constantly seek connections between illnesses and caretaking techniques. To this end, one promoted the methods of the other natural sciences, which had been tested and proved (Piquemal, 1974).

The terminologies employed by each of these two camps were both consistent, reflecting two stereotyped ways of knowing and acting. The question was: how should judgments be made? Should a doctor judge, as Louis suggested, in the light of knowledge based on the methodical recording of a large number of medical acts and results codified with the help of theories? Or should he rely, as Amador recommended, on specific intuition based on ancestral tradition, often handed down orally and followed faithfully by the practitioner—a man of experience who had seen many such cases? It was therefore less a matter of contrasting the generality of the "numerical method" with the singularity of the interaction of traditional medicine than of distinguishing between two ways of accumulating or summing up the previous cases. In both instances, the proponents proclaimed a form of generality: statistics or tradition.

The example of Claude Bernard was more complex, for he fully accepted and even advanced the scientific method, but was nonetheless hostile to the "numerical method" (Schiller, 1963). He accused it of diverting attention from the specific causes of each illness and of leaving, under the cloak of "probability," a part to uncertainty and approximation. In his view, a scientist had to try to make a full analysis, using the experimental method, of the chain of causes and effects. His criticism appears to have points in common with that of Amador. But for Claude Bernard, who in no way resembled the old type of vitalist, a doctor could not treat patients "on the average." He had to find the direct causes of the illness in order to eliminate it completely. This hostility between statistics and its regularity was linked to a deterministic concept of microcausality, according to which probability and numerical method were synonymous with vagueness and sloppy thinking. It thus appears that each of the three leading figures in these debates was right in his way and in his own world of logic, to the extent that their cognitive tools were coherent with the action they undertook: hygiene, social medicine, preventive and collective medicine for Dr. Louis (Lécuyer, 1982); the comparative, daily practice of family medicine for Dr. Amador; experimental medicine and technologically oriented clinics for Claude Bernard. Each of these three positions still exists and has its own consistency.

The use of statistical averages crops up more indirectly in another medical controversy, at the time of the cholera epidemic (Delaporte, 1990). This epidemic appeared in Asia during the 1820s and spread through western Europe, reaching France in 1832. A naval officer and former colonial, Moreau de Jonnès (1778–1870) followed every step of the disease's advance and in 1831 wrote a report on the subject, calling for strong measures: ports should be closed, ships placed in quarantine, and quarantine stations built (Bourdelais and Raulot, 1987). All to no avail: closing the borders was damaging to major economic interests. Cholera spread through France in the spring of 1832.

Doctors were of divided opinion as to possible explanations for the illness. The contagionists—among them Moreau de Jonnès—declared it was transmitted by contact with sick people. Others, more numerous, thought the disease was not contagious, but was spread by infection, encouraged by the unsanitary living conditions often found in certain parts of urban areas. They based their conviction on the statistics of death rates, tallied according to the various Paris streets and the economic level of their inhabitants. These proponents of the "miasma" theory were often found among the "hygienists," whose main organ was the review *Annales d'hygiène*. These statistical averages, characteristic of an "environment," or a "milieu," provided them with adequate information for political intervention: they called for measures improving sanitation, requiring sewers to be built, and regulating housing standards. But were these averages, which were consistent with macrosocial action, also effective in discovering the precise and direct causes of the disease? This debate is interesting because it does not lend itself to a teleological narrative of the progress of science: the story of the fight between Enlightenment and the dark night of prejudice. In this case history has, in a way, justified both camps: the contagionists—as would be shown by the discovery of the cholera bacillus in 1833; and the "infectionists," partisans of averages and public hygiene, who correctly reasoned that the spread of the bacillus was helped by the absence of drains.

As it happened, the use of averages was the subject of arguments between the same protagonists, the hygienist Villermé and the contagionist Moreau de Jonnès. Moreau was also interested in statistics, but not in the same way. Thus in 1833 (a year after the cholera epidemic) he was charged with rebuilding the bureau of administrative statistics that had been eliminated in 1812. This was the new Statistique Générale de la France (General Statistics of France, hereafter referred to as the SGF), which he directed until 1851. But his concept of statistics differed from that of the hygienists, with whom he engaged in polemics. In his *Eléments de statistique* of 1847, he presented statistics in a far more administrative and far less probabilistic light than did the "moral statisticians," the disciples of Quetelet and Villermé. For Moreau, statistics were linked to the management of the governmental administrative apparatus, in regard to both the collection of data and their use in the actual management. He insisted on the qualities of consistency, logic, and painstaking care that a statistician needed in order to obtain "real numbers," and averages seemed to him like fictions that might content some people, but that could never replace "real numbers." He criticized the calculations of life expectancy and the death rate tables used by actuaries.

We can see from this case that the connection Quetelet established between the two types of averages (objective and subjective, in Bertillon's terminology) was by no means self-evident: the realism of the aggregate was denied in the bookkeeping, administrative concept of statistics. Reproaches of this kind are often addressed to statisticians: "You're putting together things that are actually quite different." The reality invoked here is not the same as the reality revealed by statistical averages; each has its repertoire of uses. This criticism of encoding underlies numerous denunciations, whether they address the quantitative social sciences (Cicourel, 1964) or the bureaucratic, anonymous administration of large-scale societies. The irony here is that the attack was delivered by the head of the bureau of administrative statistics, the SGF. At this point, the consistency between administrative and cognitive tools was not sufficiently well established, and such a contradiction was still possible.

The hygienists, moral statisticians and lovers of averages, were doctors and eminent persons seeking to create new attitudes by militating in favor of social responses. Their macrosocial argument couched in terms of averages was adapted to promote mass hygiene and preventive medicine. This influential lobby was moreover linked to the free traders, especially merchants and importers, hostile to state regulations regarding the circulation of merchandise and closing of borders. In contrast, Moreau de Jonnès and the contagionists had a more administrative view. Statistics, they felt, had to assemble facts that were *certain* (not just *probable*), and rigorously recorded. Moreover, the treatment of cholera could not be limited to measures (in both the statutory and the scientific senses) that were uncertain, and effective only in terms of probability or generality: one had to isolate the "germ" (later known as the bacterium or vibrio) and not vague environmental miasmas. This position was close to that of Claude Bernard, described above.

If statistics is based on individual records, the interpretations associated with it at the time, by Quetelet, Villermé, or Louis, had more affinity with the political philosophy I have previously described as *realist,* in the medieval sense, later known as *universalist* or *holistic,* that is, assigning an intrinsic reality to categories or ensembles of individuals. Their adversaries, in contrast—Amador, Claude Bernard, Moreau de Jonnès—contested these classes of equivalence and were closer to the *nominalist* or *individualist* position that Occam had upheld, in which the sole reality was that of each individual. The choices between pertinent real objects, which different actors could rely and agree on, were related to the overall situation, and especially to the stature of these actors (the hygienists: the poor and poverty; Keynes: the unemployed and unemployment). But between these two extreme positions—that of the realism of the aggregate unified by its constant cause, and that of the individualistic nominalism that rejected these summations—a new space would open up. Attempts were made, first to test the homogeneity of this entirety summarized by its average, then to propose more complex statistical formalisms, enabling the two perspectives to be conceived together. From this point of view, Karl Pearson's formulation of regression and correlation would prove a decisive turning point between the nineteenth-century questions still linked to Quetelet's rhetoric and the entirely different formalizations of twentieth-century mathematical statistics.

One Urn or Several Urns?

Quetelet's intellectual edifice was supported by the probabilistic model of an urn of constant composition of black and white balls, from which a ball was drawn (and then replaced) many times. The probability of a relationship appearing between the black and white balls follows a binomial law, converging toward a normal law if the number of drawings increases indefinitely: this is Bernoulli's "law of large numbers." What would happen if the constancy or uniqueness of the composition of the urn were not self-evident? Quetelet reasoned in the opposite direction, inferring, from the normal form of distribution, the existence of a system of constant causes.

This question had already been raised when, at the beginning of the nineteenth century, certain thinkers had tried to reutilize methods of eighteenth-century political arithmetic to estimate, for example, the population of a country, in the absence of an exhaustive census based on the relationship—presumed uniform throughout the territory—between this population and the annual birth rate, a relationship measured in several parishes (Laplace's calculation: see Chapter 1 above). Around 1825 Quetelet tried to estimate the population of the Netherlands, and was tempted by this method. He was discouraged, however, by the criticism of a high official, Baron de Keverberg, and this led him to concentrate his energies into organizing exhaustive censuses. De Keverberg cast doubt on the idea that a single law could be valid for the whole territory, a law regulating the relationship between the population and its birth rate, and therefore that this coefficient, calculated on the basis of a few localities, could be extrapolated:

> The law regulating mortality is composed of a large number of elements: it is different for towns and for the flatlands, for large opulent cities and for smaller and less rich villages, and depending on whether the locality is dense or sparsely populated. This law depends on the terrain (raised or depressed), on the soil (dry or marshy), on the distance to the sea (near or far), on the comfort or distress of the people, on their diet, dress, and general manner of life, and on a multitude of local circumstances that would elude any a priori enumeration.
>
> It is nearly the same as regards the laws which regulate births.
>
> It must therefore be extremely difficult, not to say impossible, to determine in advance with any precision, based on incomplete and speculative knowledge, the combination of all of these elements that in fact exists. There would seem to be infinite variety in the nature, the number, the degree of intensity, and the relative proportion of these elements. . . .

> In my opinion, there is only one way of attaining exact knowledge of the population and the elements of which it is composed, and that is an actual and complete census, the formation of a register of the names of all the inhabitants, together with their ages and professions. (Keverberg, 1827, quoted by Stigler, 1986, p. 165)

This argument was not very different from the one later formulated by Moreau de Jonnès—who was also an important official—against Quetelet himself and the scholars who used averages. Yet it concerned a method of arriving at a totality on the basis of a few cases. As a young man Quetelet had already been intent on making a place for himself in the administrative machinery. He was sensitive to it, so much so that until the end of the century he and his successors were incessantly preaching exhaustive censuses and mistrusting sampling surveys that had been assimilated into the risky extrapolations of the political arithmeticians of the seventeenth and eighteenth centuries, carried out on the basis of incomplete readings. De Keverberg's criticism concerned the fundamental heterogeneity of the population, and the diversity and complexity of the factors determining the variables studied—factors so diverse and complex that knowledge obtained by generalization was inconceivable.

This doubt as to the homogeneity and oneness of the urn would be expressed and formalized more precisely by Poisson (1837), in regard to an entirely different question: that of the majority required in the verdicts of trial juries (Stigler, 1986; Hacking, 1990). Ever since these juries of ordinary citizens had been instituted in 1789, the problem of judicial error had disturbed lawmakers. How were they to minimize the probability of sending an innocent person to the guillotine, while not allowing the guilty to escape punishment? —while knowing that guilt was often uncertain and that, at the same time, juries could be mistaken? There was a double uncertainty, involving the reality of the fact incriminated and the fallibility of the juries' judgments. Poisson, following Condorcet and Laplace, tried to estimate the probabilities of judicial error, according to different theories of what constituted an acceptable majority in the verdict of a twelve-member jury (for example, seven against five, eight against four, and so on). This question fascinated public opinion and polarized Parliament, setting those who feared so irreversible an error against those who feared a justice that was too lax. Each camp wanted to support its position with as many arguments and objectified facts as possible.

Unlike his two predecessors, Poisson was able to support his arguments with statistical readings provided, since 1825, by the new *Compte général de l'administration de la justice criminelle* (General record of the administration of criminal justice), which published the annual number of indictments and condemnations. Moreover, there was a change in the law. Before 1831 only seven against five votes were needed for a condemnation; but in 1831 this majority changed to eight against four, allowing the effects of this legislative change to be included in Poisson's calculation. Somewhat Bayesian in inspiration, this calculation involved estimating two "probabilities of cause," both unknown a priori: the probability that the defendant was guilty and the probability that the jurors were fallible. Poisson assumed, first, that the defendants might come from different groups, for which the probability of guilt was not the same (today we would say that some groups were at "greater risk"), and second, that the jurors were not equally blessed with the ability to judge the facts correctly. This was how he came to introduce urns of *uncertain composition* into his reasoning. This led him to formulate his "strong law of large numbers," which appeared as a generalization of Bernoulli's law. In Bernoulli's law, the urn involved in the drawings had a constant composition, whereas in the case Poisson studied, the urns themselves resulted from a primary series of drawings. This new "strong law" indicated that, according to certain hypotheses in this initial distribution of urns of varying compositions, the probability of the drawings converging toward a normal law was maintained.

In Poisson's mind this result was important, for it allowed the preservation of the formulas of convergence that were the strength of Bernoulli's law, in cases when the composition of the initial urns was uncertain and even heterogeneous, which seemed to him far closer to the real situations in which this law was applied. Although he presented himself as an admirer of Poisson, Bienaymé quickly devalued the originality of this new "strong law of large numbers" in an article published in 1855, slyly entitled: "On a Principle Monsieur Poisson Thought He Discovered and Termed the Law of Large Numbers." He observed that as long as one envisaged the ensemble of these two successive drawings as a single probabilistic process, Poisson's "strong law" did not add much to Bernoulli's law. Bienaymé's reasoning hinged on the idea of *constant cause,* and on the limited sense he thought the probabilists had given the word *cause;* in his view,

> They do not mean to speak of what produces such an effect or event, or of what ensures its happening; they wish only to speak of the state

> of things, of the ensemble of circumstances during which this event has a determined probability. (Bienaymé, 1855)

He thereby made no distinction between constant cause and a cause varying according to a constant law:

> The smallness of the extent of the deviations remains the same if one no longer imagines constant probability as being absolutely fixed, but as having the average constant value of a certain number of probabilities that result from variable causes, of which each can be subjected equally to every test, according to a preassigned law of possibility. (Bienaymé, 1855)

Though published in 1855 this text was written in 1842—that is, at the moment when Quetelet was shoring up his construct of the average man by means of the "astonishing regularity of moral statistics, of marriages, suicides, and crimes." In it Bienaymé introduced a note of skepticism, which would later reappear (unrelatedly, no doubt) in the German statistician Lexis:

> When one does really serious scientific research . . . and one has to compare facts from several years, it is hard not to notice that the deviations set by Bernoulli's theorem are far from equaling the considerable differences encountered between the relationships of the numbers of natural phenomena that have been collected with the greatest accuracy. (Bienaymé, 1855)

Whatever Bienaymé's criticisms of Poisson, both of them again called into question the idea of constant cause as a principle of creating equivalences that allowed events to be combined by making them appear as contingent manifestations of a more general cause of a superior level. Their position was more nominalist than that of Quetelet. Poisson, however, tried partly to save the knowledge gained from constant cause, with its cause that varied according to a *constant law,* whereas Bienaymé more clearly perceived the conventional nature of the definition of causality. His formulation drew closer to the modern idea of a *model,* with its accompanying elements of convention and abstraction as distinct from tangible intuition. In contrast, Poisson's distinction between different moments in which uncertainty arises preserved the subjective dimension of probabilities: in eighteenth-century terms, the dimension of *reasons to believe* (in this case, the guilt of the man condemned to death). This was moreover the main difference between Poisson and Quetelet. Poisson still inhabited the mental universe of Condorcet and Laplace (even though he based his arguments on administrative statistics), for whom probabilities were degrees of belief that rational individuals attributed to their judgments. For Quetelet, however, who was interested in society as a whole and its continuity, probabilities henceforth had to do with the variability of things and with deviations from averages.

In all the formalizations and controversies just described—Quetelet and the appalling regularity of crime; doctors and their fight with cholera; Poisson and his trial juries—the theoretical formulations were closely connected to the themes under debate, and their impact was clearly linked to the burning relevancy of these topics. Conversely, the very appearance of these cognitive schemas is connected to the way these actors construct different realities corresponding to their own acts. Further on in this book we shall find an analogous correspondence in the activities of Galton and Pearson, in regard to eugenics and mathematical statistics. For some authors, however, such as Bienaymé or Cournot, this is less apparent and their activity seems internal to an autonomous epistemological reflection. For this reason few actors have based their arguments on them, and they are to some extent unknown.

The Challenge to Realism: Cournot and Lexis

Unlike Quetelet, Cournot (1801–1877)—a philosopher, mathematician, and general inspector of public education—was particularly interested in theoretical thought. The crucial questions he raised concerned the means the mind disposed of to attain to objective knowledge. The calculus of probabilities occupied a central place in this venture. In 1843 Cournot published a work on the subject, *Exposition de la théorie des chances et des probabilités* (A Statement of the theory of chance and probabilities), in which the philosophical status of probabilistic reasoning is analyzed with some finesse. He tried to subordinate the chaos of individual perceptions to the existence of a rational order of things that was superior to subjective intuitions. This was a more "realistic" position (in the medieval sense); but Cournot's refined analysis of the way in which proofs are constructed and supported, especially by means of statistics, led him to a position that is actually more nominalist. When trying to collect his thoughts into general principles, he invokes rational order as opposed to subjective errors:

> Our belief in certain truths is not, therefore, founded solely on the repetition of the same judgments, nor on unanimous or almost un-

> animous assent: it rests mainly on the perception of a rational order according to which these truths are linked together, and on the belief that the causes of error are abnormal, irregular, and subjective, and could not engender so regular and objective a coordination. (Cournot, 1843, p. 421)

This awareness of the tension between the objectivity sought in the rational order and the imperfections of subjective human judgment led Cournot to make a clear distinction (one of the first to do so) between the two possible meanings of the word "probability," and to study their connections, while vigorously criticizing the Bayesian point of view:

> A rule first stated by the Englishman Bayes, on which Condorcet and Laplace tried to build the doctrine of a posteriori probabilities, became the source of numerous ambiguities which we must first clarify and of serious mistakes which must be rectified. Indeed, they are rectified as soon as we bear in mind the fundamental distinction between probabilities that have an objective existence, that give the measure of the possibility of things, and subjective probabilities, which are partly relative to our knowledge, and partly to our ignorance, and which vary from one intellect to another, depending on their abilities and the data provided them. (Cournot, 1843, p. 421)

Cournot was convinced that an accumulation of statistics on a wide variety of subjects would profoundly change the nature and scale of knowledge of the social world. But unlike the preceding authors, he stayed within this philosophical perspective, and took no part in the broader themes being debated. Nor did he become involved in the administrative and political problems of compiling and formatting statistical results, as did Quetelet and Moreau de Jonnès. The subtle manipulation of cognitive schemes was enough for him.

Among other things, he raised the question of the meaning of deviations between measurements of the same quantity, made on several subpopulations resulting from the division of a general population. He thus introduced a way of testing the homogeneity of this population, and of answering De Keverberg's objections to the extrapolations of Laplace. But, by reasoning in probabilistic terms that the deviations would be other than zero, he was tripped up by the fact that such tests only provide answers with a certain threshold of significance, and that one can choose, for example 5 percent or 1 percent. Now, the number of possible ways of dividing up a population is a priori infinite; with a little patience and much effort, one can always find a few subdivisions for which the deviations are significant (naturally this would be more difficult with a threshold of 1 percent than with a threshold of 5 percent). The example Cournot offered had to do with the ratio between male and female birth rates, calculated for the whole of France and for each of the eighty-six departments. If one isolated the department where the deviation between this ratio and that of France in its entirety was greatest, could one legitimately test, for the department thus selected, the significance of the deviation with a threshold of 5 percent, or even of 1 percent? If nothing had previously singled this department out for attention, the answer would surely be *no*. But if this department happened to be the department of the Seine, including Paris, or Corsica, an island, doubts would arise. Why? Because people knew (or thought they knew) a priori that these places had particular features.

In reasoning thus, Cournot raised the question of the connection between forms of knowledge with heterogeneous modes of elaboration. What specific problems were posed by the creation of *composite* rhetorics of argumentation, the elements of which derived from different probative techniques? How was one to establish norms that could provide the basis of agreement as to the convincing nature of a statistical argument? The particular difficulty of this question derives from the fact that in statistics a highly formalizable, self-enclosable world comes in contact with another world, of essentially different perceptions. The construction of taxonomies, of categories of equivalence, takes place precisely in this contact zone. This is what Cournot perceived with his idea of a "previous judgment" determining the "cut" (that is, the taxonomy). He explained that the judgment made on an observed deviation depends on two elements. The first derives from a clearly formalized calculus of probabilities, whereas:

> The other element consists of a previous judgment, by virtue of which we view the cut that has given rise to the deviation observed as being one of those which it is natural to try, in the infinite multitude of possible divisions, and not as one of those that only hold our attention by virtue of the deviation observed. Now, this previous judgment, by which the statistical experiment strikes us as having to be directed at one particular cut rather than another, derives from causes whose value cannot be assessed rigorously, and may be variously assessed by various minds. It is a conjectural judgment, itself based on prob-

abilities, but on probabilities that are not resolved in an enumeration of chances, a discussion of which does not properly belong to the doctrine of mathematical probabilities. (Cournot, 1843, p. 196)

By introducing this kind of questioning, Cournot radically modified the above-mentioned problem of the "realism of the aggregates." He opened up the possibility of composite judgments, certain elements of which were objectifiable, whereas others could be "variously assessed by various minds." The "cuts" actually employed could only therefore result from a convention designed to institute common sense. This trend of thought is partly nominalist, despite the nostalgia manifest elsewhere for a "rational order pursuant to which these truths are conjoined . . . and for which the causes of error are abnormal, irregular, and subjective." This tension, and the efforts to think it through, make Cournot's work one of the richest among those quoted here, for it explicitly raises the question of the relative place of knowledge built on probabilities, and on statistics among other forms of knowledge, particularly in order to establish the realism of the aggregates. This realism was to be contested in yet another way by Lexis.

Quetelet's idea of "constant cause," supporting the reality of an object on the basis of the regularity of a statistical series, was criticized and demolished by an argument that used the same tools, but that showed its inner contradictions. This criticism was formulated during the 1870s by the German statistician Lexis (1837–1914) and the French actuary Dormoy, in fairly similar terms, although the two men worked independently of one another (Porter, 1986). Seeking to test the significance of the apparent regularity of a series, Lexis had the idea of considering each of the annual results as a lottery, in which black and white balls were drawn from an urn of a given composition. If this composition was constant, the distribution of the proportion of white balls in the yearly drawings could be described by a binomial law; if the size of the urn and the number of drawings were large, this binomial law would be close to a normal law, the *theoretical* dispersion of which—say, r—could be calculated. One then measured the dispersion *actually observed* in the series studied, or R. Last, Lexis calculated the ratio $Q = R/r$, which he compared with 1. In a case where Q was inferior to 1, ($R < r$) one could speak of a *subnormal* dispersion and deduce from this the existence of a macrosocial cause: divine providence, a certain "propensity," or a collective group effect. If $Q = 1$, one made sure that the urn was constant and that the binomial model was applicable. Finally, if $Q > 1$ ($R > r$), one could no longer state that the analyzed variable was constant; everything took place as if the composition of the urns varied.[1]

Lexis applied this method to a large number of series that had received enthusiastic comment from Quetelet and his disciples during the preceding decades. The results were devastating. The case of $Q < 1$ never arose, and the case of $Q = 1$ appeared only in the ratio of male to female births (the *sex ratio*). In every other case, the observed dispersion was superior to the theoretical dispersion, and one could therefore not conclude in favor of the constancy of the phenomenon being described. From this Lexis deduced that Quetelet's constant causes might possibly be invoked for certain *physical* observations (the *sex ratio*), but not for *moral* statistics, nor even for height. He thus changed the mechanism of the proof, the equipment needed to make things solid and stable. For Quetelet, the combination of the regularity of the average and the normal form of distribution allowed a constant cause to be inferred. For Lexis, that was no longer enough, since he showed that distributions of normal appearance could present a greater dispersion than that resulting from the binomial model, and could therefore result from a multitude of urns. It was then necessary to satisfy the condition $R = r$ to be able to speak of constant cause. Only the "sex ratio" survived this massacre.

Quetelet's average man was thus subject to vigorous attacks. Cournot mocked the idea: a man whose bodily attributes would be the averages of these same attributes, measured in many individuals, "far from being in some way a model for the human race would simply be an impossible man—at least, nothing that has happened thus far allows us to imagine him as possible" (Cournot, 1843, p. 214). Emphasizing, moreover, the conventional nature of statistical terminology, he struck a blow at the idea of constant cause, which Lexis had for his part damaged through a purely internal critique. However, the efficiency of Quetelet's reasoning, his ability to convert from volatile individuals to social solidity was so strong that the social sciences were often tempted to use his methods to justify the autonomous existence of the whole—even though, at the same time, the average man could appear as a mere sham of man in general, if not of his moral nature. This very particular contradiction in statistical rhetoric is visible, for example, in Durkheim, and in the sociology he inspired, when it resorts to the heavy artillery of large surveys, whose power to convince is often simultaneously displayed and denied. One of the aims of this book is to try to make explicit this tension, which is usually implicit.

Durkheim's Average Type and Collective Type

Since the 1830s the social sciences have largely used statistics to argue and prove things. This large-scale use raises two distinct questions: to what end (that is, to prove what?) and in what manner (what tools, what rhetoric to employ)? The answers to these questions underwent a clear change around 1900, when mathematical statistics in the English style appeared on the scene. In the nineteenth century, statistical arguments were used above all to make macrosocial entities hold, by resorting to averages in the perspective that Quetelet had opened up (even if analyses in terms of deviations, or of differences in averages, were already used, as we have seen with the "numerical method" of doctors, or with Cournot). In the twentieth century, however, statistics would be used to make relationships hold, thanks to regression lines, coefficients of correlation, and analyses of variance, with Karl Pearson and Ronald Fisher (even if the identification of the objects was increasingly instrumented, especially through the theory of tests and inferential statistics, with Jerzy Neyman and Egon Pearson, and also Fisher).

Quetelet's average type rests on the combination of two elements: temporal regularities, and distributions in Gaussian form. In actual use, the first element held up better than the second and was often deemed adequate, for example, in Durkheimian sociology. But in the nineteenth century, the second element was still a weighty argument on which to base the existence of animal or human species. This was the case, for example, with Adolphe Bertillon's famous conscripts from the department of Doubs (1876). Instead of being Gaussian and unimodal in appearance, as elsewhere, the distribution of the heights of soldiers from this department offered two curious humps (bimodal distribution). From this Bertillon deduced that this strange form resulted from the superimposition of *two* normal distributions, and that the population of the Doubs was a mixture of two distinct ethnic groups: Burgundians and Celts. Thus the form of the curve served as a classificatory argument.[2]

The average statistical type and its temporal regularity were widely used by Durkheim to support the existence of a collective type external to individuals, at least in the first two of his books: *The Division of Labor in Society* (1893), and *The Rules of Sociological Method* (1894). In contrast, in

Suicide (1897), he distances himself from Quetelet's average type, whom he carefully distinguishes from the collective type. This change in language can be linked to different constructs in each of the three cases: the effects of heredity, in *The Division of Labor in Society;* the definition of what is normal and pathological, in *The Rules of Sociological Method;* and the interpretation of a statistically rare fact in *Suicide.*

In 1893, in a discussion on the relative weight of heredity and the social milieu (based on Galton's results), he insisted on the *constancy* of the average type, whose attributes were transmitted through heredity, whereas individual traits were volatile and transitory:

> The average type of a natural group is the one which corresponds to the conditions of average life, consequently, to the most ordinary. It expresses the manner in which individuals have adapted themselves to what one may call the average environment, physical as well as social; that is to say, to the environment where the greatest number live. These average conditions were more frequent in the past for the same reason that they are most general at present. They are, then, those in which the major part of our ancestry were found situated. It is true that with time they have been able to change, but they are generally modified slowly. The average type remains, then, perceptibly the same for a long time. Consequently, it is it which is repeated most often and in a most uniform manner in the series of anterior generations, at least in those near enough to make us feel their action efficaciously. Thanks to this constancy, it acquires a fixity which makes it the center of gravity of the hereditary influence. (Durkheim, 1893, p. 325)

Durkheim thus gives the average type and the attributes that characterize a group—as a collectivity distinct from its members—a seemingly Darwinian interpretation, less apparent in his later texts (but frequent in Halbwachs). In *The Rules of Sociological Method,* in 1894, he tries to define normality, in contrast to the pathological type. He compares normality to the average type of "standard health," which merges with the "generic type." We can see how Quetelet's rhetoric permeated the social sciences of the nineteenth century:

> We shall call "normal" these social conditions that are most generally distributed, and the others "morbid" or "pathological." If we designate as average type the hypothetical that is constructed by assembling in the same individual, the most frequent forms, one may say

> that the normal type merges with the average type, and that any deviation in relation to this standard of health is a morbid phenomenon. It is true that the average type cannot be determined with the same distinctness as an individual type, since its constituent attributes are not absolutely fixed but are likely to vary. But the possibility of its constitution is beyond doubt, since, blending as it does with the generic type, it is the immediate subject matter of science. It is the functions of the average organism that the physiologist studies; and the sociologist does the same. (Durkheim, 1894, p. 56)

Three years later, in *Suicide,* Durkheim's position in regard to the average man and his regular features had changed completely. In the two preceding works, the stability of the average type, as opposed to the variability of individual cases, was invoked as a basis for the holistic position (to use Dumont's terminology [1983]), according to which the whole is external to, and previous to, individuals. In *Suicide,* in contrast, the statistical average is repatriated into the world of methodological individualism, and the "collective type" is no longer compared to the "average type." In this case, Quetelet's rhetoric produces nothing more than a holistic trumpery: the average statistical man is often a rather sorry figure, who doesn't want to pay his taxes or fight in a war. He is not a good citizen. This was perhaps the first time that a clear separation appeared between two heterogeneous discourses in regard to the support that statistics could provide the social sciences. For one such discourse, they provided incontrovertible proof. For the other, they missed the essential. The paradox here is that *Suicide*—a work generally considered to have founded quantitative sociology (mainly on account of the massive use of statistics concerning causes of death)—*also* introduced a radical condemnation of the holistic interpretation of Quetelet's average type, and thus of these same statistics. We can follow closely as this criticism unfolds, leading to an idea of the whole that was radically different from the average type.

> When Quetelet drew to the attention of philosophers the remarkable regularity with which certain social phenomena repeat themselves . . ., he thought he could account for it by his theory of the average man—a theory, moreover, which has remained the only systematic explanation of this remarkable fact. According to him, there is a definite type in each society more or less exactly reproduced by the majority, from which only the minority tends to deviate under the influence of disturbing causes . . . Quetelet gave the name *average type* to this

> general type, because it is obtained almost exactly by taking the arithmetical mean of the individual types . . . The theory seems very simple. But first, it can only be considered as an explanation if it shows how the average type is realized in the great majority of individuals. For the average type to remain constantly equal to himself while they change, it must to some extent be independent of them; and yet it must also have some way of insinuating itself into them. (Durkheim, 1897, pp. 300–301)

Durkheim tried to specify how the "moral constitution of groups" differed radically from those of individuals. A few pages after the above quotation he speaks of the "temperament" of a society, "unchangeable within brief periods." He speaks of "social coenesthesia," the coenesthesic state being, "among collective existences as well as among individuals, what is their most personal and unchangeable quality, because nothing is more fundamental" (let us remember that this construct is used to support the idea of a *suicidal inclination* inherent in a social group). He elaborates at length on how the collective moral sense can deviate a great deal, and sometimes be opposed to the individual forms of behavior of the overwhelming majority—a phenomenon that completely distances the hapless "average type" from the "collective type" in a society:

> Not many have enough respect for another's rights to stifle in the germ every wish to enrich themselves fraudulently. Not that education does not develop a certain distaste for all unjust actions. But what a distance between this vague, hesitant feeling, ever ready for a compromise, and the categorical, unreserved and open stigma with which society punishes theft in all shapes! And what of so many other duties still less rooted in the ordinary man, such as the duty that bids us contribute our just share to public expense, not to defraud the public treasury, not to try to avoid military service, to execute contracts faithfully, etc. If morality in all these respects were only guaranteed by the uncertain feelings of the average conscience, it would be extremely unprotected.
>
> So it is a profound mistake to confuse the collective type of a society, as is so often done, with the average type of its individual members. The morality of the average man is of only mediocre intensity. He possesses only the most indispensable ethical principles to any decided degree, and even then these principles are far from being as precise and authoritative as in the collective type, that is, in society as

> a whole. This, which is the very mistake committed by Quetelet, makes the origin of morality an insoluble problem. (Durkheim, 1897, p. 317)

The use of statistical data then appears ambivalent. Durkheim used them, but only with distaste. Such data implied collective tendencies existing outside individuals, but these tendencies could be observed directly (that is, without statistics):

> The regularity of statistical data, on the one hand, implies the existence of collective tendencies exterior to the individual, and on the other, we can directly establish this exterior character in a considerable number of important cases. Besides, this exteriority is not in the least surprising for anyone who knows the difference between individual and social states of consciousness . . . Hence, the proper way to measure any element of a collective type is not to measure its magnitude within individual consciences and to take the average of them all. Rather, it is their sum that must be taken. Even this method of evaluation would be much below reality, for this would give only the social sentiment reduced by all its losses through individuation. (Durkheim, 1897, pp. 318–319)

This ambiguity in relation to statistics and the interpretation of statistics is thus clearly visible in a work often presented as inaugurating the royal route of quantification later adopted by twentieth-century social sciences. It raises a basic question that is often swept under the rug, mainly as a result of the division of labor subsequently adopted among compilers of statistics, historians, economists, and sociologists: each of them manages his own segment, defined by a technology of production of pertinent objects and an ad hoc way of thinking. The inner contradiction thus designated in Durkheim's rhetoric could be used with a view to criticize or reduce either position. A statistician would smile at what seemed to him a form of metaphysical holistics, whereas a sociologist, attached to the idea that social structures condition individual behavior, would rely on statistics to describe structures and behavior, while immediately pointing out that they explain nothing. A possible way out of this debate is to unfold one by one the different ways of adding, and the more general argumentative grammars, philosophical or political, within which they are situated.

Thus Durkheim's distinction between the collective type—a reflection of the ideal of the good citizen—and the average type, an arithmetical resultant of selfish individuals, directly echoes the distinction made by Rousseau in *The Social Contract* between the *general will* and the *will of everyone* (Dumont, 1983). Whereas the second entity is only an agglomeration of individual wills, the first is anterior and qualitatively superior to these wills. The general will exists before the vote of the majority, which reveals it but does not generate it. According to Durkheim (commenting on Rousseau), it is the "fixed and constant orientation of minds and activities in a definite direction, that of the general good. It is a chronic inclination on the part of individual subjects" (in *Montesquieu et Rousseau, précurseurs de sociologie*).

The Realism of Aggregates

The same question has recurred throughout this entire chapter: how can collective objects—aggregates of individuals—be made to hold? Extending far beyond the history of statistics, it spans the history of philosophy and the social sciences, from Occam's opposition of realism and nominalism to Dumont's opposition of holism and individualism. I have chosen here to follow it through debates on the statistical mean, especially as formulated by Quetelet, centering on the all-unifying idea of *constant cause.* The use of this tool by doctors and hygienists has shown that beyond ontological controversies, the conventions of aggregation, whose various justifications and supports depend on circumstance, find their meaning within the framework of the practices they account for. When the actors can rely on objects thus constructed, and these objects resist the tests intended to destroy them, aggregates do exist—at least during the period and in the domain in which these practices and tests succeed.

Technologies develop and solidify, capable of making objects hold, but also of destroying them. A good example of this is Quetelet's model, which for a while constructed a holistic figure on basically individual recordings. It was then attacked on both sides, in the name of a stricter nominalism (Moreau de Jonnès, Amador, Lexis) or, on the contrary, in the name of a radically different holism (Durkheim, when he took up Rousseau's idea of general will). With his "strong law of large numbers," Poisson tried to extend the model's field of validity, but his perspective was not initially a macrosocial one. Each in his way, Bienaymé and Cournot diversified the rules of usage, preparing the ground for a more modern—less ontological and more conventionalist—idea of a model. These diverse possible orientations would soon be redefined by English mathematical statistics; but the question of the realism and nature of the causality exhibited by the new tools remains essential, as soon as one becomes interested in the rhetorics connecting these tools to everything else in life.

4

Correlation and the Realism of Causes

The aim of Quetelet's style of statistics was to make collective things hold through aggregates of individuals. The idea of cause was only present as an external hypothesis guaranteeing the consistency of these things (the "constant cause"); the strength of the connection between the cause and its effect was not itself measured—that is, equipped statistically. This construct was based on administrative records that were themselves previous and foreign to its own endeavor. Statistical rhetoric was still largely dependent on cognitive and social resources external to the new logic it was trying to promote. This double dependency made it vulnerable to all kinds of criticism, as we have seen with Cournot and Lexis. The feeling of strangeness that some of the foregoing controversies arouse today comes from the fact that statisticians of that time had to invent plausible junctions and translations linking their still summary tools with philosophical, political, and administrative rhetoric. To this end, they created and applied new objects. But for other actors to be able to appropriate them and include them in their own constructs, these objects not only had to be consistent in themselves, but also had to be capable of forming stable relationships with one another. Statisticians had to provide not only solid things but the equivalent of an erector set to make them hold together. This feat would be accomplished by Galton, Pearson, and English mathematical statistics.

This statistical venture can be seen a posteriori to have resulted unexpectedly from two different projects, neither of which tended a priori in this direction, since one was clearly political and the other more philosophical (MacKenzie, 1981). These projects are now largely forgotten; but correlation, regression, the chi-square test, and multivariate analysis—all of which resulted from them—have become pillars of modern statistics. What was the connection, if any, between a political belief now almost unanimously rejected (eugenics); a theory of knowledge that was refined and pertinent for its own time; and a mathematical apparatus destined to enjoy a splendid future? None of these three elements logically implied the two others: in fact, each of them was later taken up by actors who knew nothing about the two other ventures. In Pearson's mind, however, the three projects were nonetheless connected, and it is by no means certain he felt his own statistical innovations to be more important than his epistemological credo or his political activities.

I shall begin by presenting Pearson's philosophy as expressed in his book *The Grammar of Science*. I choose this work partly because in it Pearson evinces a decided position against the idea of causality, and partly because its successive editions show how he adopted Galton's idea of correlation, which was not initially present. This theory of knowledge belongs to an empirical antirealist trend, beginning with Ernst Mach and lasting until the Vienna circle. The political and scientific project Pearson developed next belongs in a line stretching from Darwin and Galton to the various attempts to improve the human race that were based on heredity, echoes of which could still be heard in the 1950s. This project had a political aspect, *eugenics,* and a scientific one, *biometrics,* both of which were connected. It was in the important laboratory devoted to biometric research that mathematical statistics was born. During the first half of the twentieth century the tools of mathematical statistics were adopted in numerous domains, especially in the human sciences, while the antirealist philosophy of knowledge and the concern with improving humanity through biological selection were both dropped.

We can follow the descent of the two major tools, *regression* and *correlation,* in the way the first was adopted in econometric models, which appeared in the 1920s and 1930s, and the way the second was used in psychometrics (to measure aptitudes, and especially intelligence), leading to multidimensional factor analysis. An important rift in this world of statistical formalisms, so abundant after Karl Pearson, was introduced by the adoption of probabilistic models, in what has sometimes been termed the "inference revolution" (Gigerenzer and Murray, 1987). This line of thinking became dominant after the 1960s, but some scholars preferred to keep away from it, attempting, by means of purely descriptive analyses without probabilistic models, to reveal an underlying order to the universe in a perspective not unlike that of Quetelet (Benzécri, 1982).

Associated with the extension of techniques designed to summarize and formulate a large number of observations was the development of procedures for recording and codifying them. These procedures involved creating contexts of common measurement: taxonomies, scales, standardized questionnaires, surveys by means of random samples, methods for identifying and correcting aberrant cases, checking and amalgamating files. This dual growth of techniques—for recording and formatting a host of new objects—had the effect of considerably extending the reality of the statistical world, and thus of pushing back the frontier zone in which statistical rhetoric confronted other rhetorics. In the nineteenth century this space was still exiguous, and its frontier quickly reached. Today it has become so enormous that only rarely do certain statisticians have the opportunity to encounter these contact zones. A language specific to this universe can be employed daily, and no longer arouses questions of the type raised by analysts or critics of Quetelet, or even by Karl Pearson in his *Grammar of Science.*

Once again, my intent here is not to denounce the falseness or artificiality of this reality in the name of other, supposedly more real constructs; rather, it is to study these contact zones, which are in general the blindest and most ill-equipped areas in the majority of sciences, and especially in the social sciences. This historical detour is interesting in that it turns a spotlight on such zones, enabling us to reexamine today's routines in a different light. Karl Pearson is exemplary from this point of view, on account of both the scope of the questions he raised and his political theories—absurd in this day and age. He has been studied and described by many authors, beginning with himself (K. Pearson, 1920) and his own son (E. Pearson, 1983).

Karl Pearson: Causality, Contingency, and Correlation

In 1912 the French publisher Alcan brought out a French translation of the third edition (1911) of *The Grammar of Science.* The first edition dated from 1892. There were nine chapters in the third edition. Chapters 1 through 4 and 6 through 9 were reproduced exactly as they had been in the 1892 edition. They dealt solely with the philosophy of science, and said not a word about statistics. The subject is broached—albeit rather briefly—only in Chapter 5, which was added in 1911, and entitled "Contingency and Correlation: Insufficiency of the Idea of Causation." The ideas of correlation and regression had, meanwhile, been formalized by Pearson and Yule, in publications never translated into French.

The French translator of this work was none other than Lucien March (1859–1933), then director of the Statistique Générale de France (SGF), ancestor of the present-day INSEE. He was helped in this task by three of his statistical collaborators, Bunle, Dugé de Bernonville, and Lenoir. In his introductory note he explains the importance, in his view, of this book and its translation. He scarcely refers to the statistical innovations for which Pearson was responsible, providing instead a seven-page summary of the theory of knowledge developed in the book, a theory he adopted for his own use. This theory was resolutely antirealist. It stated that man knows only sensations, perceptions he combines and classifies according to the analogies and continuities he observes in them, which Pearson describes as *routines of perception*. Reality itself is unknowable.

It is thus surprising to read—flowing from the pen of the director of the bureau of statistics, himself responsible for numerically describing the French population—formulations distinctly contrary to the philosophy of *practical* knowledge which, in his position, he had necessarily to develop in order to communicate with those around him and with those who used his publications. He wrote: "The man of science neither affirms nor denies the reality of the outer world." Was this a position he could maintain when negotiating the budget allocation of the SGF with his minister? In pointing to this contradiction I am not trying to suggest that March was an inconsistent scientist, or even a deceitful official, but that there are two different registers of reality involved here, which we must identify and analyze.

Pearson's theory of knowledge was marked by the thinking of the Austrian physicist Ernst Mach (1838–1916), a specialist in the psychophysiology of sensations. In his research on the subject, Mach had deduced an epistemology that laid emphasis, in the process of scientific knowledge, on the role of observation and sensations. It rejected "things in themselves" in favor of "colors, tones, pressures, durations, and what we habitually call sensations, which are the true elements of the world" (Mach, 1904). For Mach, the idea of the reality of the outer world melted away, since one could no longer distinguish between what was internal and what was external. He rejected the idea of causality as subjective, substituting in its place the ideas of the function and organicity of phenomena (Paty, 1991). Concepts were mental entities that did not exist outside the mind. This position, close to Occam's nominalism, was nonetheless partly distinguished from it by the fact that for Mach these concepts were "something other than words: they are stable and rich in content because laden with history and experience." That drew attention to social objectivity, and therefore to reality, bred of prolonged experience and use. The question of the accumulation of sensations in stable and transmissible concepts prompted the study of the processes by which a gestating reality solidified. We find the recurrent distinction between a radical relativism and a constructivist position, according to which reality exists to the extent that it has been constructed and shored up in a certain human space and for a certain period. But neither Mach nor Pearson (who adopts the essentials of this concept) developed this perspective further, a perspective admittedly more relevant to the natural sciences they were referring to.

Another important idea of Mach's was that science (the division between "knowledge and error") was constructed through a process of *selection,* of trial and error, in accordance with an adaptive model, analogous to the one Darwin constructed for the evolution of living creatures:[1] "Knowledge is invariably an experience of the mind that is directly or indirectly favorable to us" (Mach, 1904). The mutual adaptation of thoughts and facts, in which observations and theory were expressed, operated through a process of thrift and economy of thought, corresponding to a biological necessity. This idea of the *economy of thought* was essential for Mach. He remarked, "Science can be considered as a problem of the minimum, in which one expounds the facts as exactly as possible with the least intellectual expenditure"—a statement that could well serve as a definition for the program of mathematical statistics.

This view was adopted by Pearson. For him, scientific laws were only summaries, brief descriptions in mental shorthand, abbreviated formulas that synthesized routines of perception, with a view to subsequent, forward-looking uses. These formulas appeared as *limits* of observations that never fully respected the strict functional laws; the correlation coefficient allowed precise measurement of the strength of the association, between zero (independence) and one (strict dependence). The metaphysical idea of causality was renounced in favor of the idea of correlation between observed facts, the mutual connections of which, measured by this correlation, were capable of being reproduced in the future. Just as the reality of things could be invoked to solely pragmatic ends, and on condition that the routines of perception be maintained, in the same way "causation" could only be invoked as an admitted and therefore predictable correlation of likely probability. The fifth chapter of *The Grammar of Science,* which was added in 1911 and dealt with correlation, shows the link between statistical summaries and the needs of practical existence. This formulation opened up the possibility of analyzing the diversity of real worlds through reference to a theory of action, which seemed the sole possible issue to the paradoxes concerning reality elicited by statistics:

> [The] foundation of the idea of causation is in the routine of perceptions. There was no inherent necessity in the nature of this routine itself, but failing it the existence of rational beings capable of conduct became practically impossible. To think may connote existence, but to act, to conduct one's life and affairs, connote of necessity a routine of perceptions. It is this practical necessity, that we have crystallized as necessity existing in "things in themselves," and made fundamental in our conception of cause and effect. So all-important is this routine for the conduct of rational beings that we fail to comprehend a world in which the ideas of cause and effect would not apply. We have made it the dominating factor in phenomena, and most of us are firmly convinced not only of its absolute truth, but of its correspondence with some reality lying behind phenomena and at the basis of all existence itself. Yet as we have seen, even in the most purely physical phenomena, the routine is a matter of experience, and our belief in it a conviction based on probability; we can but describe experience, we never reach an "explanation," connoting necessity. (Pearson, 1912, chap. 5, p. 152)

The idea that the economy of thought resulting from the formulation of simple laws is an "economy of principles and causes, which constitutes the canon of scientific thinking" is compared by Pearson to the "law of parsimony" imputed to Occam and his razor:

> Occam was the first to recognize that knowledge of the beyond in the sphere of perceptions was simply another name given to irrational faith. Hamilton formulated Occam's canon in a more complete and adequate form: one must not recognize either more causes or more onerous causes than are needed to account for phenomena. (Pearson, 1912)

Pearson's choice of scientific method, which denied that "causation" or "the reality of things" exist prior to the routines of perception, led him to a radical solution of the old problem of free will and determinism. In Pearson's perspective *will* could not in any circumstance appear as a primary cause, but only as an intermediary link in the chain leading to the formation and stabilizing of these routines. In this chain, the determining elements could be social or cultural as well as hereditary. When he wrote the following passage (around 1890), Pearson had not yet settled into an almost complete hereditarianism, as he would subsequently do. The direction taken by his research could have led to a culturalist sociology, and even to the definition of a *habitus,* since the ensemble formed by action and routines of perception functioned as a program combining "structured structure and structuring structure"—as Bourdieu put it in 1980—but for the essential difference of the reference to biological heredity. A "construct projected outside itself," to which sensorial impressions were linked, constituted a reserve of acquired experience accumulated in various ways, which would condition what we usually call "will":

> When the exertion is at once determined by the immediate sense-impression (which we associate with a construct projected *outside* ourselves), we do not speak of will, but of reflex action, habit, instinct etc. In this case both sense-impression and exertion appear as stages in a routine of perceptions, and we do not speak of the exertion as a first cause, but as a direct effect of sense-impression; both are secondary causes in a routine of perceptions . . . None the less the inherited features of our brain, its present physical condition owing to past nurture, exercise, and general health, our past training and experience are all factors determining what sense-impresses [*sic*] will be stored, how they will be associated, and to what conceptions they will give rise . . . Science tries to describe how will is influenced by desires and passions, and how these again flow from education, experience, inheritance, physique, disease, all of which are further associated with climate, class, race, or other great factors of evolution. (Pearson, 1912, chap. 4, pp. 124–125, text written before 1892)

From this text—a surprising one, given the current image of a eugenicist and hereditarian Pearson (which he indeed became later on)—one can see how malleable are the reserves of theories and interpretations that a scholar can draw upon, and how dependent on a network of extended connections, in which elements habitually termed scientific, philosophical, social or political are all mingled: when he wrote this passage, Pearson did not yet know what path he would take. The one thing he was certain of was that he wanted to combat the idealism and metaphysical thinking of the old English style of university in favor of a scientific positivism embodied by the physicist Mach.

Pearson thought he had forged the decisive weapon for this battle by substituting the idea of contingent association for that of necessary causality. He swept aside unknowable first causes, as well as strict pin-point associations, in favor of *contingency tables,* which distributed a population according to two distinct criteria of classification. He adopted and gave precise formulation to a new object that had already been imagined by Galton: the *partial relationship* between two phenomena, midway between two limits—absolute independence and absolute dependence—or *correlation,* synonymous with association. The intersecting tables were termed "tables of *contingency,*" because "everything in the universe occurs but once, there is no absolute sameness of repetition." This statement, worthy of medieval nominalism, was an astonishing one to make just as the first stone of mathematical statistics was being put in place—statistics which in later uses, based on conventions of equivalence, could only in effect plow under this initial credo:

> [Such] absolute dependence of a phenomenon on a single measurable cause is certainly the exception, if it ever exists when the refinement of observation is intense enough. It would correspond to a true case of the conceptual limit—of whose actual existence we have our grave doubts. But between these two limits of absolute independence and absolute dependence all grades of association may occur. When we vary the cause, the phenomenon changes, but not always to the same extent; it changes, but has variation in its change. The less the variation in that change the more nearly the cause defines the phenomenon, the more closely we assert the association or the correlation to be. It is this conception of correlation between two occurrences embracing all relationships from absolute independence to complete dependence, which is the wider category by which we have to replace the old idea of causation. Everything in the universe occurs but once, there is no absolute sameness of repetition. Individual phenomena can only be classified. (Pearson, 1912, chap. 5)

Pearson was convinced that these two ideas of contingency and correlation would lead to an essential epistemological revolution: "The subsuming of all the phenomena of the universe under the category of contingency rather than that of causation is one epoch-making to the individual mind." But he also knew that the ideas of *reality, cause,* or *function,* indispensable to practical life and action, were "conceptual limits that man converts into realities exterior to himself." He thus in effect created the possibility of a register of realities different from those of contingent, a priori heterogeneous individuals. He provided tools for assessing the commensurability between individuals and opened up a new space for knowledge and action. Relationships usually interpreted as causalities were in fact "conceptual limits which man has intellectually extracted from them, and then—as his habit is—forgetful of his own creative facility, has converted into a dominant reality behind his perceptions and external to himself. All the universe provides man with is likeness in variations; he has thrust function into it, because he desired to economize his limited intellectual energy" (Pearson, 1912, chap. 5).

The justification of these passages to limit and of these too easily forgotten creations is provided by a *principle of economy,* that governs a limited intellectual energy. In this passage Pearson describes two essential aspects of the construction of an object. On the one hand, we find the exteriorization and passage "beyond perceptions," needed to engender common sense and intersubjectivity (which he analyzes in detail elsewhere); on the other, the principle of the economy of thought, which leads to summaries and mental shorthand. The link between the two aspects is obvious in what constitutes the heart of the modern procedure of defining "statistics," in the technical sense of judicious combinations of elemental data, that satisfy certain requirements of optimization. There criteria serve to create common sense, that rallying point of heterogeneous subjectivities: methods of least squares, the estimation of a probabilistic model through maximum likelihood.

This economic process can also be analyzed in terms of *investment.* Something is sacrificed (contingency, the multiplicity of singular cases) for a subsequent gain: the stabilization of standard forms that can be memorized, handed down, and used again; that can be joined and integrated into more complex machines (Thévenot, 1986). The question of realism and nominalism thus acquires a different content, once it is conceived in terms of the division of the work of constructing objects: initially the product of a mere convention, the object becomes real after being transmitted, key in hand, and used again by others. This alchemy is the bread and butter of any institute of statistics when it publishes unemployment figures or price indexes. The idea of investment is interesting in that it draws attention to the *cost* of stepping through the mirror, of passing from one world of realities to another—a cost of which the budget of the statistical institution is only a part.

Pearson's epistemology was opposed to that of Quetelet in that, after Galton, it emphasized individuals and their variations, denying aggregates and first causes (the famous "constant causes") any reality at all. But from another point of view it also inherited certain of its aspects, in not emphasizing an essential difference between errors in measurement and "real" variances. The connection between objective and subjective averages was so accepted that, as readers will have noticed from the preceding quotations, variances alone were noted, associations were measured by correlations, and regularities were defined as a basis for wagers on the future—the questions of the reality and causality of objects being metaphysical. This methodological agnosticism would become a characteristic trait of statistical practice, at least in certain of its uses.

Francis Galton: Heredity and Statistics

The nuances and interest of the history of science stem from its contingency, from the fact that it does not unfold in accordance with a purely internal cognitive logic. The first edition (1892) of *The Grammar of Science* made no reference to statistics. The third edition, published in 1911, proudly displayed correlation as "an epoch-making" event in the history of ideas. But was it simply a matter of the history *of ideas*? Didn't it also involve the history of Victorian England, of the anxiety raised by the problems of poverty, and by debates concerning explanations and solutions to them? In the interval between these two editions of the book, the road traveled by Pearson intersected with that of Francis Galton (1822–1911) and his questions on biological heredity. Before 1892 Pearson still ranked heredity after education and experience, mentioning it in passing and without dwelling on it: this had not yet become his fight. Twenty years later, to support Galton's eugenicist construct he put regression and correlation into mathematical form and added an original chapter to what had initially been a vulgarization of the ideas of Ernest Mach. He also set up a very active laboratory of biometrics and eugenics in which were forged the mathematical tools of modern statistics, and a network of political and scientific influence that remained in effect until at least 1914. He then ensured a large audience for ideas originally launched by Francis Galton.

Galton was Darwin's first cousin. The theory of evolution developed by Darwin describes the selection and adaptation of species in terms of a continual struggle for survival that allowed only the best-adapted individuals or groups to reproduce and survive. It dealt essentially with animals and vegetables. Fascinated by this theory, Galton wanted to transpose it and apply it to the human race, with a view to the biological improvement of humanity. This is where the essential difference arises between his construct and Quetelet's. Galton cast his attention on the differences between individuals, on the variability of their attributes, and on what he would later define as natural aptitudes, whereas Quetelet was interested in the average man and not in the relative distribution of nonaverage men. Galton did however retain from Quetelet, whom he greatly admired, the idea of the normal distribution of human attributes around an average value. But he used it as a *law of deviation* allowing individuals to be classified, rather than as a law of errors. The pertinent facts were henceforth deviations in regard to the mean—thus no longer parasites to be eliminated, as they had been for astronomers. It became increasingly important to classify individuals in accordance with orderly criteria. To describe these distributions, Galton constructed new objects resulting from this process of putting things in order. The *median* bisected the sample being ordered into two equal parts. *Quartiles* allowed the construction of a new dispersion indicator: the interquartile interval. Terminology and the old ideas were changing: Quetelet's probable error became standard divergence, then standard deviation or variance.

These new forms of objects emerged slowly from Galton's endeavor to create a space of common measurement for something previously seen as incommensurable: human aptitudes. This was an arduous task: how was one to elaborate categories of equivalence and scales of comparability for characteristics which, unlike size, were not easily measured? Initially (1869), Galton studied geniuses, as described by the literature devoted to great men, and the possible heredity of genius, as manifest in particular lineages such as the Bach, Bernoulli, or Darwin families. He systematically directed the interpretation of these family series toward biological heredity and not toward the effect of education and of the immediate environment during childhood, as we would tend to do today.

But this research into one pole of the scale of aptitudes was not enough to order the entire population in accordance with a criterion as natural as weight or size. Some further time was needed (after 1900) to instrument, through techniques such as Binet-Simon's intelligence quotient, or Spearman's general intelligence, measurements deemed indicative of individual aptitudes. During the 1870s and 1880s, Galton used a *social* classification developed by Charles Booth in his surveys on poverty in London (see box). In these, the surveyors (the persons responsible for applying the Poor Law) had to estimate the place of the homes that had been visited in a social and economic scale of eight positions (Hennock, 1987), according to a battery of indications reflecting the style and standard of living, from beggars, criminals, and layabouts to qualified workers and members of the intellectual professions. This scale, which Booth had used to count and compare the different categories of the poor in the various parts of London, was adopted by Galton as an indicator of natural individual aptitude, by associating "civic worth" with "genetic worth."

Like height, this aptitude was innate, programmed into the body and distributed according to a normal law. This Gaussian form of the frequency of degrees of aptitude, associated with the magnitude of the ordered categories noted by Booth's investigators, allowed the "genetic worth" corresponding to these categories to be standardized. Thus the encoding done "with the naked eye" by London social workers led to the naturalization of a scale of individual aptitude, by means of social and economic levels. Resorting to normal distribution in order to give a thing

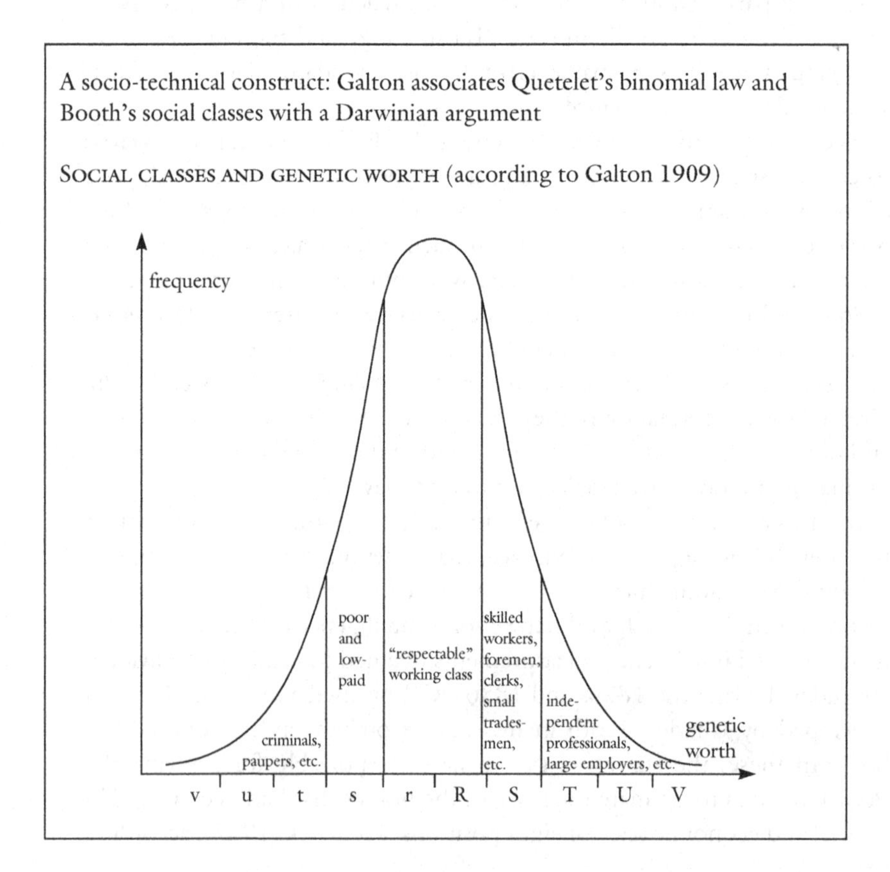

A socio-technical construct: Galton associates Quetelet's binomial law and Booth's social classes with a Darwinian argument

SOCIAL CLASSES AND GENETIC WORTH (according to Galton 1909)

consistency no longer meant the same as it did for Quetelet; for Quetelet, if such a distribution were noted, it allowed the existence of an object more general than individuals to be inferred. For Galton, on the other hand, normal distribution was presumed, by associating what was encoded in the social survey with a biological attribute comparable to height. This allowed one to infer a scale of measurement for an object attached to the individual, his natural aptitude.

But Galton's essential innovation in relation to Quetelet was that he no longer viewed the normal distribution of human attributes simply as the result of a large number of variable random causes, which were small and independent (and therefore indescribable). Rather, he sought to isolate, among these causes, the one whose influence was presumed to be enormous: heredity (Quetelet had never raised this question). The apparent paradox raised by this problem would generate a new intellectual con-

The concern with giving consistency to an inherited human aptitude, by standardizing it on a continuous scale, led Galton to a montage that combined three previously consolidated but unrelated forms: the "normal" distribution of heights, made popular by Quetelet; the division of the population into social categories, used by Booth in his survey of poverty in London; and the idea of inherited individual attributes, both physical and psychological, inspired by Darwin (Galton's cousin), with a view to a eugenicist policy (MacKenzie, 1981). Galton transformed the normal (or "seemingly normal") distribution observed in heights into the supposedly normal distribution of a hypothetical height, an aptitude that was biological in origin, described as a "genetic worth" or "civic worth." He then associated the hypothetical normality of this individual attribute (comparable to height) with the distribution in percentages of the sizes of eight population categories defined on the basis of economic, social, and cultural indices. These indices were observed and recorded during visits to individual homes for a research project on the causes of poverty organized by Booth in 1889.

In supposing that the social level encoded with the naked eye reflected the critical value to be standardized, Galton used the tabulation of the surface limited by Gauss's law to assign numerical values to boundaries between the strata, divided on the *x*-axis of the graph representing that law. This construct did not allow a direct numerical measurement to be assigned to an individual's aptitude (or to a home), but it did express the idea that such an attribute—as biologically natural as height—existed, and needed only to be measured. This measurement came later with Spearman's "general intelligence" (1904), then with intelligence quotients, or IQs (Gould, 1983).

struct: "regression," in the modern statistical sense. The paradox was as follows: attributes (which could be measured) seemed partly hereditary, but only partly. Knowing the height of the father did not automatically determine the height of the son. In addition to the massive cause (heredity), other causes came into play that were numerous, small, and unrelated. And yet, from one generation to the next, not only was the average height (more or less) constant, *but so was its dispersion*. This was the puzzle: how was it possible to reconcile heredity, the chance that the son's height would continue to be the same as the given height of the father, with the fact that there was no change in dispersion between two generations?

Galton had no mathematical training, and was not capable of formalizing the problem. But he had considerable experimental imagination and a fine intuitive grasp of geometry. He was also deeply influenced by the model of Quetelet's law of errors, which he transformed into a "law of deviation." The solution to the puzzle was therefore reached by means of two devices Galton thought up: the "quincunx" and the harvesting of peas, sown in accordance with previously established standards. The combination of these two techniques led him to a new formulation of the problem formerly discussed by Poisson, Bienaymé, and Lexis: could a random process, resulting from drawings made from urns *themselves of random composition,* lead to regular and probabilizable results?

The "quincunx" was a vertical rectangular board, studded with nails arranged at regular intervals in a quincunx pattern (that is, at the four corners of the small squares of a grid, and at the intersection of the diagonals of these squares). Shot was poured from the upper edge of the board. As it fell, the shot spread out at random, collecting in transparent vertical compartments at the bottom. The resulting outline followed a normal curve. But at a later date, Galton made his machine more complex, stopping the fall of the shot at *an intermediate level,* and collecting it in a first series of vertical tubes, describing a first normal curve, of dispersion D_1. Then he reopened the tubes separately. At this point he noted two things: on the one hand, each of these tubes produced a normal distribution, of *the same dispersion,* D_2; and on the other, the reamalgamation of all these small distributions resulted in a *large normal distribution* (of dispersion $D > D_1$). Naturally, the small distributions produced by different tubes were of different sizes, since they contained different quantities of shot; but the important result was that their *dispersions* were the same. Thus Galton dealt with a particular case of Poisson and Bienaymé's problem, in which an initial drawing made from separate urns itself followed a normal law.

Bienaymé, who was certainly a better mathematician than Galton, tended in this direction in his initial criticism of Poisson, mentioned in the previous chapter. But he was not driven, as was the Englishman, by so clearly formulated a puzzle (see box).

Galton thus took a big step toward solving the mystery. If one divided the population of fathers into sections by heights, each section produced a subpopulation of sons with its own dispersion. It was observed that the heights of the two total populations, fathers and sons, followed the same dispersion. Now the dispersion of the sons' heights (we would now say, the total variance) could be broken down into two parts: one part, resulting from the height of the fathers, was due to heredity, and the other was interior to the subpopulations. But since the *total* dispersion remained constant, the dispersion of heredity was *necessarily smaller.* This was how Galton reasoned after seeing the results of his second experiment, which involved the cultivation of standardized sweet peas.

In this culture experiment, Galton divided the "father" peas by weight into seven groups and observed the distribution of the "son" peas, according to the same division. He thus constructed a table of contingency with seven lines and seven columns, valuing each father-son couple according to this double standardization. This table resembled modern "social mobility" tables: its diagonal was heavily laden, while boxes further away from it were less laden. Moreover, the average of the weights of the sons, *whose fathers were of given weights,* varied proportionately to the weights of these fathers, but the breadth of this variation was weaker than that of the fathers: there was a reversion toward the mean. The deviate to mean of the average weight of sons whose fathers were of given weight was only *one third* of the deviate to mean of this paternal weight. In this study carried out in 1877, Galton thus observed a decomposition of variance analogous to that described by his second quincunx, constructed that same year.

The next step was taken in 1885, when Galton managed to collect measurements of the heights of various individuals at the same time as those of both their parents. To this end he set up a special stand at an international health exhibition organized in London in 1884; families passing through were measured and the measurements recorded under headings encompassing numerous physical attributes (they even paid a small sum for it, which financed the research). A complication arose in regard to the sweet peas: there were two parents. Galton thus began by calculating a "mid-parent," the average of the father's and mother's measurements. He then constructed a cross-tabulation analogous to the preceding one. This

The Quincunx: A Dramatization of Statistical Dispersion

Between 1830 and 1870, statisticians laid emphasis on the stability of statistical averages: from one year to the next, the average height of conscripts varied only slightly, and the numbers of marriages, crimes, and suicides were more or less constant. In this perspective, which was Quetelet's, the ideas of statistical dispersion and mean variance could not take shape. It was precisely these two questions that the theory of the "average man" seemed to have resolved, by treating them as contingencies, as parasites comparable to the errors in measurement that an astronomer or physicist strove to eliminate.

Unlike Quetelet, Galton was interested in differences between men rather than what they had in common. The ideal was no longer the average man, but the man of genius. The questions that the eugenicist asked were: How could the human race be improved by producing more geniuses and fewer inept fools? Was genius hereditary? The question of inherited aptitudes turned the spotlight onto two aspects that the theory of the average man eschewed: the dispersion and variation of averages. Children looked like their parents, but were not identical to them. If the father was tall, the son would probably be tall too, but this was not certain. In other words: to the initial dispersion of the heights of the fathers a further dispersion was added of the heights of the sons, for a fixed height of the father. And yet, when all was said and done, the general dispersion of the heights of all the sons was no greater than that of the fathers. Such was the puzzle, counterintuitive in nature, that Galton sought to untangle. To do so, we must imagine a formalization that analytically separated these two successive dispersions.

A posteriori, the formula for linear regression provides the solution

$$Y_i = aX_i + b + \epsilon_i$$

X_i = the height of the father
Y_i = the height of the son
ϵ_i = dispersion of the height of the son for a fixed height of the father.

This is only possible on condition that one draws in a square an ellipse symbolizing the two-dimensional normal distribution of the pairs "height of the father—height of the son." For that, the slope a of the "regression line" Δ, with an equation $Y = aX + b$ is necessarily less than 1.

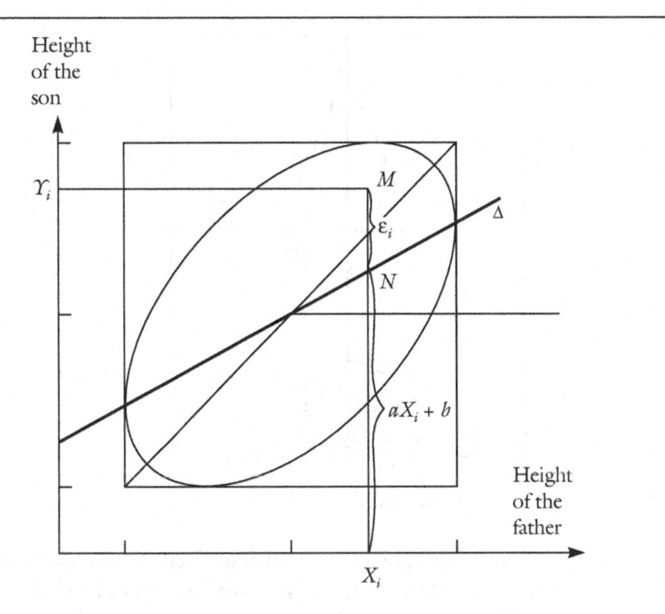

Figure A Relationships between the heights of fathers and sons

In other words, if the father was larger than the mean height, the son was probably larger too, but on average (point N), he deviated less from the mean height than his father did. Precisely for this reason Galton described this formulation as regression (toward the mean), an idea that has disappeared in the modern use of the expression linear regression (of which the slope, naturally, could be higher than 1).

But Galton was no mathematician. To have a proper understanding of (and make others understand) this new idea of *partially* transmitted inherited characteristics, he needed an intermediate apparatus that showed how two successive dispersions could be combined (here: dispersion of the heights of fathers, then dispersion of the heights of sons for a given height of the father). This was the "two-stage quincunx" that Galton imagined and described (no one knows whether he succeeded in constructing it and making it work).

The simple quincunx was a vertical board, on which nails were arranged in a quincunx pattern. Shot released from the center of the plate's upper edge dispersed at random as it fell, and collected in tubes at the bottom. If the quantity of shot was large, its dispersion in the tubes was close to "normal law," the limit of binomial dispersion resulting from the shot colliding with the nails.

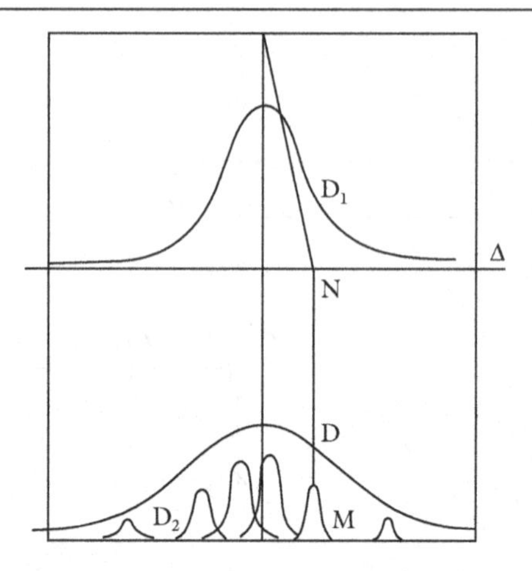

Figure B The two-stage quincunx

This first, simple experiment could have been conceived by Quetelet. Galton finished the experiment by adding a series of intermediate tubes halfway up the board, causing an initial dispersion D_1. Then these tubes—containing shot that had meanwhile been painted different colors—were opened, giving rise to a series of elementary distributions, D_2, at the bottom of the plate. The resulting mixture produced a total dispersion D, which was larger than D_1. Galton thus constructed a mechanism for deconstructing a total variance (D^2) into an interclass variance (D_1^2) and an intraclass variance (D_2^2), since $D^2 = D_1^2 + D_2^2$.

In addition the small dispersions D_2, resulting from the intermediate tubes, were all equal. Or rather: the dispersions of the heights of subpopulations of sons produced by previously sorted fathers were equal. This property was rendered visually by the two-level quincunx.

The quincunx was an intermediary machine. It allowed scientists to overcome the difficulties of adding and combining dispersions. It also showed that, the further to the right a shot fell on the first level Δ, the greater its chance of finally coming to rest on the right. But this was not entirely certain: it could sometimes end up on the left. In other words, a tall father might have a short son, but this was very unlikely (naturally, the analysis had to take into account features inherited from the *mother:* Galton conceived of a "mid-parent," the mean of both parents). But the quincunx did not permit a direct visualization of the linear relationship, of the type $Y = aX + b$, between the variables. That is why Figure A, which is more synthetic, prevailed over Figure B, although Figure B gives a good idea of the difficulties that the inventors of linear regression had to overcome.

time the two marginal distributions (parents and children) were normal. He noted the same effect of "regression toward the mean" of the heights of children in relation to those of "mid-parents," with a coefficient of two-thirds (slope of the "regression line"). On the cross-tabulation, he drew a line uniting boxes of the same value, thus tracing a series of similar concentric ellipses, centered on the point corresponding to the average heights of the two generations. Not knowing how to interpret them, he presented them to a mathematician, J. Hamilton Dickson. Dickson immediately recognized a *two-dimensional normal law.* This law of probabilities joined by two variables had already been identified in the context of the theory of errors of measurement by Bravais, a Frenchman (1846) who had already used the word "correlation" at the time. But Bravais had not drawn any conclusion from it in terms of variations united by two variables, or "co-variations," as Lucien March put it, although March also spoke of the "Bravais-Pearson coefficient" to indicate the correlation.

Calculations That Were Hard to Refute

Galton presented these results in September 1885, in the presidential address he delivered to open a meeting of the anthropological section of the British Association for the Advancement of Sciences. He thus belongs in the context of physical anthropology, in a Darwinian tradition, and not in the context of statistical methods. He was not a member of the Royal Statistical Society, and almost never wrote anything in that body's journal. In his view, and in the view of his audience, what he had to communicate concerned heredity and not the techniques of data analysis. It would later be published under the title *Regression towards Mediocrity in Hereditary Stature* (Galton, 1886). In this perspective, he drew people's attention back to what he thought was his most important discovery: the idea of reversion toward the mean (whence the term *regression,* which survived but completely lost this original idea). What was paradoxical about its formulation was that it could be read in two opposite senses, depending on whether one compared the coefficient 2/3 to 1 (which is suggested by the words "reversion toward the mean"), or to 0 (which was in fact Galton's aim, since he wished to quantify the effects of heredity).

We can see from this example how different, even contrary interpretations could be joined to formally identical statistical constructs. The reading of these constructs was rarely unequivocal, despite the increasingly rigorous purification and stylization that statisticians later engaged in, for this work often (though not always) had the effect of marginalizing an attention to interpretive rhetoric that was as rigorous as the one presiding over the actual formulations. Galton's rhetoric was still fairly simple and did not as such attract attention (at the time). But the ambiguity of the possible readings of his results would later recur, for example, in the interpretation of social mobility tables, where the occupations of fathers and sons intersect. These can be read in reference to a perfectly diagonal table (complete social heredity), and commented on in terms of the mobility and mixture of the social groups. Conversely, they can be compared with the table termed "marginal product" (complete equiprobability and zero social heredity), and support an interpretation in terms of the reproduction and transmission of social status. Whether one is dealing with the different possible readings of Galton's results, or with those of mobility tables, one would have to reconstruct, in each of these four cases, the social, philosophical, or political configurations, thus probabilizing some particular manner of deriving support from these things that have been objectified by statistical methods.

Returning to Galton, let us take two examples of readings of his results, readings that are far removed from what he himself intended. Both examples are French, but originate in the work of authors who were intellectually quite different: Cheysson and Durkheim. Emile Cheysson (1836–1910), a disciple of Frédéric Le Play, was an engineer of bridges and a member of the Société de Statistique de Paris. Quickly acquainted with Galton's paper of September 10, 1885, he reviewed it shortly thereafter in an article published in *Le Temps* (the equivalent of today's *Le Monde*) on October 23, 1885. He gave an astonishing reading of Galton's piece, interpreting the height of the son as "the average of the father, the mother, and his race." This is how he arrived at this conclusion:

> Mr. Galton's law consists of a kind of fatal, irresistible regression, of the individual type toward the average type of his race . . . If we term "deviate" the deviation between the height of an individual and the average height of his race . . ., this law states that the deviate of the stature of the product is equal, on average, to two-thirds of the deviation of the mid-parental stature. Or again, in an equivalent but perhaps more comprehensible form, that the stature of the product is equal, on average, to one-third of the sum of the height of the father, the mother, and the average height of the race. Let T, T', M, and t

represent the heights of the father, the mother, the racial average, and the product. Galton's law is expressed by the equation:

$$t - M = \frac{2}{3}\left(\frac{T \times T'}{2} - M\right)$$

from which we deduce:

$$t = \frac{1}{3}(T + T' + M)$$

(Cheysson, 1885)

Elementary, my dear Galton . . . Cheysson's formulation faithfully transmits the hereditary message that Galton had in mind, but places more emphasis on the "influence of the race," with terminology still close to that of Quetelet:

> In this form, the law clearly demonstrates the racial influence which, tending incessantly to reproduce the average type, impresses its particular stamp on a whole people despite more or less exceptional deviations. Beneath superficial accidents that cross in different directions and neutralize each other, there exists a deep and permanent cause, which acts constantly and in the same direction, suppressing individual deviations and maintaining racial genius. (Cheysson, 1885)

This reading is not very different from the one given by Durkheim who, in *The Division of Labor in Society* (1893), mentions the question of heredity, but relies on Galton to show that the "social group" (and no longer the "race") pulls the individual back toward its "average type":

> The recent researches of Galton affirm, at the same time that they enable us to explain it, this weakening of hereditary influence.
>
> According to this author, whose observations and calculations appear irrefutable, the only characters transmitted regularly and integrally by heredity in a given social group are those whose reuniting sets up the average type. Thus, a son born of exceptionally tall parents will not be of their height, but will come close to a medium height. Inversely, if they are very small, he will be taller than they. Galton has even been able to measure, at least in proximate fashion, this relation of deviation. If one agrees to call the average parent a composite being who represents the average of two real parents (the characters

> of the woman are transposed in such a way as to be able to be compared with those of the father, added and divided together), the deviation of the son, in relation to the fixed standard, will be two-thirds of that of the father.
>
> Galton has not only established this law for height, but also for the color of eyes and artistic faculties. It is true he has made these observations only as to quantitative deviations, and not as to deviations which individuals present in relation to the average type. But one cannot see why the law applies to one and not the other. (Durkheim, 1893, pp. 324–325)

Thus Durkheim bases his remarks on an "author whose observations and calculations appear irrefutable," in order to support a thesis quite removed from Galton's: "The only characters transmitted regularly by heredity in a given social group are those whose reuniting sets up the average type." That is not exactly what Galton meant, since he himself emphasized inherited *individual* attributes, rather than average attributes. It is also striking that, among the attributes transmitted by means of heredity, Durkheim lumps together "height, the color of eyes, and artistic faculties." This is not really consistent with his principle that "the social is explained by the social": the intellectual climate of the late nineteenth century was strongly permeated by Darwinism, which represented scientific modernity.

As these quotations from Cheysson and Durkheim show, statistical innovations were not initially what drew the attention of Galton's contemporaries to his work. The results of 1885 led to the publication, in 1889, of his best known book: *Natural Inheritance* (his earlier *Hereditary Genius* appeared in 1869). Among these innovations, the first to be taken up were the idea of building *a statistical scale,* based on Quetelet's normal law; the measures that derived directly from it—the median, deciles, or the interquartile interval; and, more generally, the techniques for transforming nonmetrical data into metrical data capable of being recorded on this scale. Thus Galton opened up areas of common measurement. In contrast, more time was needed for scientists to notice and reuse the intellectual breakthroughs constituted, in 1877, by the two-stage quincunx and the "reversion towards the mean" observed in the heredity of peas; similarly, in 1885, by the normal two-dimensional distribution of pairs of parent-child heights, and the distribution of variance that led to it. These results had little mathematical formulation, but implied an uncommon degree of geometric and statistical intuitiveness for that time. The regression line of the height of children (y) in relation to that of their parents (x) was drawn graphically based on concentric ellipses and the points of the ellipses with vertical tangents (whereas an inverse regression, of x on y, was possible based on points with horizontal tangents). The slope of this line was measured graphically, and no optimization, of the type of least squares, was yet in sight.

This formalization was prompted by a desire to shore up the sociopolitical eugenicist construct through measurements of the effects of heredity. That explains its dissymetrical nature: parental heights influenced the heights of children. There were "explicative" variables and "explained" variables. But Galton soon applied it to pairs of brothers, then slightly later to measurements of limbs (arms and legs) from the same human being or animal. These cases no longer justified an asymmetrical treatment, and that led to measurements of co-relationships, or *correlations*. The study of relationships between various parts of the same human being had been suggested to Galton by the anthropological writings of Alphonse Bertillon (1853–1914), an expert from the Paris police headquarters. Bertillon researched a set of measurements allowing the unique and particular identification of every individual, thus facilitating the search for delinquents and criminals. But he did not think to wonder about the possible interdependence of these measurements. The calculus of correlations was well suited to this question. Furthermore, Galton contributed another solution to the problem of identifying people, one that rivaled Bertillon's: fingerprints. Thus the two problems—how to create common measurements and unique identification of individuals—seemed like corollaries. The policeman and the new model statistician (that is, Galton rather than Quetelet) had in common that they were both interested in the relationship between each individual and his relative global distribution. The difference was that Bertillon had his eye on individuals, whereas Galton was concerned with analyzing distribution. But both of them conducted surveys, managed files, and researched good indices, with a view to constructing a reality on which a judge or politician could rely when making a decision involving widespread agreement—thanks mainly to the quantity and quality of the proof accumulated by either the policeman or the statistician.

These were the very instruments of proof that English-style mathematical statistics introduced with the tradition begun by Galton and Pearson. A proof is a complex, more or less fragile construct, which combines different things into one entity and makes it hold up against the blows that various people then try to deal it. For Quetelet and others after him, a normal distribution was proof of the link between singular individuals. Lexis destroyed this proof with his test. For Galton, however, this normal distribution was acquired knowledge, which no longer needed to be proved. It could therefore be used to standardize a scale of aptitude. In comparison, however, with the tools subsequently developed by Karl Pearson, then by his son Egon Pearson and Jerzy Neyman and finally by Ronald Fisher, the reasoning employed by Quetelet, Lexis, or Galton seemed extremely summary. During the 1890s, a discontinuity appeared in England that would radically transform the apparatus of proof in numerous domains, scientific or not.

The chronology and social and intellectual context of this decisive episode have been abundantly described by various authors, to whom I make frequent reference. These authors have tried mainly to elucidate the link between what would later appear to be simply a technology with a mathematical basis, and the social Darwinism and eugenicist activity that inspired some of the major protagonists in this story, beginning with Galton, Pearson, and Fisher. It so happens that the reference material (files, correspondence, memoirs) left by the members of the laboratory that Pearson founded in 1906 was very rich and has been extensively studied. Some scholars have examined it with a view to reconstructing a minute (and precious) internal history of the tools, giving only secondary mention to the evolutionist and eugenicist perspectives with which these techniques were originally connected (E. Pearson and Kendall, 1970; Stigler, 1986). Other scholars subsequently embraced the opposite approach and reconstituted these other aspects of the episode (Norton, 1978; MacKenzie, 1981).

These two forms of research, internalist and externalist, have often been opposed in a recurrent and inconclusive debate as to whether a science develops mainly through an internal, cognitive dynamics, or whether the historical and social context is what essentially determines its path. In the latter case, this science could conceivably be more or less damaged, through an implicit rebound, by the denunciation of its "social origins." A critical mood has doubtless existed at times, especially during the 1970s; but it does not stand much development, because the tools clearly become detached from their origins and subsequently have one or even several different lives. The response of the opposing camp—which states that science has its own logic and necessity and that the context of its birth and development is contingent on this internal necessity—raises more problems than it solves. It does not allow us to think out in detail how a scientific rhetoric and practice are in fact articulated (or *translated,* to use Michel Callon's term [1989]) into other rhetorics or practices. This is especially true of statistics, which was destined to be used as an instrument of proof in other fields, whether scientific, administrative, or political.

From this point of view historical research of any trend can prove highly illuminating, for example, in providing detailed analysis of controversies concerning specific technical points. Thus Pearson found himself caught up in such debates throughout his life. Stigler describes the discussions Pearson had with Weldon and Edgeworth, while MacKenzie studied the bitter polemics Pearson engaged in with Yule and Fisher. The typically internalist manner of interpreting controversies is to find out who, retrospectively, was right, and to pin a medal on the person already crowned by history. In statistics, this is not especially easy, because history has not in general really spoken out on it; but in other fields, it is a frequent practice. The externalist reading, in contrast, tends to ignore the contents and arguments of discussions in favor of research into the hidden interests of the participants, in terms of social groups or of positions in larger contexts overdetermining the possible choices of position.[2]

Five Englishmen and the New Continent

A classic case of controversy was the quarrel over who came first. We have already seen several examples of this, with least squares (Gauss or Legendre), and correlation (Bravais, Pearson, or Edgeworth). In this case, the obvious reading in terms of individual interests seems to suffice. It is nonetheless useful to seek why and how a finding becomes visible at one moment rather than another. We can thus analyze the translations that have allowed one of the actors to see the paternity of an object be ascribed to him: such is the case with Legendre or Pearson, in relation to Gauss or Edgeworth. Within the small group of the founders of mathematical statistics, there was no dearth of such quarrels. Each man had a particular way of articulating the new statistical language that the group fashioned collectively, with other languages that came from other sources. These particularities were linked to their training, their university positions, and their intellectual or political commitments. We need only give the examples of five Englishmen, with well-etched profiles: Galton, Edgeworth, Weldon, Pearson, and Yule. They invented a new continent of statistics, and also helped move it from Europe to England before it crossed the Atlantic.

According to Stigler, Francis Galton (1822–1911) was the "romantic figure in the history of statistics, perhaps the last of the gentleman scientists." Born into a typical upper middle-class cultivated English family (Darwin was his cousin), he studied medicine at Cambridge but did not practice it. He traveled about the world, and became interested in geography and weather forecasting: he published the first meteorological maps with a wealth of technical details and original conventions for representing weather conditions. But his main fight stemmed from the conviction that it was possible to improve the human race using Darwin's results in regard to heredity and natural selection. Ten years after Darwin's *Origin of Species* (1859), Galton published *Hereditary Genius* (1869), containing examples of whole families of exceptional men. In their time, these works were seen as participating in the great challenge issued by science and reason against the Church and traditional obscurantist religion. Galton hoped that a "scientific priesthood" would be established, replacing the old holy order, and declared that he had demonstrated the absence of any statistical proof of God's existence.

In this perspective, eugenics presented itself as a rational method for orienting the evolution of the species—a kind of scientific alternative to Christianity and its fatalism. The aim of statistical tools was to measure and undo the effects of heredity. If men could be classified in order, from the very best to the very worst, the technique of regression showed the proportion of specific qualities and defects in both good and bad that was preserved by heredity. These results were associated with the fact that the English upper classes—that is, the better, more cultivated classes—had fewer children than the lower classes, which contained the more inept specimens. This led eugenicists to prophesy, using science to support their remarks, a bleak long-term future: if nothing was done, the quality of the English nation would inexorably decline. Such was the kernel of hereditary eugenics as adopted and developed by Karl Pearson. It is of course difficult, in light of the later criminal practices which, during the 1920s, 1930s, and 1940s (at least) were based on this politico-scientific construct, to imagine that before 1914 this argument was seen by many as an element in the fight of progress and science against ignorance and darkness. But this is unfortunately the case.

Francis Edgeworth (1845–1926) cut a more discreet figure than Galton. He is best known as an economist. After studying literature, law, and mathematics, he set about applying the research done by psychophysicists on sensations and their measurements to the economic theory of utility. He was the author of one of the first works of mathematical economics, *Mathematical Physics* (1881). He reflected deeply on the "different capacity of individuals for experiencing happiness," and thus became interested in Quetelet, Galton, and the normal law. He produced a large body of theoretical work, notably in mathematical statistics, even before Pearson, but he lacked Pearson's talents in constructing a solid and coherent network. His work was not driven mainly by a political agenda, and eugenics did not interest him. Galton, aware of his own inadequacy in mathematics, took steps to associate himself with a specialist in that field: he sought the help of Edgeworth. But the two men had no common scientific project or political attachment. The statistical work undertaken by Edgeworth during the 1880s and 1890s was at least as important and innovative as that done by Pearson, but Edgeworth did not have at his disposal the channels and research institutions that Pearson was able to set up.

Edgeworth was one of the few to have recognized, before 1890, the interest posed by Galton's work in terms of cognitive schemes—not only from the point of view of heredity, which held no interest for him. In a text written in 1885, he reproduced the schema resulting from experiments with the quincunx, showing the decomposition of a large normal distribution into small conditional distributions of equal variances (which he termed "moduli"). This led him to formalize the analysis of variance (he speaks of "entangled moduli") and the tests of comparison of means (Edgeworth, 1885). He studied the general properties of the normal law with n dimensions, which he wrote in matrix form (Edgeworth, 1892). In 1893 he explained correlation as a "product-moment" (that is, the normed total of the products of two correlated variables). But none of his formulations was then taken up and retranslated by others, and the standard expression of the coefficient of correlation is attributed to Pearson, who published it three years later, in 1896. The difference between them is that Edgeworth soberly entitled his article "Exercises in the Calculation of

Errors," which was technical but established no link with anyone at all, whereas Pearson's title—"Mathematical Contributions to the Theory of Evolution: Regression, Heredity and Panmixia"—managed to combine mathematics, Darwin, Galton, and heredity (not to mention a fresh subject, panmixia, that did not survive). Evolution and heredity were then crucial points of interest for Pearson, and aroused the curiosity of a larger public than did Edgeworth's mathematical formulas.

Raphael Weldon (1860–1906) was a biologist. He was the story's "go between," the one who connected Galton with Edgeworth and Pearson. It was also Weldon who oriented Pearson's work toward a discipline that the two men would create together: biometrics. Weldon had been educated in a tradition of evolutionist biology, the aim of which was to construct genealogical trees linking the different species, based on the observation of morphological transformations. But this tradition lacked the intellectual means for studying such transformations on a large scale—nor did it even think to do so. Weldon took up Galton's statistical methods in order to describe and analyze measurements obtained in the raising of crabs and shrimp. He was initially helped in this by Galton, but the material proved so complex that he was forced to seek (as Galton had done before him) a mathematician to develop his analysis. He showed his data to Edgeworth and Pearson (in 1892). It seems that both men had the idea of studying multiple correlations, and thus of formalizing Galton's intuitive perceptions from graphs, in the light of Weldon's shrimp and crabs. Pearson seized the opportunity and transformed it in his own way, incorporating it into his preexisting philosophical and political concepts. Their collaboration gradually resulted in the institutional kernel that led to the laboratories of biometrics and eugenics. The connection with the world of pure biology was, however, broken when Weldon died in 1906 (a factor that doubtless contributed to the bitterness of subsequent debates, in which Pearson was pitted against the Mendelian biologists).

Karl Pearson (1857–1936) took the whole process one step further. He created a scientific network, institutions, and a new language. He acquired considerable repute and attracted students from other continents. He was capable of action in vastly different spheres—mathematical, philosophical, and political—and it was this combination of abilities that constituted the strength of his endeavor, the kind of strength that Edgeworth lacked: although his statistical innovations were a priori as pertinent as those of Pearson, he had neither the qualities nor the desire to be a manager. In the early part of his life, before 1892 (the year in which he published *The Grammar of Science*, and began working on Weldon's data), Pearson studied mathematics at Cambridge, history and philosophy in Germany. He was attracted to socialism, or at least to what the Germans called "professorial socialism," which was a criticism of the traditional bourgeoisie as seen by scholars and professors, rather than by the working classes. This attitude, which could be compared to that of the French Saint-Simonians, led to a rationalist and activist scientism. It encouraged the demand for an increase in social power for the more competent individuals, as selected by higher education: the English *professionals* emerging from Oxford or Cambridge, and the French engineers trained in the Grandes Écoles, as opposed to the established conservative aristocracies.

This criticism of traditional mores led Pearson to play an active part in a militant Men's and Women's Club that sought a transformation of the social role of women. From this point of view, the eugenics of the 1880s and 1890s was sometimes presented by Pearson in a feminist, working woman's light. The overfertility of working-class women created disastrous living conditions, prevented them from raising their children correctly, and provided the middle classes with a surplus of labor that enabled them to pay ridiculously low wages. The only way that working-class women could escape from this vicious circle would be to have fewer children. Obviously, this solution was more acceptable than the version of eugenics that spoke of degeneracy and the need to separate the wheat from the tares; but it is striking that the two forms could coexist, in debates that also belong to the prehistory of feminism and the movement in favor of birth control (Zucker-Rouvillois, 1986).

At the time when he met Weldon and plunged into statistics, Pearson had at his disposal three categories of intellectual and social resources: mathematics, the philosophy of science, and an ideological and political network. He would gradually make use of and reconvert all three categories for his projects, and in 1906 this resulted in a strange phenomenon: two barely distinct laboratories, one of biometrics, the other of eugenics (neither was described as "statistical"). By 1936 this combination had given rise to *three* laboratories: one in applied statistics, with Egon Pearson; one on eugenics, with Ronald Fisher; and another in genetics, with Haldane. Pearson's mathematical competence, which was immediately brought to bear in 1892, allowed him to assume a dominant position. At that time, in fact, only Edgeworth could rival him; but Edgeworth, the pure academic, thought that only scholarly questions were at stake, whereas Pearson involved himself in the theory of evolution and heredity (as can be seen from the titles of their articles, noted above).

At first, Pearson went so fast he rather forgot the epistemological creed expressed in *The Grammar of Science* and interpreted statistical distributions, both normal and asymmetrical, in a more realistic way, like Quetelet: I shall presently mention his early debates with Weldon and Edgeworth, which took place between 1892 and 1896. But around this time he succeeded in reconciling the new idea of correlation with his previous philosophical critique of causality, which descended from Mach. This gave statistical rhetoric a considerable degree of autonomy, allowing it to free itself from the constraint of having to integrate "external causes"—that is, differently constructed facts. Of the period between 1892 and 1900, the event that subsequently proved the most decisive was probably this conquest of the autonomy of statistical discourse, which was based not only on mathematics but also on a philosophy of knowledge that in large measure suspended the constraints of having to combine with other discourses. This position did not, of course, result directly from Mach's thought and Pearson's interpretation of it. But it was soon to constitute normal professional behavior. It was sufficient to make the profession of statistician autonomous, with attendant university courses and jobs in various specialized institutions.

We shall never know what Lucien March, director of the SGF, had in mind when he translated this ponderous philosophical work, which had so little to do with statistics, at a time when his institution was still very modest in size, comprising only five or six professional "statisticians." But as the reader may have noted, English administrative statistics was completely absent from the Galton and Pearson saga. Only much later, in the United States of the 1930s, was a connection established between administrative statisticians and mathematical statisticians. This makes March's translation of Pearson's *Grammar* all the more significant, but also more puzzling. It is possible that this connection with Pearson was favored by two factors: March's demographic interest in eugenics, and the personality of Yule, an essential link in the chain connecting the new English statistics and the economic and social sciences.

* * *

George Udny Yule (1871–1951) was actually the direct ancestor of all statisticians working in this field. His manual, *An Introduction to the Theory of Statistics* (1911), has been published in fourteen editions, the most recent being in 1950 (Kendall collaborated after 1937), and has helped form several generations of students of economics and sociology. Yule studied engineering in London and physics in Germany, before becoming one of Pearson's first students, then his assistant in 1893. While initially proving to be his master's faithful disciple in regard to techniques of regression and correlation, he applied them to entirely different fields and thus entered into contact with other circles. Theories of evolution and eugenics did not interest him, but in 1895 he became a member of the Royal Statistical Society, to which neither Galton nor Pearson belonged. Founded in 1836, this association brought together statisticians in the nineteenth-century sense of the term. Concerned with solving social and economic problems involving poverty and public health, these "statisticians" sought to treat these problems as distinct from passional and polemical arguments, objectifying them by means of statistics generated by administrative or private sources. Linked to churches, to philanthropy, and to hygienist movements, these "ameliorists," as they were known at the time, were often criticized by the radical eugenicists, who accused them of impeding the course of natural selection by their efforts to help the most wretched—in their view, the most inept. The debate between these two currents of thought, hereditist and environmentalist, would last from the 1870s into the 1930s and 1940s. Pearson, of course, belonged to the hereditist trend; Yule was closer to the environmentalists.

This affinity led Yule to use these new tools in the crucial political debate of the time, which involved the means of giving assistance to the poor. Should help be guaranteed according to the strict forms of *in-relief* prescribed by the Poor Law of 1834—that is, in closed asylums, or workhouses (forced labor, very badly paid); or through *out-relief*—aid administered in the home, intended more for the elderly, for families, and the sick? These two forms of assistance were guaranteed at a local level by the Poor Law Unions established in each county. The relative proportion of in- and out-relief was thought to reflect the strictness or leniency of the management style of every *local* union. The political debate was over relief administered in the home: didn't the extent of this aid help to maintain poverty, even to increase it? Yule had data on each of the 580 unions: the sum of the total number of poor people receiving aid (outdoor plus indoor) was supposed to measure pauperism, whereas the ratio between the two forms of aid (the ratio of indoor to outdoor relief) was an indicator of the leniency or strictness of the local union management. Yule first calculated the correlation between these two variables, then the regression of the first over the second. The positive correlation (0.388 with a probable error of 0.022) led him to conclude that pauperism decreased when help administered in the home was less easily granted. He then complicated his model by reasoning in *variation* (between 1871 and 1891), and by integrating other explanatory variables: the proportion of elderly persons and average salaries, according to the unions (Yule, 1899, 1909).

Yule published five articles on this subject between 1895 and 1899. Reading them, one can easily get the impression that his first concern was to demonstrate the interest offered by the new tools, while at the same time putting his own stamp on them. This is most apparent in an article of 1897, in which he uses—for the first time in a context other than the theory of errors—the method of *adjustment by the least squares:* he defines the line of regression as the one that minimizes the sum of the squares of the deviations of the observations on the adjusted line, something neither Galton nor Pearson had done before. At the same time, he also presents *regression with several variables,* since, in order to analyze the variations of pauperism he uses as explanatory variables not only the variance of out-relief but also the variance of the population and of the ratio of old people (Yule, 1897). The coefficients of this multiple regression are named "partial regression," while "partial" and "multiple" regressions are also presented. Seen in retrospect, this use of the method of least squares and formulation of multiple regression appear to have been the most notable contributions of this article. And yet its title, "On the Theory of Correlation," gives no hint of these two points (not forgetting that Pearson's article of 1896, formulating the coefficient of correlation, was entitled "Regression, Heredity, and Panmixia." An astonishing display of footwork . . .). On the other hand Yule had the feeling that his original contribution (implicitly in regard to Pearson) was in demonstrating that it is possible to free oneself from hypotheses of normalcy and adjust a cluster of points by means of linear regression. These hypotheses were closely connected with measurements of the human body, from Quetelet to Galton and Pearson. The translation of these tools with a view to subjecting (rarely normal) economical and social data to them involves this generalization. It carries the embryo of future conflicts between Yule and Pearson, linked precisely to these hypotheses of underlying normalcy and, more generally, to the nature and extent of the new realities they had constructed.

Controversies Concerning the Realism of the Models

In December 1890 Pearson found himself in charge of a geometry course at the University of London. He taught what would later constitute *The Grammar of Science*. The only mention of statistics appeared in an account of the various types of graphs and of geometric diagrams. He did not mention Galton, although he had read his *Natural Inheritance* in 1889, and had commented skeptically on it at the Men's and Women's Club, where he had already raised the question of the realism of the mathematical formulations applied to "descriptive sciences" (in 1889):

> Personally, I ought to say that there is, in my own opinion, considerable danger in applying the methods of exact science to problems in descriptive science, whether they be problems of heredity or of political economy; the grace and logical accuracy of the mathematical processes are apt to so fascinate the descriptive scientist that he seeks for sociological hypotheses which fit his mathematical reasoning and this without first ascertaining whether the basis of his hypothesis is as broad as that human life to which the theory is to be applied. (Pearson, speaking at the Men's and Women's Club on March 11, 1889; quoted by Stigler, 1986, p. 304)

We can better understand, on reading this text, how hard Pearson later had to work in order to reconcile his philosophy of science with his mathematical constructions—a labor that lead him to adopt the idea of contingency and to reject causality. He entered the arena in 1892, attacking Weldon's measurements of crabs and shrimp. He examined their distributions and normalcy, in the perspective of Quetelet and Galton. Observing clear asymmetry, he plunged into the mathematical dissection of these *skew curves* (asymmetrical curves), dividing them into two or more normal laws. His problem was not unlike the one that had confronted Adolphe Bertillon, with his bimodal distribution of conscripts from the department of Doubs. But Pearson sought a mathematical representation, which was of necessity very complicated, that would fit the observations as well as possible (something he had actually mistrusted, three years earlier). To this end he made calculations in order to estimate the parameters of these complicated laws that combined several normal laws, the *moments* of successive order (up to four or five) of the observed distributions, beyond the *mean* (first order moment) and *variance* (second order moment). This "method of moments," used in order to find fits for the *skew curves,* was his first major work in mathematical statistics. It would be used in his laboratory for roughly twenty years.

This work did not fail to raise objections on the part of those few colleagues capable of reading it: Galton, Weldon, and Edgeworth. They criticized him—rather as he himself had criticized Galton three years earlier—on the realism of his models. In addition, they raised a question of more general interest: how was one to take into account *what was already known from other sources,* for example, in order to eliminate possible aberrant observations—something Pearson emphatically refused to do? Despite his pronouncements to the contrary, his initial way of proceeding seemed in direct descent from Quetelet, for whom normal distribution attested to an underlying homogeneity and a constant cause. In dissecting the "skew curves," he seemed to be seeking several constant causes, which his analysis would attest to. But, asked Galton during the discussion, what were they to do with the little they already knew?

> The law of frequency is based on the assumption of perfect ignorance of causes, but we rarely *are* perfectly ignorant, & where we have any knowledge it ought of course to be taken into account. (November 18, 1893)
>
> [Your charts] are very interesting, but have you got the *rationale* quite clear, of your formulae, and the justification for applying them? Or are they mostly mere approximative coincidences?
>
> The estimated mean for tints [an example in Pearson] must fall into line with some other physiological processes, but which ones? & what is the common feature in their causation?
>
> Wherever the normal law is found inapplicable, it is most probable that some big dominant causes must come in, each of which might be isolated and separately discussed and the law of its frequency ascertained. (June 17, 1894) (Galton, quoted by Stigler, 1986, pp. 335–336)

Although closely involved in the undertaking, Weldon was aghast at the deluge of formulas and felt he had been stripped of his knowledge as a biologist. He wondered if he could trust Pearson, whose unmathematical reasonings seemed rather lax to him. He wrote to Galton:

> Here, as always when he emerges from his clouds of mathematical symbols, Pearson seems to me to reason loosely, and not to take any care to understand his data . . . Pearson, whom I do not trust as a clear thinker when he writes without symbols, has to be trusted implicitly when he hides in a table of Γ-functions, whatever they may be. (February 11, 1895)
>
> The more I think over Pearson's theory of skew variation, the less evidence I see that it is a real thing. (March 3, 1895)
>
> I am horribly afraid of pure mathematicians with no experimental training. Consider Pearson. He speaks of the curve of frontal breadths, tabulated in the report, as being a disgraceful approximation to a normal curve. I point out to him that I know of a few great causes (breakage and regeneration) which will account for these few abnormal observations: I suggest that these observations, because of the existence of exceptional causes, are not of the same value as the great mass of the others, and may therefore justly be disregarded. He takes the view that the curve of frequency, representing the observations, must be treated as a purely geometrical figure, all the properties of which are of equal importance; so that if the two "tails" of the curve, involving only a dozen observations, give a peculiarity to its properties, this peculiarity is of as much importance as any other property of the figure.
>
> For this reason he has fitted a "skew" curve to my "frontal breadths." This skew curve fits the dozen observations at the two ends better than a normal curve; it fits the rest of the curve, including more than 90% of the observations, *worse*. Now this sort of thing is always being done by Pearson, and by any "pure" mathematician. (March 6, 1895) (Weldon, quoted by Stigler, 1986, pp. 337–338)

These letters, which Stigler exhumed from the Galton archives, are illuminating. Here we see Weldon—whose role as an intermediary between Galton and Pearson was essential to what followed—admitting his anxiety before the unbending self-confidence of the mathematician he had introduced into the sheep pen of biologists: "But what does he know of my data?" However, these fears were not enough to disrupt their collaboration, which remained close until Weldon's death in 1906.

This was not the case with Edgeworth, whose age, character, and chief interests kept him further away from Pearson and his much-publicized ventures. In fact Edgeworth and Pearson became rivals, between 1893 and

1895, over the study of the skew curves produced by Weldon's data. Edgeworth wrote an article on the subject in 1894, but when it was rejected for publication, he suspected Pearson of having had a hand in the matter, although this was never proved (Stigler, 1978). Basically, they indulged in polemics as to the significance of the normal laws that Pearson thought were underlying the skew curves. For Edgeworth, the link between normal laws and skew curves was too fragile to support an argument. The curves were simply empirical ad hoc constructions. They could not be invoked to *infer* homogeneity. If they fitted the data well, "the question was, how much weight should be attached to this correspondence by someone who saw no theoretical reason for these formulas." On the contrary, it was necessary to submit *hypotheses* of homogeneity, based on knowledge from other sources. Thus Edgeworth imputed a kind of tardy "Queteletism" to his rival.

Sensitive, perhaps, to this criticism, Pearson was gradually incorporating his antirealist philosophy of knowledge into the interpretation of statistics. This enabled him to make an elegant escape from the questions raised by Weldon and Edgeworth. If there were no "cause" external to the statistical construction, and if mathematical formulae were simply mental shorthand, then everything would become possible. In particular, the problem of the link with other branches of knowledge (Weldon's "fractures and regenerations") became less constraining. Pearson was certainly less dogmatic, if only in establishing and maintaining ties with worlds different from his own; but this form of argument offered him an almost invulnerable position of withdrawal, in a context of objections bearing on the realism of his objects. Criticized by Edgeworth for having inferred on the basis of a good fit the presence of a normal law and a simple explanatory cause, he replied: "The question is not to know if those are really the elements, but if their total *effect* can be described so simply."

This formulation offers the advantage of allowing one to slip from one register to another, almost unconsciously, depending on the interlocutors and situations involved. In certain cases, things exist because other people need them, and ask to be provided with things that are really there and that hold. In other cases, faced with criticisms as to the hypothetical and constructed nature of things, one can describe them as mental shorthands or expedient conventions. This continual slippage is neither a deception nor a trick. Both positions are equally tenable and necessary to social life. What matters is to be aware of this and not become entrenched in one stance or the other, viewing it as the only true philosophy of knowledge. Each of these positions is accurate depending on the situation. At certain moments, it is better to be a realist. At others, a degree of nominalism can help us rediscover things long encapsulated inside larger things that take up the whole stage.

Around the turn of the century, a form of statistics shaped largely by mathematics was born of the alliance between biologists and mathematicians, giving rise to *biometrics.* Other alliances sprang up later. Yule had already begun to shift the calculations of regression and correlation in the direction of economics where, thirty years later, Irving Fisher and Ragnar Frisch would give birth to *econometrics.* Spearman, a psychologist, standardized the results and interpretation of intelligence tests by means of factor analysis—a statistical technique derived from those of the biometrics laboratory. He thus outlined the shape of psychometry. In both cases, the objects that had been created and manipulated were again called into question: those of Yule by Pearson himself and those of Spearman by Thurstone and subsequently by many others.

Yule and the Realism of Administrative Categories

When he first made use of the new tools in 1895, Yule invented a language to deal with the then burning question of poverty and relief. He compared the relative weights of the diverse possible causes of variations of pauperism in order to provide evidence to support people who were trying to reform the Poor Law of 1834. Yule offers an elaborate example of a political problem being translated into an instrument of measurement that allowed arbitration of a controversy. The problem in question had haunted English legislators for three centuries: how were they to help the poor in a way that dispelled the social danger they constituted, but in an economically rational manner? English history is punctuated with Poor Laws: there was one in 1601, another in 1795 (Speenhamland's Law), and of course the law of 1834, establishing workhouses and the distinction between in- and out-relief. Throughout the nineteenth century, debates over this law polarized philosophers, statisticians, and economists (Polanyi, 1983). Surveys designed to classify and count the poor were organized during the 1880s by Charles Booth. Galton used them in constructing his scale of aptitude.

But the only *regular* information on the evolution of pauperism was provided by the statistics of the administration of the Poor Law Unions, which made a distinction between relief afforded in the home and that provided in the asylums. In the debates of the time, "pauperism" was something that could be measured. It was the number of people receiving assistance through application of the Poor Law, just as today unemployment is measured by the number of people registered at unemployment agencies. The object exists by virtue of its social codification, through the reification of the results of an administrative process with fluctuating modalities. It is this slippage from the process to the thing that made Yule's conclusion so ticklish to interpret. Having to do with the correlations between measurements resulting from the same process, this conclusion could be read either as arithmetical evidence or as information on the effects of a social policy.

Yule studied *three* series describing, from 1850 to 1890, (1) the total number of persons receiving relief; (2) those receiving out-relief; and (3) those receiving in-relief [(1) = (2) + (3)]. The first two series were strongly correlated, showing, in particular, a marked reduction between 1871 and 1881, whereas the third series (the workhouse) had little connection with the first two. The same correlation was observed for the year 1891, among the 580 local county unions. The dominant fact was that the politics of out-relief had become more restrictive, especially between 1871 and 1881; since relief administered in the workhouse was far less variable, "pauperism" had been reduced. This seems obvious, a priori, from the arithmetic. But Yule's question was far more complex: what proportion of the reduction of pauperism was due to administrative changes in relief management, and what proportion was due to other causes—changes in the total population, or in its age structure? He calculated the multiple linear regression in the change of the pauperism figure (in each county) against changes in the three supposedly explanatory variables: out-relief, the total population of the county, and the proportion of elderly persons in that county. With the following words, he concluded the first "econometric" study ever carried out:

> Changes in rates of total pauperism always exhibit marked correlation with changes in out-relief ratio, but very little correlation with changes in population or in proportion of old in the different unions.
>
> Changes in out-relief ratio exhibit no correlation one way or the other with changes of population or proportion of old.
>
> It seems impossible to attribute the greater part, at all events, of the observed correlation between changes in pauperism and changes in

> out-relief to anything but a direct influence of change of policy on change of pauperism, the change in policy not being due to any external causes such as growth of population or economic changes.
>
> Assuming such a direct relation, it appears that some five-eighths of the decrease of pauperism during 1871–81 was due to changed policy. The decrease during 1881–91 cannot be so accounted for, policy having scarcely changed during that decade.
>
> In both decades there were considerable changes in pauperism not accounted for by changes in either out-relief ratio, population, or proportion of old. These unaccounted changes were decreases in the more rural groups, but increases in the metropolitan group in both decades. The unaccounted changes are the same in sign, and of the same order of magnitude as the changes in in-door pauperism, and so are probably due to social, economic, or moral factors. (Yule, 1899, pp. 277–278)

Neither in Yule's account nor in the commentaries that followed was the meaning of the object "pauperism," as *defined* by the number of persons receiving aid, explicitly discussed. But the demonstration he offered resulted in this meaning's being called into question, since the variable explanatory principle—the decline in out-relief—was described as a *change in policy.* Yule even specified that five-eighths of the decline in pauperism were due to it, the rest being attributable to causes connected with society itself and not with the recording process. It would seem everyone understood that this study implied that variations in pauperism measured in this way only partly reflected the changes in a possible "real poverty," which no one, however, mentioned explicitly. Was this self-evident? Was it useless to speak of something everyone had understood? At the end of the meeting of the Royal Statistical Society two economists, Robert Giffen and Edgeworth, spoke out. For Giffen, Yule's results

> confirmed in a general way the impression regarding it, vis-à-vis, that the administrative action of the local Government Board and poor law authorities counted for something in the reduction of pauperism in the last thirty years . . . Those connected with poor law administration had been able to notice in a great many cases that, if the strings were drawn tightly in the matter of out-door relief, they could immediately observe a reduction in pauperism itself. (Giffen, in Yule, 1899, p. 292)

Edgeworth observed that some people had thought they could deduce, from a strong increase of pauperism in Ireland, that this country had experienced a corresponding decline in economic health: had they read Yule's work, they would have wondered "whether administrative changes might not in large part explain this growth in pauperism." (Edgeworth also worried about the normality of the distributions of the various figures used in the calculations.)

Thus Yule, Giffen, and Edgeworth understood that the variations in pauperism had something to do with the manner in which it was managed and recorded, but none of the three went so far as to question the actual use of the word pauperism to designate this administrative recording. Thus Giffen's first sentence could be read as meaning "the administration's action has succeeded in reducing poverty"; but then doubt appears in the next sentence. It is of course standard to observe the reification of an encoding procedure, resulting in the creation of a thing that exists in itself, independently of this initial step—for example, delinquency or unemployment. The variations in these "things" are then ambiguous. They are most often read as reflections of social phenomena. But they can also be read as reflecting changes in the behavior of the police or of employment agencies. Econometric before its time, Yule's study is, however, precious in that it bears precisely on this point, although it does not reach its logical conclusion, which would be to question the actual construction process of the category of equivalence used in the model: pauperism itself.

As it happens, this question of the meaning and reality of categories was crucial to a polemical debate that set Pearson against Yule, between 1900 and 1914, to such a degree that it affected their relationship. The debate bore on a seemingly technical point: how to measure the strength of association—or "correlation"—between two variables when those variables were not measurements made on a continuous scale, but classifications in discontinuous categories, or "discrete variables"? The simplest case involves the intersection of two categories at two positions, or a table with $2 \times 2 = 4$ boxes—for example, how many people survived or died in a given year, depending on whether or not they had been vaccinated against smallpox. This case was habitual in the human sciences, where populations are sorted into categories treated as clearly discontinuous categories of equivalence: sex, marital status, activity, or inactivity. In the preceding study, Yule classified people according to whether or not they received assistance, at home or in the workhouse, in cities or in the country.

Pearson, however, worked within the framework of biometrics on continuous measurements, physical or nonphysical, the typical curve of which followed a normal law. If the variables were not continuous, his reaction was to use observed frequencies of the categories to standardize them on a continuous scale, by supposing normal distribution. This is what Galton had done in the 1880s, by using the categories of Charles Booth's surveys. Furthermore, the theory of correlation had been constructed on the basis of the two-dimensional normal law, and the coefficients of regression were parameters of that law.

From this developed their different measurements of the correlation of a table in which discrete variables crossed, and especially tables with four boxes. Yule had no reason to think that his categories reflected underlying variables that were continuous and normal. He proposed an indicator that was easy to calculate. If the table contained, on the first line, the two numbers a and b, and on the second, c and d, then the strength of the association was measured by $Q = (ad - bc)/(ad + bc)$. This expression respected several of the properties to be expected from a coefficient of association: a value $+1$ for the complete positive dependence, 0 for independence, -1 for complete negative dependence. (He nonetheless presented the drawback of the value $+1$ or -1 if *just one* of the four boxes is zero, which is hard to consider as a complete dependence.) Pearson, in contrast, had nothing but disdain for this formulation. It was arbitrary and not based on anything. It could moreover be replaced by Q^3 or Q^5 and satisfy the same necessary properties. To measure the association in a manner that seemed unequivocal, he constructed a normal two-dimensional law whose marginal distributions fitted the two marginal curves observed. He demonstrated that there was one and one alone, and that one of its parameters provided the desired correlation, which he termed "coefficient of tetrachoric correlation."

This quarrel planted the seeds of discord among the members of the biometry laboratory, who exchanged incisive articles in the *Journal of the Royal Statistical Society* (Yule's camp), and in *Biometrika*, a journal created by Pearson. It has been analyzed in detail by MacKenzie, who explicates the different rhetorical and social strategies employed by both protagonists, showing why each was attached to his own method and had difficulty understanding the other's method. Yule attacked hypotheses of normality that were "useless and unverifiable." Going back to the table illustrating the effects of vaccination he showed that in this case the conventions of equivalence were difficult to deny:

> All those who have died of small pox are all equally dead; no one of them is more dead or less dead than another, and the dead are quite distinct from the survivors. At the best the normal coefficient . . . can only be said to give us in cases like these a hypothetical correlation between the supposititious variables. (Yule 1911, quoted by MacKenzie, 1981, p. 162)

But Pearson retorted that Yule was reifying his categories, and that cases with such clear-cut categories were few and far between. He accused him of "realism" in the medieval sense of the word, and went so far as to maintain that the distinction between life and death was basically continuous (we all end up dying):

> Beneath indices of class such as "death" or "healing," "employment" or "unemployment" of the mother, we see only continuous measurements of variables which, of course, are not necessarily a priori Gaussian . . . The controversy between us is far deeper than a superficial reader might think at first sight. It's the old controversy of nominalism against realism. Mr Yule juggles with the names of categories as if they represented real entities, and his statistics are merely a form of symbolic logic. No practical knowledge ever resulted from these logical theories. They may hold some pedagogical interest as exercises for students of logic, but modern statistics will suffer great harm if Mr Yule's methods become widespread, consisting as they do of treating as identical all the individuals ranged under the same class index. This may very well occur, for his path is easy to follow and most people seek to avoid difficulties. (Pearson and Heron, 1913)

In contrast to the conventions of equivalence and discontinuity between categories postulated by Yule, Pearson posed those of continuity and normalcy, and the more realist of the two may not have been Yule. When Pearson accused him of "juggling with the names of categories as if they represented real entities," and "treating as identical all the individuals ranged under the same class index," he was raising the very problem that concerns us here. But in so doing he was not so much interested in the actual process of classification seen as an activity of administrative coding, as in substituting another form of classification that seemed to him more natural: continuity, according to normal distributions, or at least distributions that could be approached by the simplest possible mathematical laws. The form of *category of equivalence*—a convention treating all its members as identical—was very common in judicial and administrative practices, with the twin aims of accurate and economical procedures. But these imperatives are less constraining for a biologist, which explains why Yule and Pearson employed such different styles of statistical rhetoric.

A Psychometric Epilogue: Spearman and General Intelligence

The discussion on the reality of objects created by statistics was repeated almost identically, thanks to a kind of historical stammer, between 1904 and the 1950s, in relation to the interpretation of factor analyses effected in *psychometrics* (Gould, 1983). Tests played the role of Quetelet's contingent individuals. In 1904 Spearman (1863–1945), a disciple of Karl Pearson, showed that aptitude tests applied to children were highly correlated. He conceived the method of factor analysis in principal components, seeking, in the vectorial space constituted by these tests, the orthogonal axes that successively explained the maximum of the variance of the cluster of points corresponding to the results for each child. Because of the correlation between the tests, the first of these axes preserved a major part of the total variance. It was a kind of average of the different tests. Spearman called it *general intelligence,* or *g factor.* He promoted and orchestrated it—as Quetelet had done with the idea of the average man—making it into a more general object than the particular tests, which were contingent manifestations of it. This reproducible thing, which could be reused in other contexts, provided a context of common measurement for individual aptitudes, the existence of which Galton had postulated without being able to measure them directly.

Spearman's theory was completed by Cyril Burt, then later attacked and demolished by Thurstone. In its diverse twists and turns, this sequence of events greatly resembled the peripeteia by which Quetelet's concepts of the average man and constant cause were successively completed and transformed by Galton, then demolished by Lexis or Edgeworth. While holding completely to the idea of general intelligence, Burt tried, on the one hand, to explain and interpret the variance not explained by the *g factor:* were there secondary factors reflecting specific aptitudes, independent of *g*? On the other hand, he tried to demonstrate that this general intelligence was innate and inherited. He was in charge of scholarly psychology for the county of London. His work led to the establishment of a system of tests applied to children aged eleven. These tests functioned like an examination, dividing the children into two quite different educational tracks, depending on their level on the *g* scale. This system, known as the *eleven plus,* was used in England between 1944 and 1965. It was based on a phenomenon that appeared real during the first half of the century, in the sense that its existence was shared by many people; that it could be measured and could enter into vaster constructs either as an explanatory variable or one that had to be explained; and that it seemed to be consistent and capable of resisting attacks.

These attacks took several forms. First, the American Thurstone (1887–1955) conceived of a refinement of factor analysis. He divided *g,* through a clever rotation of the axes, into seven *primary mental aptitudes* all independent of one another, based on the fact that the tests could be redivided into subgroups that were especially well correlated. This division of *g* had the advantage of not classifying individuals on a single scale, and thus seemed better adapted to the American democratic ideal, as opposed to the rigid one-dimensional hierarchy of British society. But these differentiated mental aptitudes did remain innate things: we find here an equivalent of Bertillon's or Pearson's multihumped curves, divided into several normal laws. But it was rare for Spearman, Burt, or Thurstone to resort to the prudent nominalist rhetoric of Pearson, which might perhaps have made their constructs less convincing socially.

The second attack came later, from the sociologists. They observed that the hereditary correlation and transmission of a propensity to succeed could also be interpreted as effects of family and social backgrounds, and not as the result of biological heredity. In any case, the statistical studies may be reread in the perspective of a sociology of inequality that developed during the 1950s, just as psychometrics were on the decline. But psychometrics left behind complex techniques of multidimensional factor analysis that could be reused in other contexts. As for Galton and Pearson, the loop closed again: the rhetoric of eugenics and psychometrics disappeared, but a formal grammar took shape and found other uses.

5

Statistics and the State: France and Great Britain

Statistics in its oldest, eighteenth-century sense was a description of the state, by and for itself (see Chapter 1). During the early nineteenth century in France, England, and in Prussia, an administrative practice took shape around the word statistics, as did techniques of formalization centered on numbers. Specialized bureaus were charged with organizing censuses and compiling administrative records, with a view to providing descriptions of the state and society suited to their modes of reciprocal interaction. The techniques of formalization included summaries, encoding, summing, calculations, and the creation of graphs and tables. All this allowed the new objects created by this state practice to be grasped and compared at a single glance. But one cannot logically separate the state from society, and from the descriptions of both provided by the bureaus of statistics. The state was constituted into particular forms of relationships between individuals. These forms were organized and codified to varying degrees and could therefore be objectified, mainly by means of statistics. From this point of view, the state was not an abstract entity, external to particular societies and identical from one country to another. It was a particular ensemble of social ties that had solidified, and that individuals recognized as social "things." To the extent that they did so, at least during the period when this state held together, these social facts were indeed things.

Within the limits traced by this historical stabilization of state ties, the bureaus of statistics and their tabulations provide sources for historians. But historians can also view the vicissitudes and peculiarities in their gradual implementation as moments in the nineteenth-century development of modern states. The consistency of the things produced by statistics has been studied, in the two previous chapters, from the standpoint of their scientific rhetoric, from Quetelet to Pearson. But this consistency was also linked to that of the state institutions, to their solidity, to what makes individuals treat these institutions as things, without constantly calling them into question. This solidity can itself result from the arbitrariness of force, or from a carefully created legitimacy, in the states based on the rule of law that were founded during the nineteenth century, and that took various forms. This legitimacy did not fall from the sky by decree. It was shaped and woven day after day, forgotten, threatened, questioned, and rebuilt at further cost. Within the legitimacy enjoyed by state institutions, statistics occupied a particular place: that of *common reference,* doubly guaranteed by the state and by science and technology, the subtle combination of which constituted the originality and particular credibility of official statistics.

If the following chapters give the impression of tackling subjects having scant a priori connections, this is certainly because it originally took much time and money to combine these two guarantees in a way that seems natural today. Statistics is now one of the attributes of a nascent or renascent democratic state, along with other judicial or political attributes. But it also depends on the particular forms created by the histories of these states, and on the nature of the bonds existing between the various public institutions and the other parts of a society: administrative and territorial centralization; the status of officialdom; relationships with other centers of expertise such as universities or the learned and philanthropic societies so important in the nineteenth century, or the private business foundations that appear in certain countries during the twentieth century.

In Chapters 5 and 6 I shall suggest a few of the specific ties among bureaus of statistics, the structures of particular states, and other objects of social analysis for the period that lasted from the 1830s—when many of these bureaus were created—until the 1940s. At this latter date, these objects would change radically in nature and scale. This was due on the one hand to transformations in the roles of states, and on the other, to the alliance that had finally been forged between economic and social statistics and mathematicians' statistics. I shall compare four countries. In two of them, France and Great Britain, the unified state was an ancient and legitimate phenomenon, although it assumed very different forms (Chapter 5). In the other two, Germany and the United States, it was experiencing gestation or rapid growth (Chapter 6). These differences in the consistency of states may be seen in the histories of their statistical systems. In each case, I shall mention not only the conditions in which public statistics forged its legitimacy but also the public contexts of debate within which statistics found a place.

In France the state is centralized, and so were its statistics during this period, from both an administrative and a territorial standpoint. Generally speaking, expert competency was internal to the administration, through corps of engineers and officials, and the university had less influence than in the three other countries. Public statistics, of which the SGF was the principal (but not the only) element, was organized around censuses. It was most active in demographic (birth rate crises) and economic questions (industrial structures, work, salaries, cost of living); but, on account of the weight and authority of the administration, these questions were less subject than in the other countries to major public debates between technicians and laymen.

In Great Britain, the administrations were more autonomous in relation to one another, and the county and village authorities had more power than in France. Here statistics has never been centralized into one sole institution, and the national bureaus had to deal with local bureaus, such as those in charge of the civil service and the application of the Poor Law. Treated separately, the two main provinces were on the one hand, external commerce and affairs (administered by the Board of Trade), and on the other, population, poverty, hygiene and public health (overseen by the General Register Office, or GRO). The parliamentary commissions of inquiry—set up, for example, in time of grave social crises linked to rapid industrialization and sprawling urbanization—were the cause of intense debate among scholars, statisticians, economists, and political officials.

In Germany, this period was marked first by the gradual development of the Empire as it consolidated around Prussia, a development completed in 1871; then, between 1871 and 1914, by the growth of industry; and finally, by the economic and political crises of the period between the two world wars. Official statistics, which had already existed for some time in the various German kingdoms, coalesced into a unified whole after 1871 through the organization of vast surveys and censuses concerning industrial production. As for academic statisticians and economists, they produced numerous, copious descriptive and historical monographs, laden with statistical content. These works of the "German historical school" were prompted and discussed by an association called the Verein für Sozialpolitik. Highly active between 1872 and 1914 it sought, rather unsuccessfully, to influence the decisions of the imperial bureaucracy.

Last, the United States constituted a recently created federation whose population grew rapidly through successive waves of immigrants. Gathering public statistics was punctuated by the decennial censuses prescribed in the Constitution of 1787, the idea being to share financial burdens and seats in the House of Representatives among the states in proportion to an ever-changing population. Debates about statistics were linked to repeated upheavals in the demography of the country and the political and social consequences thereof (the lawful representation of the states; the integration of immigrants, unbridled urbanization, and delinquency). Statistical administration remained fragmented into different services. The most important of these was the Bureau of the Census, which did not become permanent until 1902, and there was no coordination of services until 1933. After that date the political, economic, and budgetary roles of the federal government changed radically. Official statistics were reorganized and coordinated at the prompting of a group of academics, statisticians, and economists. It greatly extended its field of action, recruiting young people with a mathematical background, implementing new techniques such as sampling surveys, and constructing, by statistical means, new objects such as unemployment or social inequality. Only at this point, in this overview of public statistics in the four countries mentioned, does the reader discover a familiar landscape: in the United States, the more typical characteristics of modern statistical duties took shape during the 1930s.

This initial summary of the situations of these four countries between 1830 and 1940 shows that a comparison of international systems of statistical description cannot be limited to institutions of official statistics, on account of the differences between their relative importance, means, administrative strengths, and, especially, their objectives. What is needed is a more complete institutional and sociological table of the places where statistical knowledge was being developed. But the historical research already undertaken on these questions—rather numerous and detailed for Great Britain and the United States, less so for France and Germany—in general bears only on partial elements of these contexts, and views them from vastly different perspectives. Accordingly, there is a risk that the comparisons will be biased, attributing much importance to a certain aspect solely because it is well documented. The following analyses rely on secondary sources for Great Britain, the United States, and Germany, and cannot therefore be protected from such errors of perspective: this risk is almost inevitable in any comparative study. It was precisely on account of this problem that German statistics—in the eighteenth-century sense of the word—failed, as did the French prefectorial statistics of 1805. A context of comparison must be established before an actual comparison is conducted, but naturally this concerns more than just description or historical research. The following remarks thus constitute more a reasoned juxtaposition of four narratives than an exhaustive comparison.

French Statistics: Discreet Legitimacy

From 1833 to 1940, the Statistique Générale de France (SGF) was still a minor service that had only taken root in Paris. Its main activity was to organize and make use of the quinquennial censuses and to analyze "population trends" on the basis of civil service records (births, marriages, and deaths). Its history reflects the gradual creation of a discreet legitimacy, based on austere technical skills—especially after the 1890s, when Lucien March radically transformed the entire process of census taking, from the questionnaires and the abstracts made of them to the actual publications. This legitimacy was not acquired right from the outset. It was linked to the establishment of administrative routines and also to the action of a small group, which was reformed after 1860 as the Société de Statistique de Paris, or SSP (Kang, 1989). But the cost, centralization, and interpretation of statistical activity were not—as was the case in Great Britain or the United States—subject to major discussions in the newspapers or in Parliament.

When first created in 1833 the SGF was attached to the Ministry of Commerce and was responsible for assembling, coordinating, and publishing statistical tables generated by other administrations. The founder of the SGF, Moreau de Jonnès (1778–1870), was its director until 1852. In 1835 he laid out a detailed plan for a publication comprising fourteen volumes, spread out until 1852, that included diverse domains of administrative action. Public health, which then occupied an essential place in British statistics, figured only slightly in this plan, through data on hospital management.[1] The "moral statistics" that formed around the *Annales d'hygiène publique* remained external to French public statistics, whereas their English counterparts in the public health movement were at the core of official statistics, in the General Register Office.

The main difficulty encountered by the budding SGF was to gain acceptance from the other ministries, by virtue of some specific technical skill. Combining the statistics of agriculture, industry, and commerce with those of "population trends" (the civil service) and censuses was not just a matter of course; the administration of the latter still depended on the Ministry of the Interior, who in 1840 created a bureau of statistics, directed by Alfred Legoyt (1815–1885). Legoyt, who criticized Moreau de Jonnès, succeeded him as director of the SGF from 1852 to 1871. The call for centralization of the numerical documents with a view to publishing them was not enough to guarantee the institution's power, unless the standard recording tools were in force: regular inventories, indexes, terminologies. The confidence (or "reliability") generated by statistics was linked to the coherence and stability of the administrative machinery.

During its initial period, the SGF published (in addition to censuses) not only regular administrative data produced by other bodies but also the results of exceptional surveys carried out at its prompting, involving agricultural (1836–1839) and industrial systems (1841, 1861). In the case of compilations produced by other administrations, the value added by the SGF was weak, and verification difficult; nonetheless, habits were created by this means: for example, criminal statistics, initiated in 1827, or statistics relating to the mineral industry (for which Le Play was responsible, in 1848). Structural surveys, in contrast, were closer to monographs in that their results—produced by ad hoc techniques that were only used once—were hard to generalize. They remained isolated points and did not help to generate a quantitative routine. Agricultural surveys were judged poor while cadastral surveys remained unfinished: statistical activity could only be operational if it dovetailed with a management practice with which it was consistent.

The verification of statistics was a central question of that time. Bertrand Gille's work (1964) on "the statistical sources of the history of France" describes the ambiguity of the requirements of the period, discussing them from a historian's viewpoint. He distinguishes between "collective, almost democratic verification," "local verification, drawing on the expert knowledge of the leading citizens," and "logical, scientific, verification." The model of locally acquired knowledge guaranteed by collective agreement was frequent in the nineteenth century, as for example in the case of epidemics or the fight for healthier living conditions. Local commissions of notable citizens were then charged with tasks which, later on, would be assumed by centralized administrations applying routine regulations. In this spirit, a vast project for setting up "cantonal statistics commissions"—begun for the agricultural survey of 1836—was continued in 1848 and systematized in 1852. It attempted to establish a system for having information verified by the interested parties. But these commissions proved unsuccessful, perhaps because they did not recognize themselves in the standardized questions sent from Paris. This combination of locally acquired knowledge and collective validation (which Gille contrasts with "logical and scientific" verification, using calculus to check internal consistency) threatened to bring out specific local characteristics as well as discord between particular and general interests. Like the prefectorial statistics of the turn of the century, it was not viable because it was not consistent with the establishment of a context of administrative management structured by equivalent rules and procedures throughout the territory. This context was certainly adequate for statistical summing. But there again, the case of France was different from that of England, where locally validated information was a necessary element in establishing the legitimacy of the General Register Office.

Social statisticians, who had some influence among these leading local citizens, criticized this attempt at a centralizing administration that could not yet justify itself by its technical prowess. For Villermé (1845), this administration accomplished nothing more than the awkward compilation of inconsistent data, which it did not verify or cross-check. He called for an organization that would be closer to the decentralized English model:

> We reproach statistical publications for their inaccuracies, their lacunae, and the lack of unity in planning and thinking that does not allow them to be compared or checked against each other . . . We emphasize the disadvantages of entrusting important publications to a ministry that is unfamiliar with the subject matter in question. We point out that analogous documents will be published concerning the economy, finances, the military, the navy, the judicial system, forms of worship, and public education. The administrations related to these objects must be charged with the work. We hope they will consider themselves honor bound to assume responsibility for it, and that if necessary the chambers of government will not allow the funds they enthusiastically approved for providing France with a general system of statistics to be put to uses so unworthy of the nation. (Villermé, 1845)

The dispersion of responsibility is clearly visible in census taking, which at that time was still organized by the bureau of statistics of the Ministry of the Interior; it was under Legoyt's direction, but its figures were published by the SGF. In 1852, however, Legoyt took over the SGF and combined census taking operations there. He introduced important innovations in regard to occupations (1851) and economic activities (1866), whereas previously there had only been a nominal list of individuals. Legoyt's position at the Ministry of Commerce then seemed more certain than that of his predecessor.

Elsewhere, the rivalries between administrative and social statisticians were lessened when a group of economists and social researchers—the most famous of which were Villermé, Michel Chevalier, and Hyppolite Passy—formed the Société de Statistique de Paris (SSP) in 1860. They asked for, and obtained, an official guarantee from the Ministry of Commerce. Legoyt was the SSP's first president. Until the 1960s, when the large administrations and statistics schools were founded, this learned society functioned as an important meeting place for statisticians (public or private) and people who used their work, and as a center for diffusing their ideas and claims. Its review, the *Journal de la Société de statistique de Paris,* provided a rostrum for a small group of untiring propagandists of statistics, the holders of professorial chairs and members of administrative commissions charged with promoting and coordinating statistical work. These were Émile Cheysson (1836–1910), a bridge engineer and disciple of Le Play; Émile Levasseur (1828–1911), a university geographer; Adolphe Bertillon (1821–1883) and his son Jacques Bertillon (1851–1922), both of whom were doctors, besides being officials of a bureau of statistics for the city of Paris, which published information on public health and causes of deaths (Alphonse Bertillon, 1853–1914—the anthropometric expert for the prefecture of police, mentioned in Chapter 4 in connection with Galton—was another of Adolphe Bertillon's sons). We find these same militant experts seated on the Conseil Supérieur de la Statistique, which was created in 1885 to support the SGF and other statistical bureaus and to direct their work. The statistics of this period, however, still had few instruments, in the sense of the tools that constitute its power today: the administrative and regulatory infrastructure needed to identify and define the objects; mechanographical machines, followed by computers; and mathematical techniques of analysis and interpretation. The continued efforts of the SGF, the SSP, and of the Conseil Supérieur de la Statistique were concerned with the infrastructure of statistics, and with affirming its importance. But that was not enough. During the 1890s the SGF was still a very small service, and the struggle had to be constantly renewed.

This ongoing effort to establish the discipline can also be observed in the international congresses of statistics (introduced by Quetelet) between 1851 and 1875, then during the early period of the International Statistical Institute (ISI), created in 1885. The ISI initially brought together administrative statisticians, who found it a suitable place for promoting their activities and endowing them with prestige. Its congresses were opened by the highest public figures of the host countries. They provided an opportunity to militate in favor of harmonizing methods and terminologies and enabled every statistician to shore up his own position by relating it to a higher order of legitimacy, that of international scientific authority (Jacques Bertillon developed and had ratified by the ISI, in 1893, two important classifications: *occupations* and *diseases*). The role of the ISI would alter radically after World War I, and change even more completely after World War II, for two reasons. The task of coordinating national statistics would be taken up again at that point by new international institutions: the League of Nations, the International Labor Office (ILO), and last, the United Nations. This would allow the ISI to devote itself henceforth to the discussion and diffusion of mathematical methods resulting from biometrics and from English statistics.

These latter disciplines received their first serious discussion at the twelfth congress of the ISI, held in Paris in 1909. Edgeworth spoke of the "application of probability calculus to statistics." Yule summarized his work on the "method of correlation as applied to social statistics." March showed how "mathematical procedures allowed the comparison of time series, with a view to researching their agreement or disagreement, and to the prediction or appreciation of the possible relationships of the sizes compared." Bowley presented an "international comparison of wages, aided by the median."

This first significant appearance of "mathematical procedures" in economical and social statistics was encouraged, in the case of France, by the transformation of the SGF during the 1880s. Whereas the English statisticians of the GRO had, in the mid-nineteenth century, established their position by making themselves indispensable in regard to problems of unsanitary urban conditions and public hygiene, their French colleagues at the SGF did so at the end of the century, by intervening in the labor problems resulting from the economic crisis that began around 1875. The efforts of the small group of statistics activists then combined with those of a network of civil servants, intellectuals, and union or political officials who wanted both a better knowledge of wage-earning labor and new laws for organizing and protecting it. Such men were Arthur Fontaine (1860–1931), a mining engineer; Isidore Finance and Auguste Keufer, trade unionists; Alexandre Millerand and Albert Thomas, who were Socialist reformers; and later on Lucien Herr and François Simiand, who were connected with Durkheim and the École Normale Supérieure.

In 1891, within the Ministry of Commerce, the Office du Travail (Labor Office) was set up, directed by Arthur Fontaine (Luciani and Salais, 1990). The SGF was attached to it, and there the young mechanical and industrial engineer Lucien March organized a centralized process of census taking in 1896. He introduced the "Hollerith machines" into France, "electrical census-taking machines" invented in the United States that used the first punched cards. He invented a more perfect machine, the "classifier-counter-printer" used until the 1940s (Huber, 1937). Finally, by skillfully sorting the census bulletins according to the address of the workplace of the persons counted, he managed to transform this a priori administrative and demographical process into a vast survey of production and the labor force, wage earners or not. It had hitherto been deemed impossible to conduct such a direct survey of the actual firms, for fear of their reticence or their hostility: the legitimacy of public statistics was not considered sufficient for such an operation to be possible. The central position of the SGF and its connection to the new office enabled March to put census taking to this double use, both demographic and economic. The policy was pursued with few changes from 1896 to 1936, providing a series greatly prized by historians.

Between 1891 and World War I, the Labor Office constituted the first scientific-administrative institute endowed with significant means (anticipating those set up after 1945). An ancestor of the Ministry of Labor, established in 1906, it combined services for conducting surveys and research into wages, unemployment, and workers' living conditions with more managerial services that dealt with the placement of labor, workers' legislation, the regulation of work, hygiene, and security, trade unions, and conciliation boards. The research was based on direct statistical surveys (of wages and hours worked), and on monographs in the LePlaysian tradition of studying workers' budgets (Du Maroussem, 1900; Savoye, 1981) or data from administrative sources, such as statistics relating to strikes (used by Michelle Perrot, 1974).

Outline and Failure of a Network of Influence

The situation that obtained in French administrative statistics between 1890 and 1920 can be compared with others. Sometimes, in particular historical circumstances—such as crises, war, or rapid changes in the concept and management of society—the statisticians in charge of their bureaus manage to enter into vaster networks, at once scientific, administrative, and political, in such a way that recourse to their tools becomes indispensable and almost obligatory. This was the case with the English GRO between 1840 and 1880, and with the U.S. Census Bureau during the 1930s and 1940s at the time of the New Deal and during World War II. The same thing happened with the French INSEE during the 1960s and 1970s, when the planning and modeling of economic growth was based largely on national accounting tables. From this point of view, though with far smaller means at their disposal, the Labor Office and the SGF both present sketches of such networks, especially during World War I.

Between 1875 and 1890 the industrialized countries suffered a grave economic crisis. Its social consequences were more serious in Great Britain and in the United States than in France. But in all these countries new laws were discussed, intended to define and institute labor legislation—novel, in regard to traditional civil law—and to establish protection in case of unemployment, work-related accidents, disability, and the destitution of the elderly. A higher Labor Council was set up in 1891 and reorganized by Millerand and Fontaine in 1899 to prepare these laws. Bringing together civil servants, heads of firms, and trade unionists, it anticipated the commissions of the Plan, the equal representation and the contractual politics of after 1945. In this connection, in 1893 the Labor Office published a major survey of "wages and work hours in French industry," and later conducted surveys of workers' household budgets and the cost of living. Elsewhere, scholars influenced by Durkheim worked on the same subjects, publishing the first theses in sociology with strong empirical and statistical content. Simiand (1873–1935) analyzed the wage variations of mine workers, and his work was discussed by March at the Société de Statistique de Paris (Simiand, 1908). In his *La Classe ouvrière et les niveaux de vie* (The working class and standards of living, 1912), Halbwachs (1877–1945) studied the budgets of working-class families compiled by German statisticians (although his further thesis, on "the theory of the average man, Quetelet, and moral statistics," adheres strictly to the line of nineteenth-century questions). Halbwachs himself conducted a survey of the budgets of workers and French peasants, publishing the results in the *Bulletin* of the SGF in 1914 (Desrosières, 1985, 1988b).

These relationships among statisticians, academics, and political officials were strengthened during World War I, especially in the cabinet of Albert Thomas, minister of armaments (Kuisel, 1984). Some intellectuals played an important part in administering the war effort. The mathematician Paul Painlevé was war minister, then president of the Council for several months. His colleague, the probabilist Emile Borel (1871–1956), was secretary general of the presidency of the Council, a coordinative and administrative post to which he tried, unsuccessfully, to attach a strengthened SGF (Borel, 1920). Simiand and Halbwachs were members of Albert Thomas's cabinet. Because of its importance in directing war-related industries, which were of necessity highly centralized, and because of the standardized procedures this entailed, this ministry was able to invent completely new forms of state management and professional relationships. Although this wartime economic structure temporarily disappeared after 1918, the memory of it remained, inspiring other, later forms established after 1940, such as the organizational committees of the Vichy government, or the Plan following the Liberation. The presence of one man was especially felt during these two periods in which the wartime economy was rationalized and reconstructed: that of Jean Monnet (1888–1979), who later created the Plan's commissionership in 1946, and initiated the European Union.

Simiand acquired considerable importance as secretary of the committee of statistics for wartime manufacturing and raw materials. Several times he sought to transform the SGF into a vast central office of statistics, attached to the presidency of the Council. In 1921 he led a delegation—consisting of Borel, the statistician Liesse, and the industrialist Gruner—to see Millerand, then president of the French Republic (formerly minister of the Commerce Department, on which the Labor Office and the SGF both depended), in order to promote this reform of the SGF. But the conditions for returning to peacetime in the 1920s were not yet those that would prevail after 1945 with the creation of INSEE and the Plan. The SGF was finally attached to the presidency of the Council in 1930, but its means and jurisdiction were not increased. It remained a small service, with about one hundred employees; directed between 1920 and 1936 by Michel Huber (1875–1947), its primary task was to carry out demographic surveys (Sauvy, 1975). In 1932 Simiand again took action, this time as a member of the national economic council, calling for the SGF to be reinforced and scientifically autonomized (Journal Officiel of April 13, 1932), but these efforts came to naught. The SGF was eventually absorbed, in an entirely different context, into the very large national statistical service set up in 1941 by René Carmille, a military engineer (Desrosières, Mairesse, and Volle, 1977).

A further consequence of the bonds forged during World War I was that the teaching of statistics—an issue on which the statistics activists of the 1880s had failed to obtain action—suddenly opened up (Morrisson, 1987). In 1920, at the instigation of March and Borel, the Institute of Statistics of the University of Paris (ISUP) was founded. The principal courses were taught by March, Huber, and the mathematician Georges Darmois (1888–1960). The creation of this institute was helped partly by the importance that statistical rationalization had assumed in managing the wartime economy, and partly by the changing relationships between practical statistics and probability theory. A good illustration of these two points is provided by the joint activities of the probabilists Fréchet (1878–1973) and Borel (also a member of a council of the SGF which, from 1907 to 1936, assisted March and then Huber), and the sociologists Simiand and Halbwachs. In 1924 Fréchet and Halbwachs coauthored a small manual entitled *Le Calcul des probabilités à la portée de tous* (Probability calculus for everyone). The connection—unlikely at the turn of the century—among administrative statisticians, mathematicians, and sociologists was facilitated by the scientific and political contingencies of the time. Neither Cheysson (although he was a polytechnician), Bertillon, nor Levasseur had tried to establish links with the probabilists of their time, and they had ignored Pearson. In contrast, March had introduced the mathematical tools of the English into France between 1900 and 1910. A new generation of probabilists had appeared with Borel, Darmois, and Fréchet. Henceforth a supply of technologies and qualified professionals existed, as well as the politico-administrative opportunity to establish this new academic discipline. However, the teaching of statistics remained small in scale and Durkheimian sociology did not become mathematical. Statistical instruction only acquired major importance during the 1950s, with the ISUP on the one hand and the École d'Application of INSEE on the other, this latter becoming the École Nationale de la Statistique et l'Administration Économique (ENSAE) in 1960.

All this should not, however, suggest that from 1890 to 1920 a vast politico-scientific network existed, within which the SGF played an essential part. This was not the case at all. The connections I have described involved only a small number of people, and their influence was only significant during the 1914–1918 war—an exceptional period. Writers such as Lévy (1975) or Le Bras (1987) who, like Sauvy, describe the SGF of the time as weak, self-involved, and ill-coordinated with other statistical services are not entirely wrong, and Simiand's numerous initiatives suggest as much. But these authors view this era from a post-1945 perspective, when the new modernizing networks outlined during the 1930s, and during the war and the Occupation, took permanent root in the institutions established at that time: the École Normale d'Administration (ENA); the commissionership to the Plan; the Centre Nationale de la Recherche Scientifique (CNRS); the Institut National de la Statistique et des Études Économiques (INSEE); and the Institut National des Études Démographiques (INED). In comparison, their predecessors of the years between 1890 and 1920 naturally appear as failures. When Alfred Sauvy (1898–1990) entered the SGF in 1923, he found a small group of scholars and technicians who were competent and efficient but introverted and "unconcerned with selling themselves," that is, with translating their work in such a way that it became necessary to others. He then set about—successfully—devising and popularizing new public images, first of statisticians and then of demographers, sweeping clean all that had gone before. He never mentioned March or the Labor Office. It was as if the 1914–1918 war had interrupted a trend. The reformers of the 1930s had almost nothing in common with those of the turn of the century, and they were the ones who invented the dominant language of the 1950s and 1960s.

The question here is not to revive forgotten precursors, but to try to understand why the network assembled in the 1890s did not hold, whereas the one outlined between 1930 and 1950 has held for almost half a century. There are, of course, macrosocial historical explanations, such as those suggested by Kuisel's analysis (1984) of the evolving relationships between "capitalism and the state in France" from 1900 to 1960. According to Kuisel, prior to 1914 a free-market concept predominated, in which the state did not allow itself to intervene at a macroeconomic level. This concept was taught at the university and promoted by the best-known economists of the time. An early form of organized economics was tested in the exceptional circumstances of World War I, with Clémentel (minister of commerce) and Albert Thomas (armaments), but was quickly dismantled in 1919. Then the crisis of the 1930s and the extreme penury of the 1940s—both before and after the Liberation—allowed more interventionist (or "controlled," as they were then termed) trends to emerge. These first took on an administrative and authoritative form under the Vichy government, with Bichelonne, then a more democratic and stimulatory form after 1945, with the concerted, statistically enlightened planning of Jean Monnet, Pierre Massé, and Claude Gruson. This history, which Kuisel relates in some detail, scarcely mentions the statisticians' work, but does cast light on the change that came about between the SGF of March and the postwar INSEE, especially in regard to the role (or absence of one) played by the economists. Before 1939, it was unusual for anyone to base arguments on statistical data, not only out of ignorance or incompetence, as Sauvy suggests, but because the overall construct did not exist, either at a political or intellectual level. This construct only appeared during the 1960s, notably with Keynes, in Great Britain and in the United States, and in France during the 1950s.

But one can also look for microsocial explanations, linked for example to a complex personality like that of March. A nineteenth-century engineer, March was an imaginative technician who introduced and transformed the mechanographical procedures invented by Hollerith in the United States. But he also shared the preoccupations of many of his contemporaries in regard to population, the decline in the birth rate, and the "quality of the race." France did not at that time have so large and radical a eugenics movement as existed in Great Britain: it would have been difficult to advocate simultaneously an increase in the birth rate and a reduction in the fertility of the poorer classes, which is what the English were doing. As a statistician, however, March knew and admired the work of Galton and Pearson, importing both their statistical and eugenicist aspects. As early as 1905 he presented research on the correlation between numerical curves to the Société de Statistique de Paris, then guided two young statisticians of the SGF, Henry Bunle (1884–1986) and Marcel Lenoir (1881–1926), in using this to analyze connections between economic series. Bunle (1911) studied correlations among marriage rates, price levels, unemployment, and external trade. Lenoir (1913) defended a university thesis on the curves of supply and demand, with a view to explaining how prices are formed and change—one of the first properly econometric pieces of research. It had little impact and was not known to the Americans (Moore, Working, Schultz) who, at about the same time or slightly later, invented econometrics (Morgan, 1990; Armatte, 1991).

March's work (and that of his students) was presented and discussed before the Société de Statistique de Paris, and published in its *Journal*. His research, however, and the positions he adopted in regard to birth and eugenics—apparently inspired by Pearson—were published elsewhere, in reviews and books issuing from vastly different spheres: *Revue d'hygiène et de médecine infantiles, Revue philanthropique, Eugénique et sélection*. (This last was a collective work published by a French eugenics society that March had helped found in 1912, on his return from a world congress on eugenics, held in London.) The combination of mathematical statistics and eugenics, then fully established in England, could not be imported directly: March was the sole French statistician to occupy the two fields. His two-pronged, purely personal interest was therefore of no help in introducing administrative statistics into a broader political and scientific network, since the selectionist eugenics trend did not have the same importance in France as on the other side of the English Channel.[2] Thus not only was the political and economic context still unpropitious for the development of a major statistical institution that would centralize and synthesize research into different demographic, social, and economic domains, but the personal ties and interests of Lucien March did not favor it. Even so, the original organization of the population censuses between 1896 and 1936, ingeniously combining demography and descriptions of the structures of businesses, could have provided an embryo for such an institution—impossible in Great Britain, where demographic and economic statistics were separate.

In short, the situation in France, on the eve of World War II, was characterized by the fact that there was little context for meeting and debate between specialists in the social sciences—statisticians, demographers, economists, or sociologists—and political or administrative officials, heads of businesses, and trade unionists. Such contexts still did not exist within the state, but were being outlined by some intellectuals and engineers. Looking back through the history summarized here we find recurrent movements of enlightened persons who, independently of the state, undertook to inquire into and analyze French society. This happened before the state—itself sometimes transformed by these reformers—integrated them one way or another into the new institutions. Thus, having become departmental prefects, the surveyors and time travelers of the eighteenth-century philosophers prepared the "statistics" of their departments. Until the 1830s, provincial middle-class doctors, magistrates, and businessmen continued to compose local monographs. Some of them competed for the Montyon prize in statistics, presented by the French Academy of Sciences (Brian, 1991). They reappear in connection with the *Annales d'hygiène*, a review founded in 1829 (Lécuyer, 1982) and active until the 1850s which criticized official statistics: such were the upholders of "moral statistics." After 1860, the two categories of statisticians, administrative and "moral," drew closer and militated together in the Société de Statistique de Paris (SSP). This alliance was symbolized by Adolphe and Jacques Bertillon, doctors who succeeded each other as officials of the bureau of statistics of the city of Paris, and by Levasseur (a university professor) and Cheysson (a bridge engineer).

The trend in surveys undertaken independently of the state took on an original form, from the 1840s to the end of the century, with the "budgets of working families" collected by Frédéric Le Play (1806–1882). Le Play was a mining engineer, and the prophet of a conservative school of social sciences that was hostile to universal suffrage, to the capitalistic wage structure, and to the egalitarian division of wealth among heirs. At the same time it was technical in bent, open to science, and attentive to workers' living conditions (Savoye, 1981). This dynamic group took on a clannish air, on account of the positions of its founder, who was hostile to the republican state. Yet some of its members played significant roles: Cheysson at the SSP and as a professor of statistics, and Du Maroussem at the Labor Office, where he conducted monographic surveys of businesses (but the surveys of workers' budgets organized by Halbwachs in 1907, or by Dugé de Bernonville at the SGF in 1913, were not directly inspired by those of Le Play: I shall discuss these surveys in greater detail in Chapter 7). The LePlaysian frame of reference, based on the triple principle of "family, work, and locality," would later exert an influence outside France, in Great Britain (Geddes) and Germany (Schnapper-Arndt). But in France this trend, cut off from university sociology, began to wilt. Its final representatives were close to the Vichy regime (Kalaora and Savoye, 1985).

Statistics and Economic Theory: A Tardy Union

Before the 1930s it was common for certain demographers, philanthropists, and social reformers to have recourse to statistical arguments. It was far less common among specialists in economic theory, whether literary or mathematical—which may seem surprising. While correlations and multiple regressions bearing on economic variables had already been calculated by statisticians (Yule, March, Bunle), they were not linked to a way of reasoning that emerged from economic theory (except, however, in Lenoir's work [1913] on the construction of supply and demand curves). The connection was not self-evident. Economists felt that the available measurements of, for example, prices, were the final products of complex chains of events, in which it was difficult to discern the economic mechanism theoretically described in their models. Such was the case most notably with Jean-Baptiste Say, Walras, and even Cournot, although Cournot was very well informed about statistics (Ménard, 1977). At a deeper level, economists still often associated the idea of probability with that of imperfect knowledge, and not with the intrinsic variability of economic phenomena. It was incompatible with the deterministic philosophy resulting from nineteenth-century physics: in their view, this latter constituted a model that allowed economics to escape from the uncertainties and indeterminations of the other social sciences and philosophy (Mirowski, 1989a). Since no one really knew what statistics reflected, they were not necessarily a good ally in the economists' fight to obtain recognition equal to that enjoyed by physics. "Statistics is not an experimental science": ever since the criticisms formulated by Claude Bernard, debates as to the nature of statistical knowledge were recurrent.

During the 1920s, however, physics definitively abandoned Laplacian determinism with the advent of quantum mechanics and Heisenberg's relationships of uncertainty. At around the same time probability theory, which had been axiomatized by Kolmgorov, became a full branch of mathematics and assumed a central place in the various natural sciences. Last, inferential mathematical statistics acquired, first with Ronald Fisher, then with Neyman and Pearson, a sufficiently general level of formalization to be used in various contexts. These diverse developments permitted the conjunction of three previously quite separate tendencies: the neoclassical economic theory of Walras and Marshall; the early econometrics of the years between 1910 and 1930, practiced without models and in the manner of Lenoir and Moore; and the new inferential probabilistic statistics that emerged from English biometrics. This convergence, which gave birth to econometrics in the modern sense, was formalized by Haavelmo (1944) and led to the program of the American Cowles Commission, which was centered on the structural estimation of models for simultaneous equations (Morgan, 1990).

This new synthesis of statistics and economic theory was disseminated in France, in the immediate postwar period, more by engineers than by university professors: Maurice Allais and René Roy, who were mining engineers, and Edmond Malinvaud of INSEE and ENSAE (Bungener and Jol, 1989). Thus a strong alliance was forged between economic theory, applied economics, and statistical apparatus, helping to endow statistics with a legitimacy and authority quite incommensurate with what it had experienced twenty years earlier. This essential transformation in the role and mode of intervention of economists—teaching in engineering schools rather than universities, or working in administrations or public businesses—had been anticipated and prepared before the war by a circle of former students from the École Polytechnique.

The Centre Polytechnicien d'Études Économiques (or X-Crise) was established in the context of the economic, political, and intellectual crisis of the 1930s (Boltanski, 1982), by a group that included Jean Ullmo (1906–1980), Alfred Sauvy, and Jean Coutrot. The conferences it organized provide a good example of the construction of a network bent on developing a common language: a new system of references in the turmoil resulting from the crisis. Statistics and economics, which were in the process of being combined in econometrics, would henceforth be a part of these common objects. Apart from the above-mentioned founders who often gave talks, presentations were made by the economists Charles Rist and Jacques Rueff; the sociologists Bouglé, Halbwachs, and Simiand (who then enjoyed considerable prestige in this circle); the historian Marc Bloch; the probabilistic mathematician Darmois; and even by the politician Paul Reynaud and the writer Paul Valéry (X-Crise, 1982). But, unlike their Saint-Simonian ancestors or the nineteenth-century LePlaysians, these polytechnicians were no longer becoming sociologists, but economists.

A series of lectures was devoted to "econometrics" (in its primary sense of the quantification of economic models). In 1933 François Divisia (1889–1964), one of the first French university economists to promote the systematic use of statistics in economics, spoke of the "work and methods of the Society of econometrics" which had been founded three years earlier. He quoted Irving Fisher, and also March and his work on the price index; he mentioned Lenoir, "the first to introduce the method of multiple correlations into the study of prices"; cited Gibrat and his "index of inequality" of income distribution; and quoted the English statistician Arthur Bowley for his "study on price dispersion as a symptom of economic crises." In 1938 Jan Tinbergen (born in 1903, winner of the Nobel Prize in Economics in 1969) described his "economic research on the importance of the Stock Exchange in the United States." In 1934 Jacques Rueff gave a presentation entitled "Why, in spite of everything, I am still a liberal" (in the European, free-market sense) while in 1937 Jean Ullmo examined "the theoretical problems of a controlled economy."

These last two titles, among others, show how much the theoretical reflections prompted by the crisis of the 1930s were linked to criticism of free-market thought and an interest in "controlled" or "planned" economies. They were quite unlike those that would be inspired by the crisis of the 1980s, based on the contrary criticism of planned economies and macroeconomic regulations that resulted from those same theories formulated during the 1930s, in circles such as the X-Crise. The ideas discussed subsequently recurred in institutions such as the Plan, the Services des Études Économiques et Financières (SEEF) founded by Claude Gruson in

1950, INSEE (directed by the same Gruson between 1961 and 1967), or the INED, founded by Sauvy in 1946. Thus, once again, themes developed independently of the state led to the transformation of the state and to the institutionalization of the public context of debate. The trend was for technical competence to be incorporated into the state. This clearly distinguished France from Great Britain or the United States. In these countries, the university centers possessed great vitality and formed relationships for exchanging information with the administration, which were close but not permanent, thereby endowing the association of science and state with an entirely different configuration.

British Statistics and Public Health

Nineteenth-century Britain was not simply the country that gave birth to biometrics and mathematical statistics. It was also the country where, in the context of industrial and urban growth and their dramatic consequences, the most varied forms of relationships were studied, involving statistics (administrative or not), social analysis, and legislation intended to treat the problems connected to this growth: poverty, public health, unemployment. British administration was not inclined to integrate the greater part of statistical knowledge and expertise into bureaus that depended on it. These bureaus existed, naturally; but they were more scattered than in France. They were subject to dispute, and their organization was often thrown into question. Beyond them, groups of social reformers, scholarly societies, and university professors maintained a strong tradition of inquiry and debate concerning the construction of directly applicable practical knowledge. The two spheres—the one administrative, the other involving social researchers and scholars—were distinct, but their interactions were constant: the second was permanently associated with the decisions of the first, as for example in the framework of parliamentary investigatory commissions. This distinguished the case of England from that of France, where the second of the two milieus was less active, or less connected with the administration (the LePlaysians), and also from the case of Germany, where the scholarly milieu existed but had more difficulty in making itself understood. If there is a figure who symbolizes this typically English liaison between the university and the administration, it is that of Arthur Bowley (1869–1957), a professor at the London School of Economics (LSE). In the early twentieth century Bowley gave clear formulation to the scientific and professional norms of a profession that did not yet exist in that form: that of the administrative statistician relying on the knowledge acquired in mathematical statistics.

The history of official British statistics is punctuated by a succession of legislative acts of great import and by the ad hoc creation of institutions, leading to a fragmented system only partly coordinated, in 1941, by the Central Statistical Office. Between 1832 and 1837, a series of political and economic reforms established the free-market framework of reference, which remained the backdrop for all these debates, at least until 1914. This framework involved, in particular, the principle of commercial free exchange; the importance of local powers (municipalities and counties); and relief for the poor provided in workhouses rather than by direct assistance (in accordance with the Poor Law of 1834, replacing Speenhamland's Law of 1795, which prescribed monetary assistance for the parishes [Polanyi, 1983]). These three factors determined the forms of the early statistical system. In 1832 a bureau of statistics was created at the Board of Trade (Ministry of Commerce), dealing with commercial trade, exports, and imports. Elsewhere, the Poor Law was locally administered, entailing the creation of Poor Law Unions in every county (1834). This new structure, extending throughout the land, made it possible for the General Registry to be accepted in 1837. This was a secular civil service, created at that time to remedy the fact that the dispersion of various religious denominations outside the official Anglican Church prevented a consistent recording of baptisms, marriages, and deaths. From its very beginnings, the secular civil service was thus associated with management of the Poor Law and the local authorities responsible for it. And from the outset British public statistics was divided into two distinct branches: the Board of Trade, which dealt with economic statistics; and the General Register Office (GRO), which dealt with social statistics. To a large extent this situation still exists: the Office of Population, Censuses and Surveys (OPCS), which succeeded the GRO in 1970, has remained an autonomous institution, while the Central Statistical Office (CSO), which theoretically coordinates the entire system, focuses more on economic statistics. In France, in contrast, from 1833 onward, the SGF had as its mission to centralize statistics (even if it did not do so completely). It was a purely Parisian bureau, without local ties: the cantonal commissions of statistics that were drawn up between 1837 and 1852 all failed.

The situation in Britain was different. The GRO, whose guiding light between 1837 and 1880 was a doctor named William Farr (1807–1883), was closely connected to the public health movement, comparable to the

French hygienists. It was involved in the administration of the Poor Law, not only through its connection with the local Poor Law Unions, but by virtue of the fact that this administration allowed it to acquire a legitimacy that could not be conferred on it from the outset by the central state (Szreter, 1991). At the time, public opinion in England was against the interventions of a national executive body that was readily presumed to be of "Bonapartist" tendencies. The public health movement could only rely on autonomous local initiatives. For its part, the GRO had no direct access to the decision-making process. It had little power, except to persuade and give advice. It forged a policy of alliances with doctors and local authorities to promote *vital statistics:* the ensemble formed by the traditional civil service (births, marriages, and deaths) and by morbidity—that is, the percentages of deaths associated with specific causes. These data were presented according to a refined geographical analysis, intended to designate with speed and accuracy the seats of epidemics and destitution. The GRO thus played an essential role in debates on the diagnosis and treatment of a problem that obsessed English society during the entire century: the problem of poverty linked to anarchic industrialization and urbanization. The GRO did so, for example, by publishing and comparing the rates of infant mortality in the large industrial cities. In Liverpool, half the children died before the age of six—whereas people had previously inferred, from the rapid growth in the population of this city between 1801 and 1831, that the climate there was doubtless particularly healthy.

With its unified data the GRO thus created a frame in which cities could compete and be compared. It stirred up interest in a national death rate competition. The various local medical services could have formatted this information themselves, since they possessed it; in simultaneously gathering and publishing such data, however, the GRO gave rise to a new need, thereby creating a market for its products. Mortality rates, whether of infants or the population in general, became pertinent indicators of municipal policy. In the absence of a national policy on poverty, it was impossible directly to arouse general interest in these statistics. But public interest was attracted by the rivalry promulgated between the cities and evinced by their death rates. These figures thus became a matter of national importance, and were even quoted in a new public health law in 1848. This law stipulated that areas with death rates of more than 23 per thousand (the national average put out by the GRO) had to establish "health tables" in support of sanitary reforms. William Farr went still further during the 1850s, calculating the average mortality rate of the *healthiest* 63 districts (one out of ten), which was 17 per thousand, and setting this rate as a goal for all the others. He thus replaced the national average with a more ambitious optimal goal, thereby anticipating Galton's subsequent shift from the notion of the ideal expressed by the average (Quetelet) to that of the optimal goal emanating from the extreme end of the curve. In referring to the death rate noted in districts scattered throughout the land, Farr was then able to calculate the thousands of deaths that could have been avoided in the other regions, if only the figure of 17 per thousand—which was not theoretical, but real—had actually existed everywhere.

In adopting this line of conduct, the GRO placed itself at the center of a more general movement whose objective was defined as *prevention*. It thus provided a specific language and instruments that differed from those of the individual doctors confronted with specific patients. It succeeded where its French counterparts had failed, managing to implant—first at a local administrative level, and then after 1900 at a national one—the already familiar tools of actuaries, of the statistical and probabilistic management of risks. It thus translated into the sphere of public action ideas already implicit in the concept of averages made popular by Quetelet, whom Farr admired. Moreover, it focused attention and debate on economic and social environment as an explanatory factor in mortality, itself perceived as a consequence of poverty. As in France during the same period, the use of averages as an analytical tool was consistent with the mode of action, with the improvement of health and sanitary conditions in an urban environment.[3] Farr's reclassification of causes of death was elaborated in a generalist perspective, with a view to action:

> The superiority of a classification can only be established by the number of facts which it generalized, or by the practical results to which it led . . . The classification of diseases should be founded upon the mode in which [they] affect the population, . . . the first class will embrace endemic or epidemic diseases; it is this class which varies to the greatest extent . . . and almost always admits of prevention or mitigation. (Farr, 1839, quoted by Szreter, 1991, p. 444)

This manner of classification, associated with a preventive or "reducing" factor, can deviate from a scientific or clinical etiology based on the examination of individual cases made by a doctor intent on curing *his* patient. In concentrating its efforts on improving the nation's health, the GRO established a position of scientific authority at the center of an informal national network for the promulgation of information. This network included the doctors and also the local and administrative authorities involved in this improvement: "The power of the messages disseminated does not depend simply on the scientific rigor of the approach used, but also, more critically, on how aptly the information is communicated, manipulated, and presented, in such a way as to exert a maximum of influence on a larger and more varied audience than that of doctors alone" (Szreter, 1991). Statistical summing elicited a more general viewpoint than that of the doctors who, seeing only patients—especially if they worked in a hospital—had a different perception of public health issues.

The form of action undertaken by the GRO helped prepare the transformation of the British state. It also helped shape public perceptions of government, ranging from the fear of an oppressive, centralized state to the welfare state of the twentieth century. Statistics played an essential part in this transformation, both as a tool (sampling surveys were born in this context) and as a symbol of a new function of the state. This shift had already been observed in 1859 by John Stuart Mill in his essay on *Representative Government*. Commenting on problems of public health, Mill remarked: "By virtue of its superior intelligence gathering and information processing capacities, there is a distinct role for the central state acquired in a liberal society over and above both local and individual claims to autonomy" (Mill, 1859, quoted by Szreter, 1991, p. 448).

The GRO's policy was simultaneously epidemiological and anticontagionist on the medical front, and environmentalist and reformist at the social level. Beginning in the 1880s, it was countered and to some extent marginalized by a dual evolution: progress in microbiology on the one hand, and the rise of social Darwinism and evolutionary theories of natural selection on the other. Bacteriological discoveries concerning the causes and direct treatments of diseases reduced the relative importance of statistical epidemiology and discredited the miasma theory. For their part, the eugenicists maintained that philanthropic measures for relieving the poor favored the reproduction of the most inept and hindered natural selection. This position was violently opposed to the policies of the public health movement, and the attention it paid to local milieus and environments. Initially (before 1910), the GRO upheld this position and continued to emphasize geographical divisions in presenting the results of censuses—divisions, it was thought, that best explained social situations. In contrast, the GRO's eugenicist adversaries spoke in terms of degeneracy and social downfall, resulting from hereditary attributes and aptitudes that could be summarized in a one-dimensional scale. This conviction led Galton to create adequate instruments for measuring these inherited characteristics (Chapter 4): the statisticians of the GRO and the founders of biometrics thus found themselves in opposite camps. The biometricians used individuals' occupational activities as indicators of their aptitudes, which led to their being ordered on a one-dimensional scale (Szreter, 1984). For some time, therefore, the debate between the two currents of thought was reflected by this contrast between two supposedly pertinent divisions: geographical in one case, occupational in the other.

But the situation changed toward 1910. Between 1906 and 1911, a series of new social laws replaced the old Poor Law of 1834. A system of pensions (1908), employment bureaus (1909), and social safety nets (1911) was set up. Unlike the relief policies of the nineteenth century, which were essentially local, this welfare state was created by national laws, in accordance with forms that were homogeneous throughout the land. This changed the terms of the debate waged between the statisticians of the GRO and the eugenicists. The GRO agreed to test the eugenicists' hypotheses by using a national classification of occupations, divided into five hierachic categories, at the time of the census of 1911. This classification subsequently constituted the original matrix of the nomenclature of social classes used by English and American demographers and sociologists throughout the twentieth century. Its confirmed one-dimensional structure, which differed greatly from that of the French socio-occupational nomenclature, originated in this debate, which had been going on for almost a century.

The public health movement and its liaison with the GRO provide an exemplary case of a concatenation observable in other cases: debates and statistical surveys in the reformist milieu independent of the state are taken up by ad hoc parliamentary commissions and result in new legislation, which itself gives rise to a specialized statistical bureau. The issue of education for the working class was raised in connection with criminality and alcoholism. Statistical surveys were done on this subject, with controversial results: some people even deduced therefrom that delinquency was no less frequent in the educated classes. A law on education was passed in 1870. A bureau of statistics devoted to this subject was created in 1876. Attention then shifted to the economic statistics produced by the "self-help" institutions within the working class: mutual insurances and savings banks. In 1886 the Labour Department was created within the Board of Trade (in France, the Labor Office was formed in 1891 within the Ministry of Commerce; analogous institutions were created in the United States in the same period). The Labour Department compiled statistics produced by trade unions, mutual insurances, and cooperatives (Metz, 1987). Until the creation of employment bureaus in 1909, data from trade unions were the only figures enabling unemployment to be measured; this appeared around 1880 as an autonomous social problem, distinct from that of poverty (Mansfield, 1988). During these thirty years there were an increasing number of taxonomical debates, seeking to distinguish the inept, the poor, and the unemployed in the context of broader debates on possible ways to reform the law of 1834 (Yule's research on pauperism, presented in Chapter 4, can only be understood in this context). A Royal Commission for the reform of the Poor Law was established in 1905. Its report was published in 1909 (with an opposing view from the minority on the commission), and employment bureaus were instituted, initiating the organization of the labor market.

The multifarious bureaus of official statistics, which were not intercoordinated, gave rise to criticisms and proposals for reform during the 1920s and 1930s. More than once the creation of a Central Statistical Office (CSO) was suggested to cover this ensemble, but in vain. In fact this only occurred in 1941, with the intense mobilization of economic resources accompanying the war (Ward and Doggett, 1991). The establishment of the first calculations of national income (Richard Stone) played a federalist role, partly unifying the perspectives of the various services. However, the CSO did not rearrange—as did its French and German counterparts—the majority of administrative statistics. It did coordinate the methods of several autonomous services, the two most important of which were the Business Statistical Office (BSO), responsible for statistics of production (this finally merged with the CSO in 1989), and the old GRO, which was integrated in 1970 into a new Office of Population, Censuses and Surveys (OPCS). At each stage in their histories, these services relied heavily on the advice and collaboration of university professors, for it was usually in universities that methodological research was conducted, leading to statistical innovations. Harry Campion, the founder of the CSO, was a professor at the University of Manchester. Arthur Bowley, who still played an important advisory role until the 1940s, taught at the London School of Economics. Richard Stone, the father of national accounts, was a faculty member of the University of Cambridge. Nothing similar, or almost nothing, existed in France, where official statisticians emanated from the world of engineering and academics only intervened (Divisia, Perroux) in statistical debates on rare occasions.

Social Surveys and Learned Societies

This involvement on the part of scientists and intellectuals in the practical problems of governing their country is a characteristic of British society. English scholars invented the techniques of objectifying scientific facts, facts that could be detached from the observer, that could be transmitted and reproduced, shielded from contradictory passions and interests. From the seventeenth and eighteenth centuries on, this model had been constructed for the natural sciences, physics, and astronomy. Whereas the word "statistics" in its (German) eighteenth-century sense denoted a general description of the states, in the nineteenth century it came to mean the numerical description of societies. This shift in meaning was linked to an intense effort to transpose the requirements and methods of the natural sciences into human society. That this transposition did not come about automatically is apparent from the interminable debates on the nature of statistics that animated the statistical societies of European countries throughout the nineteenth century. Was statistics an autonomous *science,* or simply a useful *method* for the various sciences? Could it be compared to experimental sciences? Could facts be separated from opinions about facts? In France and Germany these debates often remained academic in nature; in Britain, however, they had mainly to do with practices used in surveying, with discussions on crucial problems, and with political measures employed in dealing with the problems of free exchange, poverty, public health, education, crime, and unemployment. The work conducted by the Statistical Society of London, founded in 1834 (which became the Royal Statistical Society in 1887), shows the tension that resulted from the difficulty of pursuing two partly contradictory goals: one, to keep the establishment of raw facts completely separate from all interpretation and causal analysis; and two, to remain close to those responsible, nationally or locally, for measures designed to resolve practical problems (Abrams, 1968). In the manifesto published at its foundation in 1834, this society stated the difference between political economics and statistics. They pursued the same goal, the "science of wealth," but statistics was distinguished by the fact that it

> neither discussed causes nor reasoned upon probable effects, but sought only to collect, arrange and compare the class of facts which can alone form the basis of correct conclusions with respect to social and political government . . . [a difference which] absolutely excluded

> all forms of speculation . . . The Statistical Society will consider it to be the first and most essential rule of its conduct to exclude carefully all Opinions from its transactions and publications—to confine its attention rigourously to facts—and, as far as may be found possible, to facts which can be stated numerically and arranged in tables. (London Statistical Society, 1837, quoted by Abrams, 1968, pp. 14–15)

The founders and driving forces of this society were very close to the government, so much so that its board members could be described as the "subcommission of the liberal cabinet." Linked to civil service, serving and advising a government they tried to influence and rationalize, they invented and developed the rules for dividing labor between politicians and technicians. The elimination of opinions and interpretations was the price they had to pay, the sacrifice they had to make, in order for their objectives to acquire the credibility and universality required by politicians. In *Hard Times* Dickens mocked the certainty of his fictional Mr. Gradgrind, an austere and implacable figure who brought death to innocent people while calling for "the Facts, Sir; nothing but Facts." A member of the statistical society of Ulster declared that "the study of statistics will in the long term preserve political economics from the uncertainty that currently envelops it." This link between statistics and objective certainty, opposed to speculation and individual opinion, helps us understand why probabilities—originally associated with the idea of uncertainty—were so tardily incorporated into the statistician's box of tools (using the word statistician in its nineteenth-century sense), the role of which was clearly defined in 1837 by this manifesto.

But so drastic a sacrifice contained its own limitations, in confining the statistician to a narrow role, and in forbidding him to deploy his energy and talents contingently in two distinct directions, politics and science. He could not enter the field of the political decision maker without immediately forfeiting his credo of fact without opinion. Nor could he play an active part in the social sciences, economics or sociology, for he would then have to combine his facts with more general explanatory and interpretive systems which he could not account for simply by reason of the duly noted fact. And so he was completely stymied. Thanks to such questions "statistical enthusiasm," which was very pronounced during the 1830s and 1840s, ran out of steam for the next thirty or so years. Initially, ambitious survey projects were conceived by the various statistical societies, not only in London but also in Manchester and in several other industrial towns. They dealt with savings institutions, crime, rent, workers' living conditions, and strikes. The questionnaires were addressed not to a general population sample, but to competent authorities: police, hospitals, school councils, insurance companies, Poor Law administrations, factory directors, property owners, magazine editors, magistrates, prison governors, and above all to the local members of statistical societies.

But in the absence of a set of standardized definitions and a homogeneous network of fact gathering—and also because certain results were unexpected or unintelligible—these experiments came up short, rather as, forty years earlier (though within the administration itself), the survey conducted by French prefects had done (Chapter 1). The technical apparatus, administrative infrastructure, and tools for formalizing data were still too limited to allow a statistician to sustain the principles abruptly proclaimed in 1836, by means of sophisticated work. He could not clearly impose his specific legitimacy on politicians, economists, nor even on mathematicians and probabilistic philosophers, who were also active in Great Britain, but in a different sphere (George Boole, John Venn) that had no connection with statistics.

Between 1850 and 1890, the work of the Statistical Society was fueled by aggregate data provided by the official bureaus of statistics. These data were subproducts of administrative management: customs, the Poor Law, declarations of deaths, statistics from trade unions and mutual insurances. The tradition of conducting surveys was reborn around 1890 (Hennock, 1987). Charles Booth (1840–1916), a rich reformist bourgeois, devoted his fortune to surveying poverty in London. Rowntree (1871–1954), a chocolate manufacturer, adopted Booth's methods, studying other English towns and comparing them with London. Bowley, an economist and statistician with a background in mathematics, codified and standardized the techniques of sampling surveys and, more generally, the gathering, formatting, and interpretation of social statistics, in accordance with norms subsequently adopted by the professionals, administrative statisticians, economists, and sociologists (Bowley, 1906, 1908). Until the 1940s, the successive editions of Bowley's manual provided a standard for recording and formatting economic and social statistics comparable to that of Yule's book in regard to the mathematical techniques used in analyzing these data.[4]

This work was first undertaken by prominent middle-class citizens with few technical qualifications, especially in mathematics (Booth and Rowntree). They used nineteenth-century methods and vocabulary, although the strict separation between "facts, interpretations, and recommendations" was no longer apparent. One of Booth's objectives was to distinguish those who were irreparably poor and responsible for their own downfall from victims of structural causes linked to the economic crisis. He evaluated the relative proportion of each category, so as to deduce the appropriate forms of correction or assistance. For Bowley, in contrast, the formulation of sophisticated technical norms allowed him to assert—with more likelihood of success than in the 1830s—a clear separation between the expert and the decision maker. He thus made way for the autonomy of a new figure: that of the state statistician with a particular form of professional competence, a figure distinct from politicians, from high management officials, and from university professors or researchers enrolled in an academic discipline. Though still a rarity at the turn of the century, this figure became common in official statistics after the 1940s.

In 1908 Bowley delivered a lecture before the Royal Statistical Society entitled "The Improvement of Official Statistics." He analyzed the seven conditions needed for a statistical operation to produce consistent objects: (1) defined unities; (2) homogeneity in the populations being studied; (3) exhaustiveness (or adequate sampling); (4) relative stability (or repeated measurements as dictated by the instability); (5) comparability (an isolated number meant nothing); (6) relativity (the numerator and the denominator of a quotient must be evaluated in a coherent manner); and (7) accuracy, which resulted from a combination of the six other conditions. He observed that "the accuracy of official statistics was, despite the attention and systematic verifications to which they are subject, purely superficial." Bowley presented a series of concrete examples to illustrate his remarks and made several proposals for improving these measurements, particularly through the use of "scientifically chosen samples" with obligatory answers to avoid the bias resulting from the subjective nature of responses. Last, in the absence of a central office of statistics, he suggested the creation of a Central Thinking Office of Statistics, recalling the advice given by another member of the society in regard to statisticians: "Publish less and think more."

In the debate that followed, Yule endorsed Bowley but made two significant remarks. In regard to the *homogeneity* of populations, he observed that this was not necessary, and that recent statistical methods (those of Pearson) were intended precisely to analyze *heterogeneous* populations (Bowley replied that there had been a misunderstanding). Moreover, Yule was not convinced that "a sampling can be truly random and representative of the total population, even with an obligation to respond analogous to the obligatory participation of trial juries" (an interesting and seldom made comparison). The debate continued with a seemingly practical question which in fact brought up the problem of the nature and realistic status of official published statistics. Bowley had maintained that these publications should always be accompanied by full details as to the methods and conventions used in defining, recording, coding, and tabulating. Someone objected that this "threatened to make the publications so voluminous and cumbersome that it would diminish rather than increase their usefulness." Ever the professional, Bowley suggested at least leaving the lid of the mysterious black box slightly ajar, so that the process of fabricating objects would remain partly visible. He received a utilitarian sort of answer: what was the point of this major investment if one could not use it blindfold, and readily integrate it, as such, into other constructions? This seemingly insignificant tension was central to the statistical construct of reality. Arbitration between these two claims depended largely on the legitimacy and credibility of the publishing institution. The norms determining the critical apparatus deemed necessary were not the same in a thesis, a scientific journal, or an administrative publication. In this regard, the choices made by the different organs of diffusion of a statistical bureau indicate whether it based its legitimacy on the universe of science or that of the state.[5]

The paradoxical characteristic of the English configuration was that this new professional profile had been codified in this way by an academic, which allowed the foregoing questions to be raised very early on. In Britain, abstract theoretical reflection was less prevalent in universities than in France and especially in Germany (at that time). On the other hand, close interaction between universities and the administrative, political world was a common tradition, whereas in Germany attempts made along these lines between 1871 and 1914 aroused little response among the administrative and political personnel of the Empire.

6

Statistics and the State: Germany and the United States

At the beginning of the nineteenth century in France, Great Britain, and the United States, the existence of a national state was guaranteed and no longer subject to question. This was not, however, the case in Germany, which was still divided into different states of vastly uneven importance. The successive stages in the construction of the German nation punctuated the next two centuries: first, the kingdom of Prussia, then the Bismarckian Empire, followed by the Weimar Republic, the Nazi Third Reich, the separate Federal and Democratic Republics, and finally the unified Republic of Germany. This uncertainty in regard to the state, its consistency and legitimacy, left a mark not only on the structures of statistical institutions but also on modes of thought and reasoning. It gave German statistics—and more generally the social sciences and their relationship with power in its successive forms—a particular coloration, quite different from that of the other three countries, and this lasted until at least 1945. The wish to maintain the particular features of local states (and of the *Länder*) manifest in historical and religious traditions was offset by legalism and organization, guaranteeing general cohesion in a way that was once authoritarian, but is nowadays more democratic. Philosophical debate had long occupied a significant place in the German social sciences, including statistics. It was marked by a holistic and historicist form of thought, hostile to the individualism and universalisms—which were deemed reductive—of French rationalism and English economism. The relationships that academics and statisticians entertained with the state were complex, often filled with conflict, and rarely attained the suppleness and flexibility observed in Great Britain. These characteristics were typical of nineteenth- and early twentieth-century Germany. Since 1945, several such tendencies have survived: federalism, legalism, and organization. In contrast, technical statistics and the uses they are put to in administration and the social sciences have drawn closer to those in the Anglo-American world.

German Statistics: Building a Nation

The word "statistics" originated in eighteenth-century Germany and designated a "science of the state" *(Staatwissenschaft),* a descriptive and non-quantitative framework of reference and terminology offered by university professors to princes of the numerous German states (Chapter 1). Of these, Prussia emerged as the dominant power, already using methods of quantification that were quite different from the statistics of professors. As in France at the same time, these practices obtained in two distinct milieus, the administration and the world of enlightened amateurs (Hacking, 1990). The royal government and its bureaucracy gathered information, which was secret and for their use only, in order to raise an army and levy taxes. In Prussian statistics of the nineteenth century, a fundamental separation between civilians and soldiers remained. This structured first the tabulations, then later on, in the German censuses, the distinction of officials *(beamte)* among occupational groups that is still visible today: service to the state was an important element in defining individuals. Independently, moreover, of the administration and the universities, there were "amateurs"—geographers and travelers—who produced syntheses based on numbers, closer to statistics in the sense of the next centuries. The best known was the pastor Sussmilch (1708–1767), whose *Divine Order* was a basic text in demography (Hecht, 1979).

Following its defeat at the hands of Napoleon's armies, the Prussian state was reorganized and equipped with a statistical service. This existed continuously from 1805 to 1934 and was the most important service of its kind in the German Empire proclaimed in 1871 (Saenger, 1935). The other German states—Saxony, Wurtemberg, and Bavaria—were also equipped with bureaus of statistics during the first half of the nineteenth century. An imperial service was created in 1871, but the offices of the diverse states remained autonomous until 1934, at which date they were absorbed into the unified statistical office of the Nazi state. (In 1949 the new Federal Republic restored the system that had existed prior to 1934: the *Länder* had statistical offices distinct from the federal office in Wiesbaden, but their activities were closely coordinated.) Beginning with the Prussian bureau, the nineteenth-century German bureaus of statistics inherited and amalgamated the three traditions of the eighteenth century: the political, historical, and geographical descriptions furnished by university professors; the administrative records kept by officials; and the numerical tables established by scholarly amateurs. The directors of these bureaus were often university professors who taught "sciences of the state." In these two simultaneous activities, they compiled vast amounts of information on the various aspects of a territory, with the region's historical, religious, cultural, and economic identity providing a descriptive and explanatory guiding thread. Unlike their eighteenth-century predecessors, however, these "statistics" increasingly integrated tables calculated on the basis of demography and on administrative activity. The close ties with the administration were marked by the fact that these bureaus were attached to the Ministry of the Interior, a ministry of direct political management, whereas their French or English counterparts depended more on ministries of economics (Commerce, Labor, Finance). This difference has endured: statistics is one of the mechanisms that sustains a state when its stability is more problematical than in other countries.

The Prussian bureau of statistics displayed remarkable durability. In the one hundred and twenty-nine years of its existence, it had only six directors. Two of them exercised a particularly lasting and profound influence: Hoffmann, between 1810 and 1845; and above all Ernst Engel, between 1860 and 1882. The first directors were senior officials who held other positions at the same time, not only in the university but in diplomacy or at the Staatsrat (State Council), where they participated in legislation. Before Engel's nomination in 1860 the activity of this bureau, which resembled that of the SGF under Moreau de Jonnès, consisted of gathering and publishing a vast amount of data that had been recorded by other administrations, devoid of any form of technical verification or central coordination. The published tables reflected "population trends" (from the registry) and also prices, "means of existence," financial statistics, buildings, and schools for the entire kingdom and its provinces: accurate geographical descriptions were of great importance. Taken as a whole, these elements formed an administrative and territorial patchwork rather lacking in cohesion, and seen retrospectively their reputation as a source of data is rather weak. But in all probability the existence and volume of these publications signified as much—as symbols of the power of the state and its administration—as did the detail and precision of their imposing charts. Unlike England, Germany was still not very industrialized, and its social problems were not yet as manifest. On the other hand, the questions raised by the political and military relationships among states, by the emerging Zollverein (customs union), and by the growing power of Prussia were crucial during the period before 1860. After that date the situation changed, not only on account of Engel's strong personality, but also because the economic and political context altered with the rapid industrialization and subsequent unification of the Empire around Prussia.

Ernst Engel (1821–1896) is particularly known for his work on family budgets and his formulation of a ratio of elasticity, "Engel's Law," which stated that the proportion of food consumption in a given budget diminishes when income increases. Initially, however, he was a typical nineteenth-century statistician, an activist and organizer, though still not very sensitive to the refinements of mathematics. The role of such a statistician was to create or modify administrative mechanisms; to unify or coordinate bureaus; and to subject their activities to a common logic, according to more or less centralized and hierarchic forms. His role was also to interest other actors, and to integrate the bureau of statistics into vaster scientific and political networks: we have already seen how the SGF under Lucien March and the British GRO under William Farr more or less achieved these two objectives. The case of Engel and the Prussian bureau of statistics differed, however. Engel managed, on the whole, to achieve the technical reorganization and the administrative centralization that his predecessors had not been able to accomplish. But he partly failed in his second task: the construction of a politico-scientific network, even though he had made a major contribution by helping to found the Verein für Sozialpolitik (Social Policy Association) in 1872. This failure was not his alone. It also belonged to the association of liberal moderates—professors and leading citizens—who did not succeed in influencing Bismarck's authoritarian policies. The history of Engel and his statistics bureau cannot be separated from this larger endeavor, which half succeeded with the social laws, and half failed too, since the political alliance eventually fell apart.

In 1861 Engel set about completely reorganizing the Prussian bureau of statistics, extending and unifying the statistics of the state in such a way that the greater part of their construction was entrusted to him. For censuses, he created *individual bulletins,* so that the primary data were recorded directly from the persons queried rather than from leading citizens (mayors and priests). These bulletins were conceived and analyzed by the bureau itself (this central analysis was only introduced into France in 1896, by March). He increased the number and variety of the publications. He also created a central commission on statistics that served as a liaison between ministries and bureau. Last, following a model that was typical of the German university of the time, he also created a seminar on statistics in 1870, to train other administrations or states soon incorporated into the new empire. This seminar was attended by several economists and economic historians who, after the Verein, subsequently became the best-known representatives of the current of economic thought known as the *German historical school.* Opposed to abstract and deductive economics, whether English or Austrian, these economists attributed great value to empirical monographs with a historical or statistical basis, according to methods resulting from Engel's teaching.[1]

At the time when Engel was director of the bureau of statistics, Germany was embarking on a period of rapid industrial growth that enabled it to catch up with England and France, where industrialization had got off to a quicker start. Industrial statistics was ready to be created. Engel imagined a unified reckoning of individuals, jobs, and establishments based on individual bulletins (this was also the system obtaining in France in 1896). The first census of industrial plants took place in 1876, and a complete industrial census was undertaken in 1882 (these feats were the source of an important historical series of statistics on trades and establishments since 1875 [Stockmann and Willms-Herget, 1985]). In this way the statistics of production were linked with those of employment. Industrialization brought about a rapid increase in the working classes. They were highly organized, both politically and in terms of unions, by the Social Democratic movement. Engel and the economists of the Verein militated in favor of the creation of mutualist systems of insurance, based on the idea of *self-assistance.* As early as 1858 he had imagined and set up a new kind of insurance society based on mortgages, to prevent foreclosure on workers' lodgings—a frequent occurrence at the time (Hacking, 1987). He then helped write the social laws that Bismarck's government passed in answer to the demands made by the workers' movement. In 1883 disability insurance was created; in 1885, compensation for work-related accidents; and in 1891, retirement pay. These first forms of social protection owed much to the statisticians and economists associated with Engel within the network of the Verein.

By the time these laws were passed, however, Engel had already been obliged to resign from his position, in 1882, on account of his open disagreement with Bismarck's protectionist policies. (Bismarck, who had ties with the property owners of East Prussia, was opposed to customs tariffs being levied on corn imports, which weighted down food prices and dragged down salaries. This hurt both workers and industrialists, heightening the tensions between them.) Even before this, Engel's project had run out of steam. In 1873 his seminar on statistics was becoming rather aimless and Prussian officials were showing less interest in statistics. He managed neither to sustain relationships between the bureau of statistics and local administrations (which he had initially tried to promote by creating ninety-seven local offices) nor completely to unify the statistical studies undertaken in these other ministries. After Engel's forced departure in 1882, the bond of trust between the government and the Prussian bureau of statistics was broken. Reconstructing the history of this bureau in 1935 (at which time it was absorbed into the centralized office), Saenger describes how signs of weakening appeared in the scientific and political network during the period that followed:

> Unlike Frederick the Great, Bismarck did not like statistics and even thought one could do without them. It is therefore understandable that the lightning bolt that unexpectedly struck the bureau's leader paralyzed its activity, necessarily giving rise to prudence and great caution in regard to all external intervention . . . The central statistical commission which, in Engel's time, had exercised a very stimulating activity, gradually fell asleep . . . Almost imperceptibly, the Prussian bureau of statistics was transformed from being an institution intended as an organ of assistance for legislation, the administration, and the economy, to being a scientific institute that published research according to the gifts or interests of its members. (Saenger, 1935)

This retrospective judgment was made fifty years later by an official of this same bureau, himself the recent recipient of an even more violent lightning bolt than the one sent by Bismarck. Perhaps Saenger wanted to suggest that Hitler could no more claim authority from Frederick than could Bismarck, since he manhandled the Prussian bureau of statistics. Between 1882 and 1914, however, the bureau's activities retained their importance and were not those of an "institute that published research according to the whims of its members." Demographic censuses were taken every five years. Two major censuses, of jobs and institutions, were carried out in 1895 and 1907. Statistical surveys were launched, the subjects of which clearly show how the bureau conformed to the preoccupations of the administration of the time: additional taxes; eligibility for military service, and the ratio of recruits from town and countryside; the school and university system; nationalities; and the finances of communes. But the Prussian bureau of statistics began to experience competition on two fronts, resulting from economic growth and the unification of Germany. The rapid industrial development had given rise to the creation of very large firms, cartels, and professional trade unions; these bodies kept their own records of data that had previously devolved upon official statistics, now causing it to lose its monopoly. Saenger describes the consequences of German unification in terms reminiscent of European unification a century later—but for the fact that today there is no state that clearly dominates the ensemble, as Prussia dominated the Empire:

> To the extent that a unified German economy was being created, data that concerned only Prussia lost their value. Prussia was still a state, but it was no longer an entirely separate economic entity. The farther the laws of the Empire extended, the stronger the need for data gathered everywhere on a common basis . . . Individual states could no longer be considered, and Prussia even less than the other states in the heart of Germany, which still formed more or less closed economic entities . . . The statistics of the Empire became increasingly important. The central Prussian commission was gradually replaced by labor commissions that recombined the state statistical bureaus with that of the Empire. The commission's advisory role was henceforth played by the federal parliament *(Bundesrat)*. (Saenger, 1935)

Increasingly, however, the Prussian bureau did work for other states, and the work was gradually divided between Prussians statistics and statistics for the whole of Germany. This empirical balance survived until the two services were fused in 1934. The process described is significant: unification took place a step at a time, and several services could coexist simultaneously, while rivalries and complementary relationships were regulated according to current political relationships, through negotiation and compromise. In contrast the French state, which had long been unified politically, did not experience such changing relationships. Contemporary German statistics still rely on a negotiated balance between the federation and the *Länder,* in which the federal parliament (and especially the second assembly, with the Bundesrat representing the *Länder*) plays an important role in controlling the activity of the Statistisches Bundesamt (Federal Office of Statistics). Thanks to its federal character, as negotiated between independent states, the European statistics now being created by Eurostat (the statistical office of the European communities, based in Luxembourg) may well have more points in common with the old, historically evolved German statistics than with the territorially centralized French system. The same problems of federalized states are occurring in other countries of central and eastern Europe.

The Historical School and Philosophical Tradition

The difficulties experienced after 1880 by Engel and the Prussian bureau of statistics in regard to political power cannot be reduced to the economic and administrative problems described by Saenger. The chill in relations between statisticians and the imperial government was more general, involving the entire modern liberal current that Engel belonged to, through the Verein für Sozialpolitik. In the years immediately following its establishment in 1872, this association initially played—thanks to the feebleness of Parliament—the role of advisor to the prince, preparing legislative texts: it was comparable in this respect to the London Statistical Society. But this agglomeration of nationalist free-market tendencies subsequently collided with Bismarckian authoritarianism on numerous points, and finally split into several different groups, depending on the forms of political organization advocated in the circumstance. In this new context, in which advice and direct influence were no longer possible, the Verein generated—mainly as a means for reconciling its diverse factions—an activity of economical and social surveys with a strong statistical content.

One of the most important of these surveys was the one carried out by Max Weber in 1891, on the agricultural workers of East Prussia.[2] Despite its seemingly technical and limited subject matter, this research was prompted by an economical and political problem crucial for Germany at that time: how was it possible to maintain and strengthen the identity of a national state that had only just been born and was still developing, when at the same time the development of industry upset the social balance between landowners, industrial bosses, and agricultural or factory workers? In England the problem of balance between classes was posed in terms of poverty, and of the dangers raised by the underprivileged lumpenproletariat. In Germany it was posed in terms of national identity, which was thought to be threatened by non-German contributions. Industrialization involved important shifts in population within the Empire, from Northeast Prussia toward the southwest and the Rhine. A labor force of Slavic origin (Polish and Russian) came to fill the agricultural jobs left vacant in East Prussia in the vast estates of the Prussian *Junkers,* who supported Bismarck's regime. Traditional patriarchal bonds were gradually replaced by anonymous capitalist relationships.

The aim of the surveys was to provide a description of these new relationships in economic terms, and to evaluate their impact on social and national cohesion. Two separate questionnaires were used. The first, intended for the landowners, covered factual questions: the number of salaried persons; the ratio of remuneration in money and in kind; the social characteristics of workers; forms of labor contracts; and the possibility of access to schools and libraries. Out of 3,100 questionnaires that were sent out, 2,277 (or 73 percent) were returned. The second questionnaire was an evaluation. It was addressed to teachers, pastors, notaries, or officials, who were thought to know the values and opinions of the rural milieu. Sent to 562 persons, it was filled out by 291 of them, or 52 percent. Weber drew up a report of 900 pages from this survey, which comprised a large number of statistical tables describing the economic situation of Prussian agriculture. He recommended the development of a small independent agriculture, to replace the vast capitalistic estates of the absenteeist *Junkers* living in Berlin; his aim was to establish a German work force that preferred a free lifestyle to higher wages, and to avoid the influx of the Slavic wage earners. The goal and conclusions of this survey were primarily political, even if the demonstration was based on economic arguments shored up by statistical tables. How could and should the German nation best evolve, given the social upheavals induced by industrial growth and migratory movements? The survey was well conceived technically. The questionnaires were used by Grohmann, a statistician and colleague of Weber. However, this methodological aspect was not what was most important. In Germany of that time the most convincing arguments, around which the "methodological debate" *(Methodenstreit)* was raging, were philosophical and historical. They were not yet statistical and mathematical like those being developed, during the same period, by the English school of biometrics.

In the second half of the nineteenth century in Germany, Quetelet's intellectual construct—with its statistical regularity and "average man"—had a philosophical impact unparalleled in France and England. This impact, moreover, was not direct. A series of translations (not only linguistic ones) were required, capable of making this fairly simple schema fit into the subtleties of German intellectual debate. The intermediary was the English historian Henry Buckle, whose monumental *History of England* (1857) was based on the idea that the macrosocial regularities discerned by statistics made it possible to explain the protracted, ineluctable trends in a nation's destiny. Thus expressed in terms of historical determinism and national identity, Quetelet's statistical construct could be reinjected, transformed, and criticized in a German debate that revolved around these very questions. Analogous philosophical retranslations took place at a later date with systems as varied as relativity and probabilistic models during the 1920s and 1930s or, more recently, catastrophe and chaos theory.[3]

The Germans who reinterpreted Quetelet were statisticians, economists, and historians. Primarily academics, they were steeped in an ancient philosophical tradition in which the oppositions between freedom and determinism were essential, as were those between a supposedly German holism and a reductive individualism, whether rationalist in the French manner or economical in the English way. In this context, a statistical system that claimed to display ineluctable laws analogous to those of physics appeared mechanistic and distorting. It negated the particularities and specific traits of individuals who were free and incommensurable, and also the original cultural traditions.[4] This critique of creating equivalence before statistical construction (in the modern sense) was already present in "statistics" in the eighteenth-century German sense (Chapter 1). But the idea of the *average man*—expressing, through his regular features, a reality of a higher order than that of contingent and unpredictable individuals—could also be used in an opposite sense. It could be adopted as an argument in a holistic perspective, in which the whole had more reality than the individuals who constituted it. This was the sense that Durkheim gave the idea in his early works, incorporating it into his construct of the social group as existing prior to individuals (Chapter 3). In this German context, Quetelet can thus be stigmatized as a mechanistic individualist, a product of the arid rationalism of the Enlightenment; or, depending on the case, he can be used as a modernist reference in a form of argument that lays much emphasis on traditional cultures and national communities.

The statisticians and economists of the historical school debated a great deal—especially within the Verein—about the status and methods of the social sciences as compared with the natural sciences, and in terms of a philosophy of history opposed to that of the French or English. Statistics was often used, but it was termed a *descriptive method* rather than a method of discerning *laws*. Thus Engel recognized the idea that statistics could conceivably show empirical regularities, but not that it could under any circumstances claim to establish laws analogous to those of physics, since the laws of physics entailed knowledge of underlying elementary causes.[5] German economists used the abundant data published by official statisticians, who were often close to them both intellectually and politically, to fuel descriptive monographs dealing with precise, localized themes: Weber's survey is a sterling example of this genre. For these economists, statistics appeared as one descriptive element *among other* elements, whether historical, institutional, or sociological. The fact that these various modes of knowledge were not yet classified under separate disciplines was responsible for the "methodological debate" among the academic members of the Verein, and for the philosophical nature of this debate.

Seen retrospectively, the accumulation of this fragmented knowledge and the rejection of any formalizations other than those offered by traditional German philosophy may give the impression of a scientific trend that had no future, especially when compared with all that would soon follow: the birth of econometrics and sociology based on surveys. Nevertheless, the tradition had its descendants, mainly because a number of French, English, or American academics had studied or traveled in Germany at the time and were familiar with this intellectual milieu. Halbwachs used German statistics for his thesis on the standard of living of workers. Karl Pearson, who had studied in Heidelberg as a young man, gained from that experience a philosophy of science and causality close to that of Engel, excluding "laws" in favor of observed regularities. The institutionalist American economists, such as the sociologists of the Chicago school of the early twentieth century, were well acquainted with the works and debates of the German historical school. That this intellectual trend actually ran out of steam in Germany was largely due to the fact that it did not manage to make itself an indispensable part of a radical transformation in macroeconomical and macrosocial policies—as would later be the case for the quantitative social sciences in the United States during the 1930s and 1940s, then in France and Great Britain after the war.

Census Taking in American Political History

Census taking is implicit in the very mechanism that generates and organizes the federal American state, in accordance with modes that have no equivalent in Europe. Every ten years, the population census provides a basis for the apportionment of political representation and fiscal burdens among the states that make up the Union. The principle of this regular demographic census taking, which is written into the Constitution, has been respected and applied twenty-one times between 1790 and 1990, mainly to apportion seats in the House of Representatives. The process is commonplace and takes place almost unnoticed in European countries, where the population develops only slowly. In contrast this process is extremely visible, and the subject of intense debate, in a country whose number of inhabitants grows rapidly between each census: the average decennial growth in the United States was 35 percent between 1790 and 1860, 24 percent between 1860 and 1910, and 13 percent between 1910 and 1990. The successive waves of immigrants, the expansion of boundaries westward, urbanization, economic crises, ethnic conflicts: all this has contributed to a constant, radical modification in the political balance between regions and parties.

The techniques for measuring the populations of individual states and converting them into congressional mandates have prompted recurring debates in Congress. How were slaves and foreigners to be counted? What arithmetical procedures should be used to allot whole numbers of seats in proportion to the population? Other major questions concerning censuses have been discussed, as have possible methods and conventions. How were the social and economic effects of slavery (during the 1850s) to be assessed? Could the flow of immigration (during the 1920s) be influenced or limited? How was unemployment to be measured, and social inequalities (during the 1930s)? In each of these debates statistics were quoted abundantly, criticized, and made to compete with one another. They were both ubiquitous and relative. They were used in negotiations, in the compromises that temporarily confirm power relationships and allow society to forge ahead until the next crisis. They reflected the image of a state that did not embody, as in France, a general interest superior to individual interests; rather, through a balance among the various powers codified by the Constitution and its successive amendments, they reflected compromises that allowed individuals to exercise and uphold their rights.

Thus the United States is the country where statistics has displayed the most plentiful developments. But it is also the country where the apparatus of public statistics has never experienced the integration and legitimacy that its French, British, and German counterparts—though in different forms—have managed to acquire. Conversely, in both sociology and economics, universities, research centers, and private foundations have produced numerous surveys, compilations, and analyses of statistics of very varied origins that no institution has had a mandate to centralize. As early as the 1930s some of the major technical innovations were used in experiments, innovations that, after 1945, would radically transform statistical activity and the profession of statistician: sampling surveys, national accounts, econometrics and then, during the 1940s, computers.[6]

The constitutional principle by which the population of the individual states serves as a basis for apportioning both taxes and political representation is an ingenious one, in that it prevents the states from trying to manipulate their population statistics: they would lose on the one hand what they would gain on the other. This mechanism encourages compromise between contradictory goals. But in fact it has seldom been a factor, for only rarely did the federal government use its fiscal branch during the nineteenth century. For a long time this state governed extremely lightly: Jefferson spoke of a "wise and frugal government." Customs receipts were sufficient for its budget, except during wartime. The administration of the census was reconstituted every ten years—enough time to carry out the survey and then use it to calculate the congressional representation of each state. It was dissolved three or four years later. A permanent Census Bureau was not established until 1902. Each time, the organization of the census and the creation of this ad hoc service were the subject of a special law, preceded by an intense debate in Congress. Topics raised in the debate included what conventions to use in counting and apportionment; what questions should be asked; and how to recruit the surveyors and the personnel charged with carrying out the operation.

From the very beginning, and until the Civil War, the question of how to treat slaves in the process of apportionment was raised. The states in the North and South did, of course, have opposite points of view. But unless the Union were to burst asunder, something neither side wished, there had to be a compromise, the only merit of which was that both parties accepted it: this was the "three-fifths rule," according to which a slave was counted as three-fifths of a free man. This convention strikes us as particularly shocking now, for it cannot be justified by any objective reality external to the actual conflict. It implies simultaneously that the slave was *comparable* to a free man, but also *inferior* to him. So paradoxical a combination can only be extremely unstable. It was demolished after the North defeated the South in 1865. The debate then assumed another form as the southern states managed, by various means, to continue to deprive blacks of the right to vote. The North attempted to exclude from the apportionment pool all adults unduly deprived of the vote. The South replied by proposing to exclude foreigners (that is, recent immigrants), which would have damaged the North, where they had settled. As a compromise, both sides then agreed to count all adults.

Another seemingly technical question raised further interminable problems until the 1920s: what arithmetical convention could be used to transform the apportionment of the population into whole numbers of House seats? Several possible solutions were adopted in turn (Balinski and Young, 1982), giving different results ("major fractions" and "equal proportions"). They did not benefit the same states. Statisticians representing the rival states were summoned before Congress to argue with one another between 1923 and 1929, in front of congressmen who understood nothing of their harangues, so dry was this subject. Finally, since they still had to pass the law authorizing the census of 1930, they split the problem down the middle, and a spurious system was set up, halfway between the two solutions. When the census had been carried out it was observed—quite by chance—that the two methods would have given exactly the same results (Anderson, 1988). In the meantime the Depression had begun, and congressmen and statisticians had new preoccupations. In these debates, the statisticians were called upon not so much *to tell the truth* as to *provide solutions* allowing some way out of the political impasse. They were closer to the jurists helping the decision makers than to the scholars perched above the fray that they would later become. The fixed, incontrovertible point was the constitutional principle of apportionment. The debate hinged on its modes of application.

Statistics was also invoked in regard to polemics concerning the crucial problems of American society—namely, depending on the period, slavery, poverty, immigration, unemployment, and racial integration. In 1857 Helper, an economist from the North, systematically compared the indicators of production, wealth, and social and cultural development as calculated for the northern and southern states, and imputed to slavery the fact that the indicators for the North always exceeded those for the South. His work caused a great deal of commotion and prompted a counterattack from the southerners. A journalist named Gordon Bennett wrote that "slavery was only an evil because work itself was an evil . . ., the free-work system was in reality the white slave trade of the North." This pro-South argument was based on statistics produced by the welfare administrations (managed by the states), showing that "poor people, deaf people, dumb people, blind people, and idiots" were more numerous in the New England states than in the South. These two opposing statistical arguments were contested in their turn on the grounds of the construction and pertinence of the data used. The southerners showed that Helper had not related his wealth indicators to the population of the states; this, in their view, stripped them of all value of comparison. For their part the northerners criticized Gordon Bennett's use of data on the poor who received relief, as explicitly prescribed by the laws of the various states. These statistics reflected the policies particular to the states, and not the absolute number of poor people: "Any comparison must be preceded by an examination of the laws and customs of poor relief based on public funds" (Jarvis, quoted by Anderson, 1988).

As early as the 1850s therefore, the fire of controversy led economists explicitly to raise the question of the construction of reality based on administrative records—something that, forty years later in Great Britain, Yule did not formulate as clearly in his mathematical analysis of poverty (Chapter 4). The fact that, in the United States, statistics was so closely and so rapidly involved in an arena of contradictory debates stimulated critical thinking and encouraged diversity in the uses and interpretations of this tool. More than in other countries, statistical references were linked to the process of argumentation rather than to some truth presumed to be superior to the diverse camps facing off. This way of thinking seems typical of American democracy, based more on debate and the quest for compromise than on an affirmation of general public interest and single truth: in any case, we find here a path leading to a better understanding of the significances and different roles of statistical argument, according to political and cultural traditions on both sides of the Atlantic.

Throughout the nineteenth century the United States welcomed ever more numerous waves of immigrants, multiplying its population by 23 between 1790 and 1910. Arriving first from the British Isles (England and Ireland), then from northern Europe (Germany, Scandinavia), these immigrants tended increasingly at the turn of the century to come from southern and eastern Europe (Italy, Poland, Russia, and the Balkans). Whereas the first arrivals (from western and northern Europe) were deemed more easily assimilable into the lifestyles and ideals of the initial kernel of the Union on account of affinities in language or religion, the second group was increasingly suspected, during the 1920s, of introducing cultural elements that were unassimilable and incompatible with liberal democracy, notably on account of their religion (Catholic, Jewish, and Orthodox). For the first time, the dual question was raised of restricting immigration and allotting different quotas to the countries of origin. An intense political and statistical debate was waged between 1920 and 1929 on the appropriate criteria by which to justify and fix these quotas. Not only did the congressmen and statisticians charged with providing these criteria take part in the debate, but representatives of groups with contrary interests also participated: industrialists opposed to restricting immigration for fear of finding themselves without a work force; representatives of the diverse national groups; various scientists; economists assessing the effects of slowing down the flux of immigration; psychologists comparing the intellectual quotients of immigrants according to their origins; sociologists investigating immigrants' socialization in mushrooming towns such as Chicago; and historians reconstructing the ethnic composition of populations prior to Independence, in 1776, and its later transformations.

Census taking allows the population to be sorted according to its place of birth. In 1920 a proposal was made to Congress to limit the annual immigration for each country of origin to 5 percent of the persons counted in 1910 as having been born in the corresponding country. Congress reduced the proposed quota to 3 percent. President Wilson vetoed the law, but the following year his successor, President Harding, signed it. This constituted a decisive turning point in American population policies, and was perceived as a wish to limit the irresistible growth of urban centers that were gradually nibbling away at the power of rural states. The stakes were similar in the previously mentioned, more technical debate over the means of calculating congressional seats, for it seemed that the one method benefited the urbanized states and the other, the rural states. All these debates were fueled by an anxiety as to the political balance between two different Americas: the one industrial, urban, and cosmopolitan, and the other rural and traditional.

The debate on ethnic quotas lasted through the 1920s, and the statisticians of the Census Bureau were deeply involved. It first appeared that the decision to use the most recent census—that of 1910—to settle quotas raised problems for those who favored immigration from western and northern Europe rather than from southern or eastern Europe. Since there was a majority of southern or eastern Europeans in the waves of new arrivals since 1900, the quotas allotted these countries were likely to be higher than desired—especially if, for subsequent years, it was decided to use the following census, that of 1920. To avoid this danger they proposed keeping the census of 1890 as the basis for calculating the quotas. But this proposal proved immediately and strongly unfavorable to the Italians, Poles, and Russians who were the most recent arrivals. To choose the 1890 census seemed arbitrary and unjustifiable, as was pointed out by the opponents of these measures, themselves either natives of these countries, or else industrialists who used this labor force. The discussion then broadened to include the idea of a balance among the original nationalities of the *ancestors* of present-day Americans, before their immigration. The Census Bureau was asked to resolve this extremely delicate problem. A first attempt in this direction had been made in 1909 by Rossiter, a statistician from the bureau. Rossiter's goal had been different: he compared the fertility of Americans classified by their country of origin with the fertility subsequently observed in these same countries, thus demonstrating the beneficial effects and the vitality of American democracy. In 1927 Hill, the adjunct director of the census, was charged with continuing this work, so that a new immigration law could be based on this breakdown of the nationalities of Americans' ancestors.

Hill soon observed the flaws in Rossiter's calculation: a slight variation in the estimates of the scant eighteenth-century population led to substantial modifications in the quotas calculated a century and a half later. Rossiter had used patronyms as an indication of a person's origins, without taking into consideration the fact that numerous immigrants had anglicized their names when they arrived. This resulted in an overestimation of immigrants of British origin, to the special detriment of the Irish and the Germans—two of the western and northern European countries for which higher quotas were desired. Whereas preceding laws on immigration had divided the immigrants' lobbies, this new formulation threatened to reunite them in a hostile community. A commission of historians and genealogists was then charged with studying the problem more closely, and with correcting the embarrassing results obtained by Rossiter's method. Thus large numbers of administrators and academics were mobilized between 1927 and 1929 to deal with this problem, leading to a heavy task for the Census Bureau. At the beginning of 1929 figures calculated by the bureau were ratified by historical experts. That spring, President Hoover was able to announce the "official, scientific" breakdown of the national origins of the American populace, providing a basis for immigration quotas. These quotas were then used until the 1960s, having meanwhile constituted a formidable obstacle for Jews and political refugees fleeing from Nazism between 1933 and 1945. Moreover, a few months after the conclusion of this work, the Census Bureau found itself facing an entirely different problem, a consequence of the crisis of October 1929: a new battle of figures had erupted, this time over unemployment figures. This crisis, and the political, economic, and administrative responses that Roosevelt's government supplied after 1933, involved the construction of an entirely new public statistical system, relying mainly and for the first time on probabilistic techniques and on the sampling method.

The Census Bureau: How to Build an Institution

Throughout the nineteenth century, unlike the European states, the American state did not possess a permanent bureau of statistical administration. At the time of each census a superintendent was appointed and provisional personnel were recruited, but the group was disbanded once the work was finished. This recurrent discontinuity was linked to the constitutional definition of these censuses, the role of which was to reevaluate, every ten years, the balance between the states and their representation in Congress. But since the population and its territory grew very quickly, the operation assumed, as the century went by, proportions inconceivable in 1790. With every census the organization of ground work, the contents of the questionnaire, the methods of recruiting personnel, the means and time allowed for its use were all again debated by Congress, the Budget Office, and the superintendent in charge. Inevitably, the superintendent tried to increase his mastery of the various political, administrative, and technical components of this long and complicated process by playing on the opportunities and circumstances particular to each decade. Thus the history of the census and its bureau, recounted in detail by Margo Anderson (1988), can be read as that of the lengthy construction of a political and scientific mechanism. The solidity of its products cannot be judged absolutely, but only according to the constraints and exigencies of the moment.

In his perpetually renewed negotiations with Congress the superintendent was caught in a contradiction quite familiar to any public statistics official, but which was even more pointed in the American context. On the one hand, he tried to make autonomous the circuits for collecting and treating the raw statistical material in relation to chance and external constraints, the better to stabilize an industrial routine for the methodology. On the other hand, however, he had to take part in the normal interplay of a political and administrative life, in which temporary balances were constantly challenged by lobbying groups, by changes in the congressional majority, and by upheavals in its political agenda as Congress itself was partly renewed every two years. And yet this seeming instability was itself prescribed in the immutable framework of the Constitution, respect for which hindered compromise. The decennial census and the mechanism of apportionment were part of these intangible constraints, which provided the statistician appointed to carry out this task with important bargaining power.

Political patronage offers a good example of this perpetual play of forces. The recruitment of temporary personnel, charged with collecting information in the field and putting it to use in Washington, enabled the elite to press for certain experts to be hired whose opinions they discreetly wished to guide. This is a recurrent obstacle in the formation of a stable, professional body of statistical officials, which would thus deprive congressmen of this means of barter. Linked to a long-standing reluctance to increase the number of federal agencies, this mechanism explains why a permanent Census Bureau was not created until 1902. The preceding period, lasting from the 1880s, had been marked by an intense lobbying effort, which favored the creation of this institution. It brought together university professors, businessmen, and trade unionists, all of whom in their different capacities wanted to see public statistics exceed its traditional constitutional role and produce economic and social data that reflected the industrial and urban "boom" of the moment. In this favorable context, the superintendent was able to turn the mechanism of political patronage to his own advantage by encouraging his employees to lobby their representatives in order to obtain job stability.

The successive stages through which the Census Bureau increased its control over the production chain of results were both administrative and technical. Until 1840 information was gathered by local officers (marshals) in accordance with territorial divisions in no way dependent on the federal bureau. Its temporary personnel comprised less than fifty people. It was content to recapitulate findings on states that had already been examined at the local level, and which were impossible to verify. After 1850 individual bulletins were collected and tabulated manually in Washington. The necessary personnel then increased rapidly, from 160 in 1850 to 500 in 1880, for this manual tabulation involved an enormous amount of work. Francis Walker (1840–1896), superintendent of the census between 1870 and 1880, fought to extend his control over the process and also his network of alliances. In 1880 Congress authorized him to verify fieldwork. The bureau was then able to define the survey zones itself, recruiting and paying the necessary agents and supervisors: this was both an advantage and an disadvantage, in that it increased the risks of pressure being exerted on local elected representatives to influence the hiring. In 1888 Walker proposed attaching the census to the newly created Labor Office, which had major resources at its disposal; but an unexpected change in the presidential majority caused this project to abort, for the new president wanted his party to benefit from the uses of the census. (An identical attachment occurred in France in 1891, where the SGF was included in the new Labor Office. The development of labor statistics was taking place at this time in Great Britain and in Germany.) In 1890, the enormous amount of manual work involved began to decrease with the use of the first mechanographic machines invented and constructed by an employee of the bureau, Herman Hollerith (1860–1929). But the task was still huge, and more than three thousand people had to be recruited for the censuses of 1890 and 1900.

The next problem was to convince Congress to approve funding for the establishment of a permanent statistics bureau by demonstrating the need for regular annual surveys—especially for agricultural and industrial production—as well as for decennial censuses. The opportunity to do so arose in 1899 with a slump in cotton prices, which was disastrous for small farmers. English buyers predicted increased harvests, anticipating lower prices. Cotton producers needed to be able to respond with guaranteed statistics of production, in order to prevent such speculation. From the early nineteenth century, the population census had been completed (irregularly) by a census of factories, thanks to which it was possible to identify and localize machines for separating the cotton. Established from the census of 1900, the list of factories allowed an annual survey of harvests to be organized; this in turn provided an excellent argument for persuading Congress to accept the institution of a permanent bureau, responsible for these regular surveys which were so indispensable in regulating the markets. This institutional innovation, approved in 1902, was itself part of a more general administrative reorganization involving the creation of a new Department of Commerce and Labor. Combining numerous services that had previously been scattered, the new department aimed at promoting and developing external and domestic trade, mining industries, manufacturing and naval industries, fishing, labor-related interests, and transportation. In addition to the census it included other statistical bureaus: those of labor, the Treasury, and external trade. One of the goals in creating this department had been to reunite—or at least coordinate—these bureaus, thereby avoiding pointless duplication of work, and to disseminate common methods and terminologies.

The activity of the new permanent Census Bureau then evolved in the direction of frequent and regular economic statistics, instead of limiting itself to decennial censuses. Statistics from the business world began to be widely disseminated and discussed. But the census failed in its goal of covering the whole of the statistical system. The products of other specialized services were closely linked to the current management of the administrations they depended on. These bureaus saw no interest in weakening these bonds, or in accepting the tutelage and language of newcomers whose interests were less connected to this management, and whose legitimacy was not yet manifest. The problem of the census's place in the new department is signaled by the squabble over its title: the ministry wanted to call it the Census Bureau, in keeping with the other administrative bodies, whereas the Census preferred to keep its previous name of Census Office—a term that implied greater autonomy (comparable to the French word *institut*). Such tension was typical: statistical bureaus were not able fully to justify the originality of their position, a mixture of traditional administration and research laboratory, until thirty years later, when they were able to claim a specific technical prowess, in addition to a purely administrative competence.

The irregular progress of this professionalization of public statistics was marked, in the United States, by exchanges with academics or with associations such as the American Statistical Association (ASA) or the American Economic Association (AEA), which brought together scientists militating for the development of statistical products. In the census these scientists found not only sources for their work, but also a training place for students and young researchers. Therefore they tried systematically to attract this still young institution, of ill-defined status, and to introduce some of their own people into it. But they were not the only ones. Political patronage continued to exercise considerable influence. The executive framework was still of a weak professional level and offered no prospect of advancement to young people graduating from universities. Once they had acquired a thorough knowledge of the current statistical system and its possibilities, these graduates would leave in order to pursue research in new institutions such as the National Bureau for Economic Research (NBER), the Brookings Institution, or the Carnegie Foundation.

As in the European countries, the United States' entry into the war in 1917 led to the intensive mobilization and rigorous planning of all economic resources. This enhanced the role and importance of public statistics and, in interrupting the routines of peacetime, facilitated its coordination. For a brief period, the new bonds then forged recall those of the Albert Thomas Cabinet in France. In June 1918 Wilson created a central planning and statistics bureau, coordinating all statistical work done by the various governmental agencies and consolidating the efforts of academics and former census statisticians, such as Wesley Mitchell. But this central bureau was dissolved shortly thereafter in June 1919. Nevertheless, in November 1918 these renewed relations between the administration and the academic community led to the joint creation of a committee of advisors to the census; formed by the statisticians of the ASA and the economists of the AEA, it included the best available specialists. During the 1920s this committee played the role of brain trust and public statistics lobby, guaranteeing intellectual continuity in an administrative world lacking in memory. Like its prewar predecessors, however, it failed to coordinate and inwardly professionalize these bureaus, which were subject to the hazards and debates of current politics already described—the fixing of ethnic quotas, for example, or discussions on how to apportion the congressional mandates (university professors, moreover, were not loath to intervene in these debates). It took the economic collapse of the 1930s for the landscape of public statistics and its relations with the academic world to be completely transformed.

Unemployment and Inequality: How to Create New Objects

In the space of a few years, between 1933 and 1940, the terms of the social debate and statistical tools informing it were radically altered. The two transformations were closely connected, for the same period witnessed both the birth of a new manner of conceiving and managing the political, economic, and social imbalances of the federal American state, and a language in which to express this action. Unemployment, viewed on a national scale; inequality among classes, races, or regions; the very fact of equipping these objects with statistical tools in order to debate them—all this entered into the new language, which became commonplace in all Western countries after 1945. A comparison of the ways in which these problems were outlined and treated in three successive periods, from 1920 to 1929, 1930 to 1932, and 1933 to 1940, clearly illustrates how political schemes and techniques were simultaneously affected. The two traditions, administrative and mathematical, described separately in previous chapters, converged in a construct henceforth endowed with two-fold legitimacy: that of the state and of science.

After 1929 the Republican Herbert Hoover, president from 1929 to 1933, found himself facing first the collapse of the Stock Exchange and then of the economy, together with a rapid rise in unemployment. Hoover has often been described as a fierce partisan of a free market economy and consequently hostile to any form of federal intervention, whether macroeconomic (boosting demand) or social (unemployment compensation). According to this description, he would have been content to wait passively for the ineluctable return of the business cycle, which did not require a detailed statistical analysis of the causes and effects of the crisis and the necessary remedies. But things were not so simple: paradoxically, during the 1920s, Hoover actively promoted the strengthening of federal statistics, especially in regard to the activity of businesses. As early as 1921, as secretary of commerce, he had launched a permanent survey of this activity, the Survey of Current Business. At this point, the country was already in a state of crisis and had rising unemployment. Hoover persuaded President Harding to sponsor a national conference on the subject, in order to analyze the causes of this crisis and to propose measures for reducing it. The explanations offered, the methods of observation, and the solutions envisaged were coherent and entirely different from those that would predominate during the 1930s. The crisis essentially resulted from production, from ineffectual business management practices, from waste, and from extravagances linked to speculation and inflation of the growth phase. The conference on unemployment organized by Hoover in 1921 gathered detailed reports on the situation of firms in a hundred or so cities and on the steps taken, at a local level, by bosses and political officials to make the markets stronger.

These reports were neither quantified nor combined in national measures. They were minutely indexed and analyzed, to detect the zones most affected and the local actions that proved most effective. Government and businesses had to encourage such actions in order to attenuate the eccentricities of the ascendant phases of cycles. The conference even proposed ordering civil engineering during the low phases, and restructuring unhealthy industries around the leading firms. But these actions could only accompany an assumption of responsibility at the local level for the functioning of the employment market, a responsibility the federal government could not itself assume. The main responsibility for taking action fell to the business leaders, but they had to be informed about the cycles and conditions of the markets, and this information, with its local variations and sectors, could be elaborated and disseminated effectively by federal agencies. This orientation of statistics that Hoover upheld during the 1920s did not prompt a national evaluation of the global volume of business and unemployment, nor an analysis of the living conditions of the unemployed, since no public policies to help them were envisaged. So great was the experts' reticence in measuring unemployment that in 1920 the committee of advisors to the census (from the ASA and AEA) deleted a question on this subject—a topic they discussed between 1880 and 1910—"because it was impossible to determine for what reasons someone was not working at the moment of the survey: business cycle, seasonal variations, illness . . ." A further proposal to delete the question was made in December 1928 for the census of 1930, but Senator Wagner, a Democrat who was very well informed, succeeded in having an amendment passed to reintroduce it in June 1929, four months before the collapse of the Stock Exchange.

As chance had it the decennial census took place on April 1, 1930, when unemployment had already reached critical levels and, for the first time, a nationwide measurement had become an essential political issue in the debate between the government and the Democrats in opposition. Strong pressure was exerted on the Census Bureau to provide rapid unemployment figures; but in the agenda conceived before the Depression, the number of unemployed had only a very low priority. Moreover, the information appeared on a different questionnaire, analyzed separately. To match it with the questions appearing in the main questionnaire, such as race, sex, age, or profession, was a monumental task at a time when electronic tabulation did not yet exist. But with public opinion pressuring it to provide a very quick response to a question included almost by chance in a census not conceived for this purpose, the Census Bureau announced in late June that an initial counting recorded a figure of 2.4 million unemployed. The conventions for defining and counting these "unemployed" were immediately contested, and some people suggested that the real number was between 4 and 6.6 million. Unemployment had never been clearly defined before, and in the emergency the bureau had retained restrictive conventions: it counted neither workers who still had jobs, but who had already been informed they would be fired, nor those young people who had never worked and were looking for a job. For his part Hoover maintained that unemployment was not as severe as people claimed, and that the census figure was far too high, for "many people recorded as unemployed were not really looking for work."

All the elements of the modern debate on unemployment had thus arisen in the space of a few months. What was an unemployed person? If he or she was defined as a person *without employment, looking for a job,* and *immediately available,* each of these three conditions was problematic and led to discussion, since dubious cases could abound: people who worked intermittently, for want of anything better; people who had grown discouraged and were no longer trying hard to find a job; people in serious difficulties, in weak physical or mental health, as was frequent among the poor. If in addition one measured not only the *number* of unemployed but also the unemployment *rate,* the definition of the denominator posed a further problem: should one relate unemployment to the total population, or to a potentially active population? In this case, the boundary between the potentially active and inactive populations was strewn with doubtful cases. These questions were seldom raised before 1930, and only took on meaning because the Democratic opposition was demanding a nationally organized policy to fight against unemployment. No one had raised them in 1920–1921, because the idea that health in business depended on local circumstances and initiatives was then most prevalent. This was the position Hoover continued to maintain in 1930. He did however suggest that local help be organized, that part-time employment and the sharing of tasks should be encouraged, and that illegal aliens be expelled. But he rejected proposals to modify the historical balance between the power of local councils, states, and the federal government. For this reason, statistics on agriculture, labor, unemployment, relief, education, and health were still within the competence of local authorities. Federal statistics were hit hard by the restrictions decided on by Hoover's administration. His budget was cut by 27 percent in 1932. But the situation was to change, from every point of view, when Roosevelt came to power in March 1933.

In wartime the general mobilization of resources had created unusual alliances and means of organization, the effects of which were felt in public statistics. Similarly, in 1933 when the crisis was at its height, the effort made by the new administration to correct the economic situation and help the millions of ruined farmers and unemployed led it radically to transform the wheels and role of the federal government and, accordingly, the place of statisticians in these mechanisms. The changes had to do with the goals of public policy, with the technical and administrative processes put in place to reach them, and with the language used to express these actions, so that they had a common meaning for all those who took part in them. Far more than before, statistics played a decisive role in giving consistency to things affected by collective action. These things—unemployment, social security, inequalities between different groups or races, national income—were henceforth formulated on the basis of their definitions and statistical measurements. It was not the first time this had happened: even in nineteenth-century England, public health policies referred to local death rate indicators, and the GRO derived its importance and legitimacy from this. But it was no longer a matter of expressing a particular action, even a very important one; it was a matter, rather, of coordinating and informing the activities undertaken by the sum of the administrative and social forces on which that action was based—the firms and their associated foundations, the universities and their experts, the trade unions, the charitable societies. In a time when other means of expressing collective action, such as those related to the languages of market economy or local solidarity, no longer seemed able to account for a critical situation, it became more plausible to resort to new forms of rationalism and universality as formulated in scientific, notably statistical terms, whereas this had not been plausible a few years before.

During the first three years of the crisis bitter arguments had raged, creating opposition between members of the Republican administration, the statisticians of the ASA, and the economists of the AEA. They disagreed not only on the measurement of unemployment but also on the way statistics was organized, deeming it archaic. They further disagreed on the remedial actions to be taken, so that government and public opinion could count on reliable descriptions of so novel a situation, since previous cyclical crises had never been of such magnitude. During the first weeks of the new administration the directors of the principal statistical bureaus, Republican in sympathy, were replaced by members of the ASA who had taken part in these debates. The new secretaries of labor (on whose department the Bureau of Labor Statistics, or BLS, depended) and commerce (on which the Census Bureau depended) instigated the establishment of two committees charged with studying the workings of all the statistics bureaus, and with proposing how to reorganize them. The most important of these groups of experts, the Committee on Government Statistics and Information Services (COGSIS), would play a decisive role in coordinating and rationalizing the administrative channels that lead to statistical measurements and in making the bureaus more professional. It was financed by the Rockefeller Foundation and presided over first by Edmund Day—director of social sciences of this foundation—then by Frederick Mills, professor at Columbia. Day introduced to the Census Bureau young academics of a very high caliber, with degrees in mathematics and economics. Gradually these young professionals replaced the aged executives with primarily administrative backgrounds, men marked by the political quarrels of the 1920s which were now quite irrelevant. Whereas in normal circumstances official bureaus attracted few brilliant students, this was not the case in a period when new business ventures had dried up. For some thirty years this generation, which remained active until the 1970s, lent a very original profile to the census, the BLS, the bureau of agricultural statistics, and to several other federal services that dealt with statistics and economic studies. It transformed them into places of innovation and experimentation in the new branches of knowledge that derived from the mathematicization of statistics and economics, such as sampling theory, national accountancy, and econometric models (Chapter 9).

The technique of surveys conducted by sampling, already imagined by Laplace in the eighteenth century and subsequently forgotten, had been readopted in around 1900, first by Kiaer in Norway, then by Bowley in Great Britain (Chapter 7). The conjunction of three separate events led to its being routinized and popularized in the United States of the 1930s. In 1934 Jerzy Neyman (1894–1981)—a statistician of Polish origin who had collaborated with Egon Pearson (1895–1980) in London—formalized methods of sampling and stratification. He thus opened a path that the freshly recruited young men at the Census Bureau—Dedrick, Hansen, Stephan, and Stouffer—subsequently broadened considerably, applying it to *regular* surveys. This latter technique occurred most appositely in the context of the new social and political economics, and of the efforts made by COGSIS to build a permanent statistical system allowing this policy to be pursued, which the decennial census could not do. It was used successfully by Gallup in 1936 in predicting electoral results, thereby alleviating doubts on the part of public opinion as to the consistency of measurements taken from only a tiny fraction of the population being described. Thus, after a few hiccups, this alliance among mathematicians, statisticians, political officials, and journalists met with dazzling success.

Far less costly than the census, sample surveys could be repeated with a permanent questionnaire, allowing objections as to the conventions and arbitrariness of the coding to be elegantly phrased: if the realism of the measurement *at a level* was debatable, that of its *variations* was less so, once the conventions of recording were stabilized. Thus routine use of statistics became possible, including in part a criticism of its own realism. But before it could be definitively adopted, the new reality created by sample surveys was still questioned, by the policies and even the management of the census. The young statisticians had to prove beyond doubt that their objects held and were better than those based on classical administrative records. In the history of sample surveys I am about to recount, two conclusive proofs play the part of this founding deed. One concerned *unemployment*, and convinced political and administrative officials. The other concerned the *votes* people would cast, and convinced the press and public opinion.

Unemployment remained as crucial a question for the new administration as it had been for the previous one. Toward 1935, persistent demands for better means of describing and measuring it encountered the same reticence as before. Like his predecessor, Roosevelt questioned the reality of unemployment, and even the possibility of establishing statistics for it. Nobody, he said, had "been able to define an unemployed person . . . Some girls work for pin money . . ." (Anderson, 1988, p. 177). Carpenters stop working when the weather turns bad. He thought a census of the unemployed would unduly confuse such cases with those of people who were "truly needy." But two years later, political pressure became so strong that Congress approved funding for a national survey of unemployment. This survey took the form of an exhaustive census: a form questionnaire was sent through the mail, and the unemployed responded on a voluntary basis. Statisticians disagreed with this method: they maintained that numerous people would not reply, for they viewed unemployment as a humiliating circumstance. A survey conducted by sampling would be less expensive and would allow investigators to be sent directly into 2 percent of the population's homes; this direct contact would establish a relationship of trust, thus allowing unemployment to be more realistically assessed. Dedrick and his young colleagues persuaded their director to conduct this experimental survey at the same time as the postal census, so as to compare the results for people participating in both operations. The proof was conclusive: 7.8 million unemployed responded to the exhaustive postal survey whereas, of those who declared themselves unemployed in direct interviews with sample population, only 71 percent had answered the postal survey (although this percentage was far higher for unemployed persons receiving help from special agencies, and therefore already known: 98 percent of them responded). For the first time, therefore, it was possible to compare the two standard ways of counting the unemployed, through direct survey and through the agencies responsible for helping them. Comparison of the postal census and the sample survey thus allowed a higher unemployment figure to be estimated: 11 million instead of 7.8 million.

The other conclusive proof for the method of random sampling, as opposed to spontaneous sampling, occurred in 1936 with Gallup's experimental poll: Gallup predicted Roosevelt would be reelected, whereas the *Literary Digest* magazine, which questioned readers on a voluntary basis, had predicted a Republican victory and was badly mistaken. Gallup and the upholders of the new method did not fail to point out that their random sample was far smaller than the sample taken by the magazine, but that it had nonetheless produced the right result. Thus the idea of *representativeness* was made widely popular, strengthening the arguments of statisticians calling for regular sample surveys of economic and social questions. A fresh opportunity to test their methods and prove their effectiveness was offered them shortly thereafter, with the preparation of the normal census of 1940. The numerous demands being made for the addition of new questions would have resulted in an excessively long questionnaire. To the exhaustive regular census, the statisticians of the new generation proposed adding a complementary questionnaire that concerned only 5 percent of individuals. Thanks to its association with the exhaustive survey, this operation allowed essential theoretical and practical questions to be asked and resolved concerning the trustworthiness of the sampling process. It also encouraged the testing and stabilizing of the formulation of questions and the standardizing of certain measurements destined for widespread use, such as the labor force: the total number of persons having a job or looking for one during the week in which the census took place. This measurement would henceforth serve as a denominator in the standard assessment of the *unemployment rate*. These various experiments were sufficiently persuasive for the results of a survey by sampling to be henceforth presented as the "official unemployment figure," a development which would have been inconceivable ten years earlier. After 1940, a survey of this kind was organized every month, first under the name of a "sample survey of unemployment," then in 1942 as a "monthly report on the labor force," and finally in 1947 as a "current survey of the population."

A dramatic, simultaneous change also occurred in the goals and tools of statistics, between the 1920s and 1930s, in relation to population questions and inequalities between social and ethnic groups. Before 1930 these differences were generally described in terms of innate aptitudes, or the inability to be integrated into American society for cultural and religious reasons. These analyses used intellectual constructs deriving from eugenics or from a form of culturalism emphasizing the immutability of characteristics linked to national origin. The debates of the 1920s on ethnic quotas were strongly marked by such arguments. Ten years later, the vocabulary and the questions raised were no longer the same. Committees of academics and officials, supported by the large private foundations, were instructed by the government to analyze the problems and to come up with proposals. A "liberal" policy was outlined and formulated, in the American sense of the word liberal (which is different from the European sense): that is, of "progress." In 1934 one of these committees studied the "improvement of national resources," with the conviction that sciences and techniques should be brought to bear as such to resolve social problems:

> The application of engineering and technological knowledge to the reorganization of natural resources of the Nation . . . is to be conceived as a means of progressively decreasing the burdens imposed upon labor, raising the standard of living and enhancing the well-being of the people. (National Resources Committee, 1934, quoted by Anderson, 1988, p. 180)

A further committee, on "population problems," was appointed by the first, to study more particularly the questions of social and ethnic differences. Its report, published in 1938, took exactly the opposite tack from the theories of the 1920s. In no way, it opined, had the immigration of the various national minorities diminished the vitality of American society. It emphasized the positive aspects of the immigrants' diverse cultural traditions, and those of blacks and Indians. As Anderson explains, the problems encountered by the more underprivileged groups resulted from the difficulties they had in obtaining an education, from an improper use of land and subsoil resources, and from inadequate planning for technical changes. It was the federal government's responsibility to resolve these problems, chiefly by providing the framework and funding to guarantee equal access to schooling and normal health for all Americans; by restoring forests and agricultural land that had been overcultivated by preceding generations; and by anticipating technological changes through growth planning. To thus assign such a body of problems to federal competence was a major innovation in the American political tradition, one that marked the various administrations that followed, whether Democratic or Republican, until the beginning of the 1980s. The statistical system developed in parallel with this policy (experiencing drastic budget cuts after 1981). The problem of how to describe and measure initial inequities of chance (expressed, in American culture, by such strong words as *opportunities* and *birthright*) among social groups, regions, races, and sexes was at the core of the numerous forms of political legislation subsequently tested to counterbalance handicaps of every kind. The philosophy of these attempts is well expressed by the conclusion of the committee's report, in 1938:

> It cannot be too strongly emphasized that this report deals not merely with problems regarding the quantity, quality, and distribution of population, important as they are, but also with the widening of opportunities for the individuals making up this population, no matter how many or where they are. In our democratic system, we must progressively make available to all groups what we assume to be American standards of life. The gains of the Nation are essentially mass gains, and the birthright of the American citizen should not be lost through indifference or neglect. (National Resources Committee, 1938, quoted by Anderson, 1988, p. 181)

As early as 1935, the application of this policy led to the passing of Social Security, which was federally financed from funds shared among and managed by state and local authorities. The question then arose of how to divide *grants in aid* among the diverse intermediary bodies. To this end the old constitutional process of *apportionment* gained a new lease on life, and the Census Bureau was mobilized to provide a basis from which to calculate and apportion the aid to the states. But the principles of justice that inspired these new policies were not only based on apportionment and the division of burdens among the states, as were those of the founding fathers. Henceforth they also concerned individual inequalities linked to race, occupation, and income. A new type of information was thus required in order to organize these policies. "Individual-national" statistics could not be provided by censuses, which were both costly and infrequent. But regular sample surveys, conducted at the federal level, provided adequate information for action prompted by a new concept of individual inequalities perceived on a national scale. National space was no longer simply political and judicial space. It had also become a statistical sphere of equivalence and comparability, vindicating the process of random sampling—the equivalent of drawing colored balls from an urn, except that those balls were individuals henceforth endowed with both political and social rights.

The balance sheet of American statistics during the decades of the 1920s and 1930s is a study in contrast. It was the source of two partly different modes for introducing statistics into society. It may be useful to classify these modes, at risk of exaggerating their characteristics, in the knowledge that systems that really exist combine them in various measures. The 1920s were not only years of debates on immigration and ethnic quotas. Hoover, who was secretary of commerce in 1921, took a lively interest in economic statistics concerning business and firms and even in unemployment, with the idea that the administration could guarantee and improve the rules and the environment in which firms embarked on their activities. From this point of view, information on the running of the business cycle, analyzed by sector and by region, could be very helpful in avoiding the management errors that resulted in bankruptcy and unemployment. Thus, in 1921, Hoover prodded the census into organizing a permanent "Survey of Current Business" and set up a committee to study the causes of unemployment and its remedies. But this information was more microeconomic in nature, or at least sectorial or local. In no event could the state act directly on macroeconomic balances, nor conduct a general policy of unemployment relief, for these interventions distorted the free play of the market. It was this latter aspect of Hoover's policy that was subsequently retained after the collapse of the economy, when he was president, between 1929 and 1933. But this line of policy, which opposed federal economic intervention in macroeconomic regulation, did not contradict a keen interest in statistical information designed to facilitate the smooth running of the markets.

After the economic disaster of the early 1930s had resulted in the near-disintegration of society, it became possible for the new administration to break with some of the more essential American dogmas concerning the balance of power between the federal government, states, counties, and firms freely conducting their business. The establishment of federal systems of monetary, banking, budgetary, or social regulation gave the government in Washington an entirely new role. To fulfil this new role the administration relied ever more heavily on experts in the social sciences, economics, demographics, sociology, and law, mainly through the agency of wealthy private foundations: Carnegie, Ford, and Rockefeller. During this period, close ties were formed between bureaus of official statistics run by statisticians and economists of a high caliber, and an academic world itself mobilized by the administration's demands for expertise. This period witnessed the development not only of quantitative sociology based on sampling surveys, in Chicago or at Columbia (Converse, 1987; Bulmer, Bales, and Sklar, 1991), but also of the national accounts (Duncan and Shelton, 1978) on which, between the 1940s and the 1970s, policies inspired by Keynesianism were later based. Last, in both the United States and Great Britain, the 1980s saw the return of a philosophy of economic information reminiscent, in certain respects, of that of the 1920s.

7

The Part for the Whole: Monographs or Representative Samplings

The technique of conducting surveys by means of probability samples did not appear until the late nineteenth century with the work of the Norwegian Kiaer, and even then only in a rudimentary form, more intuitive than formalized. The first calculations of confidence intervals were made by the Englishman Bowley in 1906, and the detailed formalization of methods of stratification was presented by Neyman in 1934. Surveys based on a small number of individuals had been conducted for much longer, throughout the nineteenth century in particular, often by persons with a high degree of scientific knowledge (engineers from the École Polytechnique, in the Corps of Mining or Bridges), for whom the elements of probability calculus necessary for the "intuition" of the sampling method ought not to have constituted an insurmountable problem. Laplace, moreover, had used them as early as the late eighteenth century to estimate the French population, but this experiment had not been followed up for more than a hundred years. In 1827 Keverberg had criticized this method for implicitly presuming that the "birth multiplier" was the same throughout the land (the ratio of population to annual number of births, discussed in Chapter 1). This criticism of the evaluation techniques of eighteenth-century political arithmeticians had so vividly impressed Quetelet and the statisticians of the time that, in the seventy years that followed, sample surveys appeared as poor substitutes for exhaustive censuses, which symbolized rigorous statistics.

The fact that today the formalizations and systematic uses of methods of probabilistic sampling are scarcely more than half a century old shows that to invent and apply a technology presupposes conditions that are inseparably cognitive and social. Before inventing a solution for the problem, it was necessary to invent the problem itself—that is, the constraints of representativeness, in the sense that statisticians would thenceforth give this word. One could not fail to know that this concern, expressed in similar terms for certain precisely defined elements of the part and the whole, is a recent one— subsequent, in any case, to the development of large censuses (Quetelet, beginning in the 1840s) and monographic surveys (Le Play, at about the same time). The history of empirical social sciences, of statistics, and more precisely of techniques used in sampling (Seng, 1951; Hansen, 1987) gives the impression that we have passed directly from an era in which the question of representativeness was almost never raised (see, for example, reports published between 1914 and 1916 by the Statistique Générale de la France, of surveys of household budgets conducted by the statistician Dugé de Bergonville or the sociologist Halbwachs) into another era in which it became patently obvious (debates that took place at the International Statistical Institute, first between 1895 and 1903, then between 1925 and 1934).

The debate, when it did occur, did not concern the constraints of representativeness as such. It unfolded in two stages. Between 1895 and 1903, the question to know first, whether one could legitimately replace the whole by a part (comparison with exhaustive censuses), and second, whether one did "better" in this way than with LePlaysian monographs, was still held in great respect. As we shall see, this notion of "better" did not bear directly on the constraint of representativeness in the sense of the precision of the measurement, but on the possibility of accounting for a diversified space. Then, between 1925 and 1934, the debate hinged on the choice between the methods of "random sampling" and those known as "purposive selection." Neyman's development of the theory of stratification would strike a fatal blow to the second of these choices.

Emerging from the history of sample surveys, this chronology describes how between 1895 and 1935 commonly admitted social norms were transformed for the requirements expected from descriptions of the social world intended to *generalize* for an entire society observations based on part of it. How was one to proceed from the "part" to the "whole"? The two ways of generalizing successively (and simultaneously) at work in the social surveys that had been carried out for a hundred and fifty years seemed heterogeneous and not comparable, as if each had its domain of validity and its own logic, and as if each could only confront the other in a mode of mutual denunciation. This seeming incompatibility can be better understood if we reset it in a larger opposition between the various ways of conceiving the links between the parts and the whole of a society, ways that had been in competition since the early nineteenth century, after the two social upheavals constituted by the French Revolution and English economic liberalism. We find such analyses of the transformations in the relationships between the parts and the whole in the research of Polanyi (1944), Nisbet (1984), and Dumont (1983).

The opposition—familiar to anthropologists and historians—between "holism" and "individualism" provides a definition of the "whole" that is inadequate for our purpose. In the "holistic" vision, which for Dumont was that of traditional societies before revolutionary upheavals in politics and economics, the social whole had an existence that preceded and was superior to its parts (and especially to individuals). But in the "individualist" vision, which is that of modern societies, the individuals, whether citizens or economic agents, are grouped together in various ways, but without being exceeded or totally encompassed by these groupings. But this manner of conceiving an all-encompassing whole, as opposed to the atomized individuals of modern societies (which recurs in Tönnies's "community-society" opposition) does not account for another, criterial manner of constructing the whole: and this is precisely the manner adopted by statisticians intent on creating a *representative sample.* The "social whole" of Dumont's holism and the "exhaustiveness" of statistics constitute two different ways of conceiving totality, and the opposition between them helps us understand what distinguishes the two implicit ways of generalizing underlying monographs and sample surveys.

These schematically stylized intellectual configurations were variously compared and combined during the nineteenth century in the work of the founding fathers of the social sciences: Quetelet, Tocqueville, Marx, Tönnies, Durkheim, and Pareto (Nisbet, 1966). But they did not function as a *deus ex machina,* pulling alternately at the threads of this or that practice of empirical research. Rather, they formed relatively heterogeneous constellations, each with its own coherence, which can be followed in the debates on survey methods in the period under consideration. Each of them simultaneously implied different ways of conceiving the management of the social world, the place of social sciences in this management, and the place of probabilistic schemes in these sciences, from Quetelet to Ronald Fisher. This provides a guiding line as we examine a history that described social surveys long before anyone spoke of "representative method," together with the evolution in the use of probabilistic arguments during the period, the debates of the International Statistical Institute between 1895 and

1934, the first uses of the method, and discussions between "random choice" and "purposive selection."

The Rhetoric of Example

It is possible to reconstruct the philosophy of surveys designed to make generalizations from observations carried out without the modern constraint of representativeness, on the basis of three seemingly quite different cases: the monographs of Le Play and those of his followers, written between 1830 and 1900; the surveys on poverty conducted by the Englishmen Booth and Rowntree between 1880 and 1900; and finally the studies made by the French Durkheimian sociologist Halbwachs, on workers' budgets between 1900 and 1940. The common point in these surveys is that the persons questioned were chosen on the basis of networks of familiarity: in Le Play's case, families designated as "typical" by village notables; in Booth's case, persons habitually known to the "schoolboard visitors"; and in Halbwachs's case, workers who volunteered through the trade unions. These methods of selection would later be stigmatized for generating "bias," but in the context in which they were used they were consistent with the goals of these surveys, which were to describe the functioning (or malfunctioning) of workers' communities subject to the upheavals accompanying the first phases of industrialization. The intent was not yet to take measurements in order to prepare for the measures to be taken, as would be the case when the welfare state developed. It was to assemble elements capable of grounding the characters in the story to be told or organized, through, in particular, typological works: classification, which created collective actors, was one of the products of these surveys. In the next phase, however, the atomized individuals had become essential actors (for example, in the act of voting or of purchasing in market studies), and it was important to count them accurately.

Yet it cannot be said that all probabilistic elements are absent from these surveys and their interpretations. But it is a question of the holistic version of probabilities, handed down by Quetelet. This version emphasized the regularity of averages calculated from large populations, as opposed to the dispersion and unpredictable nature of individual behavior. This regularity provided sturdy support for a concept of the social whole as exceeding and encompassing the parts that composed it (Chapter 3). The application of the law of large numbers to the stability of calculated averages (for example, birth rates, marriages, crime rates, suicides) impressed Quetelet's contemporaries and formed the framework of a macrosociology in which the "social" had a reality external and superior to individuals. This was the central idea of Durkheim's *Suicide,* and of Halbwachs's analyses of "workers' consciousness."

But it was because the probabilistic model of the law of large numbers was drawn toward the regularity of averages and not toward distributions, character frequencies, or dispersions that it could not be the basis of an application in terms of random sampling. Indeed, if the averages were stable, one only had to find cases that were close to these averages, and these "typical" cases would automatically embody the entirety. Not only did they represent but they literally were this totality since, in holistic systems, the totality was what came first, whereas individuals were simply contingent manifestations of it. The intellectual model of this concept of the average was provided by the theory of errors in the measurements made by astronomers and artillerymen. The altitude of a star was known only through a series of contingent measurements distributed according to a normal law around an average, which constituted the best estimate. Similarly, in Quetelet's view, contingent individuals were random manifestations of a "divine intent" that constituted a superior reality.

Toward the end of the nineteenth century, this concept of statistics was still dominant (at least in France, for in England this was no longer true, with the early work of Galton and Pearson). Knowing this, we can better understand how in 1890 Émile Cheysson, a disciple of Le Play, described and justified the "monograph method," shortly before the Norwegian Kiaer presented his "representative method," which differed radically from the preceding one:

> Monographs carefully avoid particular cases and pursue general ones; they neglect accidents, exceptions, and anomalies, pursuing instead the average, or type. The type is what constitutes the true essence of the monograph. Outside the type, it finds no salvation; but with the type, the monograph really acquires the privilege of brightly illuminating economic and social surveys. The observer is guided in his choices by vast synthetic statistics and by administrative surveys, which form a network covering the entire country and clear the ground, as it were, on which the author of the monographs intends to work. Thanks to the data made available to him, the author knows in advance which population he wants to study and selects his type with accuracy, and no fear of mistake. Official statistics therefore lead the way like an advance guard, detecting the averages that lead the mo-

nographer to his type. In its turn, monography performs a service for statistics, checking in detail the general results of the survey. These two processes thus audit each other, although each retains its distinctive characteristics.

> While the method of administrative surveys covers the surface of a subject, monographs delve into it deeply. Official statistics mobilizes a whole army of more or less zealous and experienced agents, accumulating a mass of rather superficially perceived facts on a single topic; it grinds them together all in a jumble, relying on the law of large numbers to eliminate mistakes in fundamental observations. Monographs, in contrast, aim at quality rather than quantity of observations; they employ only select observations, which are both artistic and scholarly, and show a powerful grasp of a typical, unique fact, and doggedly dissect it to the very marrow of its bones. (Cheysson, 1890, pp. 2, 3)

Records of expenditure and receipts of family budgets constituted the core of these monographs, noted in the course of varying amounts of time spent in family homes by the investigator. But even if such studies were composed in accordance with a "uniform frame"—that of Le Play—they were not really designed to be *compared* with a view to defining budget structures typical of different *milieus,* as Halbwachs was to do shortly thereafter, and this was indeed a criticism that Halbwachs addressed to the LePlaysian monographs (Halbwachs, 1912). The question asked was therefore: what purpose did they serve? These surveys seemed essentially designed to defend and illustrate a certain concept of the family and social relations. The arguments set forth by Cheysson had to do with the needs of administrators and lawmakers intent on assessing the effects that their general, abstract measurements had on singular, concrete cases of individual families. But this managerial concern was also linked with a moral preoccupation:

> This knowledge is indispensable both to the moralist who wishes to act upon mores and to the statesman who has the power of public action. The law is a double-edged sword: while it has great power to do good, it can also do great harm in inexperienced hands. (Cheysson, 1890, p. 6)

In actual fact, possible uses were mentioned, for studying the division of fiscal burdens among farmers, tradespeople, and industrialists, or the effects of regulations prohibiting women and children from working. But even in these cases, there was an explicit moral concern. On the other hand, no technique was manifest or even suggested allowing practical mediation between the few available monographs and the state sums: the question of a possible categorial framework within which individual cases could fit was never mentioned. The monographs of Le Play and his followers were criticized and then forgotten for reasons both technical and political. On the one hand, they offered no methodological guarantee as to the selection of the sample. On the other, they served to shore up a discourse that was hostile to the French Revolution, to universal suffrage, and to the Civil Code while seeking to reestablish the social relations of the Ancien Régime. And yet their cognitive and political plan was coherent. Knowledge resulted from long-standing familiarity on the part of the surveyor with the working family. It was explicitly presented as being useful not only in producing learning but in establishing and maintaining personal relationships of trust between members of the upper and lower classes.

Such knowledge, moreover, centered on direct contact and the value of particular examples, did not entirely preclude comparative treatments. Particular attention was given to the nonmonetary part of income, recorded as "supplements" or "alms": the right to use communal land; family gardens; benefits in kind. Not being dependent on the market, these forms of recompense maintained direct and personal links between members of the various social classes. Frequently included in advocacies of traditional family values, monographs were rarely used in structural comparisons except in the case of the proportion of nonmonetary income, which was thought to reflect the persistence of patriarchal social bonds. Thus in eastern Europe—a largely rural society with hardly any industry—this proportion was larger than in the West, where towns and a market economy took a greater place.

The idea that traditional social bonds disrupted by economic and political transformations in society could only be understood through long proximity between observers and observed—and by taking into account the universality of the meanings of acts that the researcher could not break down and codify into a priori items—is found in other modes of knowledge. These modes involve other means of generalization than those of statistical representativeness. Ethnology describes non-European societies on the basis of researchers' long, patient stays in a community. Psychoanalysis devises a model of the structures of the subconscious based on particular case material, gathered in the course of personal interactions of very long duration. This same cognitive procedure—according to which a case studied in depth can itself enable us to describe a "typical man"—and the mechanisms of a "common humanity" recur during the same era in the works of experimental psychologists such as Wundt, a German who tried to delimit such general characteristics in his laboratory. The idea of the dispersion of human psychic traits did not appear until later, with the work of Pearson and Spearman.

It is clear, from Cheysson's justification of his choice of "typical cases" in reference to averages calculated from large synthetic surveys, how for almost a century Quetelet's theory of the average provided an intellectual scheme that allowed people to conceive both the diversity of individual cases and the unity of a species or a social group at one and the same time:

> Thanks to statistical data, the monographer knows in advance which population to study, and chooses his type accurately and without fear of error. Official statistics goes on ahead like an advance guard and detects the averages that lead him to his type.

This manner of proceeding anticipates, in a way, the method known as "purposive selection" which, between 1900 and 1930, constituted a sort of link between the two types of methods, through the expedient of *territory.*

Halbwachs: The Social Group and Its Members

Halbwachs was a student of Durkheim who, to a greater extent than his master, was concerned with techniques for surveying and observing facts, as can be seen from his thesis on *La Classe ouvrière et les niveaux de vie* (The working class and standards of living) (1912). Far more than the LePlaysians, whose work he criticizes keenly, he was sensitive to the diversity of the cases observed and sought means for interpreting it, as Durkheim had done in *Suicide,* inventing modern quantitative sociology. Moreover, he was aware of certain of the probabilists' works: his complementary thesis (1913) dealt with *La théorie de l'homme moyen; essai sur Quetelet et la statistique morale* (The theory of the average man: an essay on Quetelet and moral statistics), and he even coauthored a small manual on probability with Maurice Fréchet in 1924. However, his closely reasoned argument on the problems of sampling and surveying was entitled "The Number of Budgets" and concerned the "economic" balance to be found between the number of persons surveyed and the more or less thorough nature of the observations; that is, the comparison between methods termed "intensive" and "extensive," and not the problems of selecting a sample, which he does not mention.

What Halbwachs retained from the "law of large numbers" was that a sum of small causes—numerous, random, and of different meanings—compensate for one another and produce an "average" that supposedly reveals the essential truth according to Quetelet's schema. This was especially true with the so-called extensive method, then much used in the United States, where certain samples involved more than 10,000 people. But just as he rejected the intensive LePlaysian method that provided no indication of diversity and therefore did not permit cross-tabulations suggesting explanations, so too he rejected the extensive method of the Americans; like a good Durkheimian sociologist, he mistrusted a microsociological interpretation of the law of large numbers. Indeed, he said, the defenders of extensive methods were certainly aware that the answers obtained could involve errors, inaccuracies, or slips of memory; but, he added:

> They believed that by increasing the number of cases they would obtain, through the effect of the law of large numbers, a compensation for and a growing attenuation of these errors. (Halbwachs, 1912, p. 152)

But Halbwachs supposed that, like every other social fact, these lapses or imperfections had systematic macrosocial causes and were not random in the probabilistic sense. He intuitively grasped the "systematic bias":

> It has been observed that lapses of memory do not occur by chance . . . A memory lapse is the concomitance of an individual's inattentive state and some relatively important social duty he must carry out . . . If such lapses are periodical and regular, then the causes of his inattentiveness and the causes that explain these duties must be constant forces . . . But the periodicity of these lapses suggests that the forces that explain them, by virtue of the way social life is ordered, exert themselves in turn, and their effect is in exact proportion to their constant intensity. (Halbwachs, 1912, pp. 152–153)

Halbwachs then engages in a detailed analysis of the "survey effects" linked to interaction and the role of the researcher, showing that in all these cases the errors could not be random and independent of one another. Indeed, in these extensive American surveys there was no record of a budget diary: the interviewer simply asked the interviewee retrospectively to assess his expenditures during a certain space of time. Because of this, therefore, the measurements really relied on "opinions":

> Here as elsewhere, the goal of science is to replace principles, ready-made opinions, and vague and contradictory ideas with precise knowledge that is based on the facts. But in this case the facts are distorted in advance, seen through an opinion that dims and swamps their contours. We go round in circles made up of averages. (Halbwachs, 1912, p. 157)

This criticism of the American method is based on the idea that the "law of large numbers" was supposed to cancel numerous, small, independent errors—and not on the fact that this same "law" could justify a technique of random sampling in a population that was diverse *in reality,* not on account of errors in observations. We are still close to Quetelet's model, rather than the model of Galton and Pearson. However, Halbwachs explicitly mentions the diversity of the working-class population, and it is precisely on this point that he opposes Le Play and his choice of "typical cases" guided by averages. Halbwachs asks by what visible external traits one might characterize an average working family, at least in regard to the structure of its revenue and expenditure:

> Many households don't know themselves how their budget is or is not balanced. In any case, they don't tell anyone. Now, this is something that cannot be seen, felt, or guessed at—as, if need be, is the case with the regularity or solidity of the family bond. Where are we to find and how are we to seek average cases in this field? (Halbwachs, 1912, p. 159)

It was therefore necessary to make direct observations (keeping records of accounts) of a certain diversity of families, in order to study the effects of the variations introduced by various factors: the size of a family, whether it had children, and so on. But this diversity and these variations were still macrosocial, in the Durkheimian mold of *Suicide* and Quetelet's "constant causes." The German surveys used by Halbwachs struck him as having none of the drawbacks of the others: they were small enough to allow a scrupulously maintained record of accounts, and large enough for variations to be studied.

Nevertheless, the goal of this research was to discern the characteristics of a common "worker's consciousness," the relatively homogeneous character of which did not result from some divine essence, as in Quetelet (Halbwachs was more of a materialist), but from a group of material existential conditions. So Darwinian an adaptation led to homogeneous forms of behavior, both in practice and in consciousness. In the final analysis, it was certainly because this working-class consciousness was what interested Halbwachs that he did not pose problems of sampling in the same terms as did those who, fifty years later, used such surveys to create national accounts. Thus congratulating himself on the fact that the union of German metalworkers had managed to collect 400 budgets, he noted:

> Clearly, the workers' solidarity, and the increasing influence exerted by trade unions on their members, are responsible for our having obtained results. (Halbwachs, p. 138)

Further on, commenting on the budgets of some very poor families in London as recorded by Booth, Halbwachs wondered how rigorously the accounts were kept in these extreme cases and concluded, rather icily, by asking if these most underprivileged fractions were indeed part of the working class, since they did not "rise to the level of a common consciousness":

> Because of the state of penury of these households, a brief observation may give an accurate picture of their chronic destitution: but one cannot be sure of this. In the study of social classes, moreover, this inferior social class, which does not rise to the level of a common consciousness, is not the most interesting class and can, if necessary, be studied only superficially. (Halbwachs, 1912, p. 470)

Despite all that distinguishes the LePlaysians from the Durkheimians in both their scientific and political projects, we may note a few common points in their manner of proceeding. Intent, first and foremost, on rethinking (admittedly, with more subtlety in the case of the latter) the nature of social bonds that had been riven by revolutionary changes, they employed empirical procedures designed to indicate the persistence of old forms or the birth of new ones in such bonds by assessing the moral import in each case. This manner of using an empirical procedure for purposes of social reconstruction was common to the entire nineteenth century, including Quetelet. But it did not yet accompany any direct design on social and political action, and consequently did not require the survey to be exhaustive in a territorial or national sense. The degree of generality supposed in the cases described was enough to support political and moral developments that implied no form of territorial insertion (except where Le Play made general comparisons between eastern and western Europe).

The Poor: How to Describe Them and What to Do with Them

The history of the gradual emergence of the idea of representativeness in the modern sense can be read parallel to that of the extension and transformation of political and economic tools for treating problems of poverty, from the seignorial or parochial charity of the eighteenth century until the forms of the welfare state created in the late nineteenth century. At this point, the function of social statistics changed. Whereas previously it had illustrated comprehensive analyses of a social world conceived through holistic (the Quetelet model) or organic schemes (the Auguste Comte model), it now came gradually to constitute an essential element in various policies, designed to affect individuals. The idea of representativeness then acquired decisive importance in evaluating the costs and benefits of the policies applied.

Three of these would play an essential role in asserting the constraints of representativeness in the modern sense of the word: the establishment of the first laws of social protection, in northern Europe, beginning in the 1890s; the development of national consumer markets (through the railroads) and of market studies; and last, the possibility of nationwide electoral campaigns (through the radio), which arose in the United States between the two world wars. The common point in these transformations was that *local* modes of management centered on personal relationships (charitable work, small businesses, artisanship, rural markets, electoral goodwill) were succeeded by other, *national* ones in which the land lost some of its importance as the scene of the daily reproduction of social bonds. In France, at least, this nationwide establishment of equivalence and uniformity had been prepared by the organization of the country into departments; the spread of the Civil Code, universal suffrage; obligatory military service; obligatory secular education; and the disappearance of local languages—not to mention the metric system and railway timetables. These were the preliminary conditions in which to conceive the two related ideas of *exhaustiveness* and *representativeness*, which were missing in the LePlaysian and Durkheimian procedures.

The connection between these three transformations—the delocalization of social statistics, the spread of the "representative method" and random sampling, and the creation of the welfare state—is illustrated in two studies (1976, 1987) by E. P. Hennock on the successive series of surveys on poverty conducted by Charles Booth, Seebown Rowntree, and Arthur Bowley. An analysis of the debates of the International Statistical Institute after 1895 amply confirms this hypothesis.

Throughout the nineteenth century, the history of English surveys on poverty was linked to successive interpretations and explanations of the phenomenon and to proposed methods of treatment (Abrams, 1968). Initially, a reformist tendency animated local surveys generally accompanied by recommendations for improving the morality of the working classes. The perspective differed from the LePlaysian viewpoint only in that the condemnation of a market economy was clearly less virulent. During the 1880s an economic crisis was raging and the situation seemed particularly dramatic in London, especially in the East End. Discussion arose as to what proportion of the working class lived below the extreme poverty line. This constantly recurring and ubiquitous debate—insoluble, in that the definitions of the poverty line and the means of measuring it were so conventional—is significant because it was the source of a typology of poverty at once descriptive, explanatory, and operative. From his measurements of poverty, Booth deduced measures for resolving the problem: he suggested expelling the truly poor from London—people who were poor essentially for moral reasons (alcoholism, improvidence)—in order to relieve the burden on those who were slightly less poor, and who were themselves poor for economic or macrosocial reasons (the economic crisis). The actual breakdown was in fact more complicated (there were eight categories) and had to do with both the level and regularity of income (Chapter 8). The counting was carried out on the basis of the "impressions" of "school-board visitors" and yielded detailed statistics.

What was important, however, was the connection between taxonomy and geographical results. Previously, problems of poverty had been dealt with at the municipal level. But the case of London was especially severe and it appeared that, contrary to general opinion, the proportion of "very poor" people was scarcely lower in the whole of London than in the East End considered by itself: the question of the geographical representativeness of the results analyzed by districts gradually led to political conclusions. The proposal to expel the very poor was based on the belief that the situation was most serious in London. But the survey had only been conducted in London, and it was no longer possible to generalize in the manner of Le Play or even Halbwachs. In order for action to be taken, a *maquette* or demonstration model would soon be required.

A few years later Rowntree conducted surveys in other English cities (York, in particular), surveys comparable to those of Booth in London. He sensed that Booth's methods were questionable, and paid more attention to the techniques of collecting data. He could not, however, change them entirely, for his goal was to compare the proportion of poor people in the two cities. It became apparent that the percentage of poor in York was scarcely lower than in London. This finding supported the argument that poverty could not be treated locally. The passing of a new Poor Law in 1908 took place in this context of a national assumption of responsibility for incipient social protection, thus providing macrosocial responses to problems that could no longer be presented as the result of individual morality (working-class sobriety and bourgeois good deeds). Between Booth in the 1880s and Rowntree in the 1900s, problems initially posed in local terms would thenceforth be posed in national terms. But there was still no permanent tool with which to ground this new national social statistics. Such a tool would be introduced by Bowley.

The relationship between the technique of surveying, the use expected of it, and the actual means employed changed radically, not only because of the new social laws, but also by virtue of the growing importance, in the England of that time, of debates on the rivalry between the large industrialized towns and on the question of free trade. Several surveys comparing different countries were then conducted. Rowntree made contact with Halbwachs, who organized some research according to the Englishman's methods, the results of which were published by the SGF (Halbwachs, 1914). Above all, the English Board of Trade mounted a large survey operation in several countries; while it did not yet employ probabilistic methods, it was the first survey of such scope to be carried out in Europe and dealt with several countries in particular. In France, it covered 5,606 working-class families, and the questionnaires were distributed by unions of salaried workers in thirty or so towns (Board of Trade, 1909).

Because the free market–oriented English government needed arguments in its battle against protectionism, significant public funds were made available by the Board of Trade. This made it possible to conduct a survey of a large number of towns and to pose problems crucial to the subsequent establishment of the infrastructure necessary for sample surveys: a network of homogeneous surveyors had to be organized, and differences due to "local circumstances" had to be taken into account (types of lodging, consumer habits, employment structures). The use of such surveys to compare towns *within* the country was an accidental by-product of the larger, international comparison (Hennock, 1987).

The next stage made possible by this large-scale operation was that Bowley was able to formulate in a plausible manner the conditions that would permit "representative" surveys (in the vocabulary of the period). He did so by organizing one, with a rate of 1/20, in four carefully selected towns: two were to be "mono-industrial" and two others "pluri-industrial." He also perceived that the conditions of exhaustiveness and representativeness presupposed another: the obligation to answer, a theme that had previously been absent. In so doing, he shifted the process of interaction between surveyor and surveyed from the sociable model of trust and familiarity of previous surveys, comparing it instead with a general form of civic duty analogous to universal suffrage or obligatory military service.

Similarly, there was a change in the nature of error and accuracy. Whereas Rowntree was especially punctilious about the means of recording information—though ignorant in regard to questions of sampling—Bowley was less rigorous in regard to the former. Rowntree reproached him, for example, for accepting answers about salary levels provided by the wives of salaried employees when their husbands were absent—something he himself refused to do. Above all, Bowley made imprecision and margin of error into a respectable object, something proper (confidence interval) that was no longer shamefully dissembled in the modest silence of inaccuracy. Technique and the law of large numbers replaced attempts to moralize the individual. Nor did Bowley try to identify poverty on the basis of visual impressions received during visits, as Booth used to do, but instead based it on quantifiable and constant variables. Nor did he attempt to distinguish between poverty linked to "bad habits" and poverty resulting from economic causes, thus banishing all trace of moral judgment from his investigations.

Finally—and this was a translation of all that went before in terms of professional identity—Bowley claimed that he himself did not have to offer a solution to the problems of poverty. Similarly, at about the same time, Max Weber declared that a distinction existed between "the scholar and the politician." This was an entirely new position in relation to the events of the nineteenth century:

> As economists and statisticians, we are not concerned with palliatives or expedients (for reducing poverty), but are concerned with correct knowledge and an exact diagnosis of the extent of these evils, for which reasoned and permanent remedies can then be developed. (Bowley, 1906, p. 554)

The comparison of surveys conducted by Booth, Rowntree, and Bowley, who knew and referred to one another, shows the consistency between the diverse cognitive, technical, and political aspects of the revolution in means of acquiring knowledge and generalizing that took place around 1900. The establishment of the welfare state at a national level; the nationalization of the production and interpretation of data; the replacement of moral judgment by neutral technical mechanisms; and last, the appearance of a new professional figure, the government statistician—who differed both from the literary scholar of the nineteenth century, tormented by the dissolution of social bonds, and from the managerial official directly responsible for treating social problems—all these things went together.

From Exemplary Monographs to Well-Regulated Sampling Surveys

This new figure became ever more apparent in national statistical societies and above all at the International Statistical Institute (ISI), founded in 1885, which brought together the most important of the government statisticians. In these various societies the enlightened, eclectic amateurs of the nineteenth century were gradually replaced, between 1900 and 1940, by professional technicians of statistics whose background was increasingly mathematical and less and less historical or political. In this context the "representative method" was debated twice after 1895. The initial impetus was provided by the Norwegian Kiaer, who organized the first "representative enumeration" in his country in 1894. It involved the successive drawings of localities and of persons questioned in these localities, and focused on occupations, income, expenditures, days missed from work, marriage, and the number of children.

Kiaer's initiative was widely discussed in the course of four successive congresses of the ISI held between 1895 and 1903. The congress of Berlin then adopted a motion favorable to this method, on the condition that it be clearly specified "in what circumstances the unities observed were chosen." The report called for at the time was not, however, presented until 1925 by Jensen, a Dane. A motion was adopted that did not decide between the two methods of "random sampling" and "purposive selection." The latter method was not eliminated until after Neyman's work in 1934.

During the initial phase (1895–1903), the probabilistic aspect of the new method and the need for the random nature of the samples were scarcely perceived. Moreover, during the first survey of 1894, Kiaer was not very exacting on this point, the importance of which he did not yet grasp. For example, after carefully drawing localities and streets, he allowed the researchers to choose which houses they would visit:

> They had to take care to visit not only the houses that were average from a social point of view, but houses in general that represented the different social or economic conditions present in the commune. (Kiaer, 1895)

Actually, in placing strong emphasis—for the first time in this connection—on the idea of representativeness, Kiaer wanted to show that, by taking a few (still rudimentary) precautions in the choice of sample, one obtained from a sample sufficiently good results with a few controllable variables (already present in exhaustive enumerations) to suppose that these results were "good enough" for the other variables, without having to be too precise about the meaning of this expression. Here we find the essence of the idea of representativeness: *the part can replace the whole,* whereas in previous surveys no one thought to compare the part and the whole, for the *whole* was not conceived in the same terms. Thus, for Quetelet, the average man in himself summed up an entire population; but attention was focused on these average traits and the regularity of their occurrence, and not on the actual population with its limits, structures, and exhaustiveness.

For Kiaer, in contrast, this concern with exhaustive and representative description in this new sense was in fact present, even if the technical equipment was not yet available. This equipment, which was needed to construct a random sample and to calculate confidence intervals, was not presented until 1906, when Bowley did so, outside the ISI (although he played an active part in the ISI congress of 1925 that triggered the debate again, with Jensen, March, and Gini). The technical debates of these gatherings of statisticians have been described in major research (Seng, 1951; Kruskal and Mosteller, 1980) from the standpoint of the progressive integration of the results of probability calculus and mathematical statistics into the theory of sampling. The decisive stage from this point of view was Neyman's work (1934) on stratification, sending "purposive selection" back down into the dungeons. I shall not discuss these debates here, but instead will examine how Kiaer introduced his method, and how he felt the need to compare it to LePlaysian monographs—an idea that would no longer occur to twentieth-century statisticians.

The justifications he gave from the outset for his survey are indicative of the end of a period in which relationships between classes were still thought out in terms of order and place. They were thus incommensurate with another period in which individuals of the various classes could be compared to some common yardstick, a period in which the theme of inequality—inconceivable in the other system—became fundamental, in which problems of poverty were no longer conceived in terms of good deeds and neighborliness, but in terms of social laws enacted by parliaments. Kiaer did indeed observe that previous surveys dealt only with workers (or the poor), since it was not yet conceivable for the various classes to be made equivalent within a superior whole. He was thus one of the first to raise, in such terms, the problem of "social inequalities." It is striking that the idea appears at the beginning of the first text by a government statistician dealing with representativeness, in 1895:

> One thing in particular struck me, which is that detailed surveys on income, dwelling-places, and other economic or social conditions conducted in regard to the working classes have not been extended in an analogous manner to all classes of society. It seems obvious to me that even if considering only the properly so-called working-class question, one must compare the workers' economic, social, and moral situation with that of the middle and affluent classes. In a country where the upper classes are very rich and the middle classes very comfortable, the aspirations of the working classes relative to their salaries and dwellings are measured by a different scale than in a country (or a region) where most of the people who belong to the upper classes are not rich and where the middle classes are financially embarrassed. From this proposition it follows that, in order properly to assess the conditions of the working class, one must also, in addition, be familiar with analogous conditions in the other classes. But we must take a step further and state that, since society does not consist only of the working classes, we must not neglect any class of society in social surveys. (Kiaer, 1895, p. 177)

Immediately after this, Kiaer explains that this survey would be useful in creating a fund for retirement and social security, guaranteeing social standardization and a statistical way of treating various risks:

> Since the beginning of this year a representative census has been and is still being conducted in our country, the goal of which is to elucidate various questions concerning the plan to create a general fund providing for retirement and insurance in case of invalidism and old age.

> This census is being conducted under the patronage of a parliamentary committee whose task is to examine these questions, and of which I am a member. (Kiaer, 1895, p. 177)

Two years later, in 1897, in the course of a fresh discussion at the ISI, the debate hinged on what the "representative method" provided in relation to the "typological method" then recommended at the ISI by LePlaysian statisticians such as Cheysson. Kiaer emphasized the territorial aspect, presenting a picture of the total territory in miniature, showing not only types but also the "variety of cases that occur in life." He did not yet broach the subject of random selection, but emphasized the verification of results by means of general statistics:

> I do not find that the terminology used in our program—that is, "procedures in typological studies"—is consistent with my ideas. I shall have occasion to demonstrate the difference that exists between investigations by types and representative investigations. By representative investigations, I mean a partial exploration in which observations are made over a large number of scattered localities, distributed throughout the land in such a way that the ensemble of the localities observed forms a miniature of the total territory. These localities must not be chosen arbitrarily, but according to rational grouping based on the general results of statistics; the individual bulletins we use must be arranged in such a way that the results can be controlled in several respects with the help of general statistics. (Kiaer, 1897, p. 180)

In opposing his method, which allowed a "variety of cases" to be described, to the method which only showed "typical cases," Kiaer emphasized a mutation parallel to the one that Galton and Pearson had just wrought in relation to Quetelet's old statistics of averages. In henceforth focusing attention on the variability of individual cases with the ideas of variance, correlation, and regression, the English eugenicists took statistics from the level of the examination of wholes, summarized by averages (holism), to the level of the analysis of distributions of individuals being compared:

> The Institute has recommended investigation by means of selected types. Without contesting the usefulness of this form of partial investigation, I think it presents certain disadvantages, compared to representative investigations. Even if one knows in what proportions the different types enter into the whole, one is far from reaching a plausi-

ble result for the entirety; for the whole includes not only the types—that is, the average relationships—but the full variety of cases that occur in life. In order for a partial investigation to give a true miniature of the entirety, it is therefore necessary for us to observe not only the types, but any species of phenomena. And this is what will be possible, if not completely, with the help of a good representative method that neglects neither the types nor the variations. (Kiaer, 1897, p. 181)

Then, attempting to position himself between those two fundamentally different modes of knowledge constituted by individual monographs and complete enumeration, Kiaer surprisingly insists that he could do as well as the monographs on their own ground ("blood, flesh, nerves"). In fact, however, samples would subsequently be compared with exhaustive censuses (in regard to cost and accuracy)—but certainly not with monographs. This shows the dominance of a mode of knowledge based on an intuitive grasp of the whole person:

In discussing the reciprocal roles of monographs and partial statistics, we have said that the monograph is concerned with objects that cannot be counted, weighed, or measured, whereas partial statistics is concerned with "objects which in themselves could be counted in their entirety but which, deliberately, are only counted in part." . . . In general, I think it possible to apply to partial investigations, and especially to representative ones, the eloquent words uttered by our honored colleague Mr. Bodio, in regard to our late lamented Dr. Engel's work on the budgets of working-class families: "statistical monographs and enumeration are two complementary ways of investigating social facts. In itself, enumeration can provide only the general outlines of phenomena—the silhouette, as it were, of the figures. Monographs"—and to this I add partial investigation in general—"allow us to extend the analysis to every detail of the economic and moral life of the people. They give blood, flesh, and nerves to the skeleton formed by general statistics, while enumeration in its turn completes the ideas provided by monographs." As long as we insert the words "partial investigation," I find these words of Luigi Bodio to be an excellent demonstration of the reciprocal roles of partial investigation and general statistics. (Kiaer, 1897, pp. 182–183)

Kiaer then describes his ideal tool, as rich as the monograph and as precise as complete enumeration—if we but respect the constraint of representativeness (something Kiaer had an accurate intuitive perception of, but still lacked the tools for):

> The scientific value of partial investigations depends far more on their representative character than on the quantity of data. It often happens that data that are easy to obtain represent an elite rather than ordinary types. (Kiaer, 1897, p. 183)

One can retrospectively verify that the procedure followed is a good one if the controllable variables do not differ too much between the sample and the census:

> To the extent that partial investigation has proven correct regarding the points it has been possible to verify, it is probably also correct in regard to the points that cannot be verified with the help of general statistics. (Kiaer, 1897, p. 183)

The idea that the probabilistic theorems formulated at the beginning of the nineteenth century could allow more to be said on the "probable errors" involved in random drawings of samples (and therefore on the significance of deviations observed by Kiaer) had not yet occurred to anyone. His method was rooted in a solid knowledge of the *ground,* with verification only taking place afterward. The deterritorialization and mathematization of the procedures did not happen until later.

The introduction of probabilistic systems only occurred rather timidly in 1901, in a new debate on Kiaer's method. The German economist Bortkiewicz then declared he had used "formulas deduced for analogous cases by Poisson, to find out if the difference between two numbers was fortuitous or not"; he had noted that this was not the case in the examples presented by the Norwegian, and that the deviations were significant. Thus Kiaer's sampling was not as representative as he thought. It seemed that Bortkiewicz had dealt Kiaer a deadly blow. Oddly enough, however, in the wake of the debate no one took up Bortkiewicz's words and we do not even know what Kiaer's reaction was. Perhaps Bowley had wind of it, for five years later, in 1906, he presented the first calculations of confidence intervals before the Royal Statistical Society (Bowley, 1906).

How to Connect "What We Already Know" with Chance

The idea that the representativeness of a sampling could be guaranteed through "control variables" nonetheless endured for about thirty years through the method known as *purposive selection*. This method continued to favor the territorial division (inherited from the previous era) of national space into an ensemble of districts. From this a subwhole was selected, not at random, but in such a way that a certain number of essential variables (the control variables) had the same values for the entire territory. A remarkable use of this method was presented in 1928 by the Italian Corrado Gini. After having to get rid of the individual census bulletins of 1921 because they were so cumbersome, he had the idea of keeping some that dealt with 29 regional districts (out of the 214 that constituted Italy). In these districts the averages of seven variables were close to those of the entire country (birth, death, marriages, proportion of agricultural population, proportion of agglomerated population, average income, mean height above sea level). The choice of the 29 districts that best respected these limitations was the result of a laborious and tentative process that Gini himself criticized, showing that, barring very particular hypotheses of linearity of correlation between controlled and noncontrolled variables, there was no reason to suppose this sample would be a good substitute for the whole of Italy.

This entire discussion, which lasted from Jensen's report of 1925 to Neyman's article of 1934, bore upon the question of the connection between pure random selection and "what was already known from other sources" (for example, censuses). This led first to Kiaer's methods of "control variables," then to that of "purposive selection," each being successively rejected in favor of the technique of *stratified* sampling. This was based on a priori divisions of the population, thought to summarize what was already known—that significant differences in averages existed between the classes, and that the accuracy of the general estimates was improved by a priori stratification. This therefore implied that such nomenclatures—a true repository of previous knowledge—existed, had a certain durability, and inspired confidence: after 1950 socio-occupational categories, levels of education, categories of communes, and family types would all play such a role. The construction of the mechanism of representativeness thus passed, on the one hand, through the edifice of mathematics, gradually purged of its old "control variables" (in the case of Neyman), and on the other hand, through a system of nomenclatures that recorded personal qualities within frameworks guaranteed by the state, as trustee of the general interest, and elaborated by an institution that inspired confidence.

This point was highlighted in Jensen's report of 1925 on the representative method. Observing that this method aroused mistrust because it touched on only a part of the population, he wondered whether the fact that the statistical administration inspired confidence was enough to obviate this type of criticism. That the two categories of problems—technical and sociopolitical—were mentioned at the same time is material for an answer to the initial question (why wasn't the representative method used sooner?):

> This objection contains the real kernel, that the greatest importance must be attached to the existence of a state of mutual trust between the official statistical service and the population which both supplies the material for the statistics and for whose sake all this work is done. The official statistics ought of course to be exceedingly cautious of its reputation—"it is not sufficient that Caesar's wife is virtuous, all the world must be convinced of her virtue." But it would hardly be warrantable purely out of regard to prestige, to prevent a development which in itself is acknowledged to be really justified. One does not omit to build a bridge with a special design because the public, in its ignorance, distrusts the design; one builds the bridge when the engineer can guarantee its bearing strength,—people will then use the bridge in due course and rely on its solidity. (Jensen, 1925a, p. 333)

Jensen's problem in this text was to connect the technical solidness of the object and its reputation: this connection was what constituted the strength of government statistics.

The Welfare State, the National Market, and Forecasting Election Results

Beginning in the 1930s this typically European instance of the use of the representative method in the state's management of social problems was no longer the only one. In at least two other cases, especially in the United States, certain totals affecting the entire national territory became directly pertinent: studies of consumer goods and markets, and electoral forecasts. In both cases, prior, nationwide standardization and categorizing of the products were necessary. For consumer goods, large-scale businesses had to distribute standard products regularly throughout the whole country by means of a national transportation network, and these products had to be clearly identified (Eymard-Duvernay, 1986). It then became possible to conduct a nationwide opinion poll to see if consumers preferred drinking Coca-Cola or Pepsi-Cola. As for electoral forecasts, it helped if the candidates were the same for the entire country (as was the case in American presidential elections, but not in the French district election system), and if their images were relatively widespread and unified—which was beginning to be the case, through radio. It also helped if the sampling frame was as close as possible to the electorate: everyone knows about the disastrous U.S. telephone polls which, in 1936, surveying only the wealthy persons who possessed this equipment, wrongly announced a Republican victory.

But all these surveys had one thing in common: their results were turned over to sponsors—administrations, big businesses, radio networks, or newspapers—who used them for operational purposes. The idea of representativeness made limiting the cost of this knowledge consistent with its relevance, which was both technical and recognized socially. All these cases concerned individuals (people receiving aid, consumers, and electors), rather than whole entities such as Quetelet's divine order, Le Play's lineage, or Halbwachs's working-class consciousness.

I have tried here to interpret the appearance, at the turn of the century, of a new concept of representativeness, which accompanied the passage from one mode of thinking to another. This transformation had been presented in the past in many ways. Louis Dumont echoes one of them in his *Essai sur l'individualisme* (Essay on individualism). He opposes "national cultures" deemed incommensurate with each other (the German tradition) to a universal—or at least universalist—"civilization" (the English tradition for the economist version, the French for the political version). He indicated that, to resolve this contradiction, Leibniz had imagined a "monadic system" in which "every culture expresses the universal *in its own way*"—a clever way of generalizing:

> A German thinker presents us with a model that fits our needs: I refer to Leibniz's monadic system. Every culture (or society) expresses the universal in its own way, like each of Leibniz's monads. And it is not impossible to imagine a procedure (a complex and laborious one, admittedly) that enables us to pass from one monad or culture to another, through the medium of the universal taken as the integral of all known cultures, the monad-of-monads found on the horizon of each. Let us, in passing, salute genius: it is from the mid-seventeenth century that we inherit what is probably the sole serious attempt to

> reconcile individualism and holism. Leibniz's monad is both a whole in itself and an individual within a system unified even in its differences: the universal Whole, as it were. (Dumont, 1983, pp. 195–196)

The comparison of monographic and statistical methods poses quite differently the question of the reconciliation of modes of knowledge resulting from holistic and individualist perspectives. The history of the idea of representativeness has referred us back to the idea of generalization: What is a part? What is the whole? As we have seen, the two processes—political and cognitive—of defining a pertinent whole (the nation or the market, depending on the case) were indispensable in order for the modern idea of representativeness and the necessary tools to appear.

The problem raised by the transformations of the idea of representativeness, and in particular by the debates of the period between 1895 and 1935, was to be able to link together pieces of knowledge produced according to different registers. This was the problem that Kiaer's "control variables," Jensen and Gini's "purposive selection," and Neyman's "stratification" respectively attempted to resolve (O'Muircheartaigh and Wong, 1981). This also refers back to the choice between the acceptance or rejection of "a priori laws," between subjective and objective probabilities (Chapter 2).

During the debate of 1925 Lucien March, then aged sixty-six, who had watched the whole of statistical thinking change direction, raised this problem in terms that seem both refined and archaic, for all these things have subsequently been formalized—in other words, frozen in some way:

> The system that consists in approximating as closely as possible the random selection of the unities of the survey is not necessarily the one that best corresponds to the idea of representation. As we have observed, it presupposes that the unities present no differences among themselves. Now, the goal of the representative method is to throw differences into sharper relief. The hypothesis thus seems rather to contradict the objective. This explains the preferences for a better understood and more intelligent choice. For example, some people put aside extreme cases that are judged abnormal, or else they effect the choice of specimen by basing themselves on some criterion that is deemed essential. (March, 1925)

When March speaks of "differences," emphasizing the seeming contradiction between the need for creating equivalent categories of unities and the search for their differences, we could point out that later formalizations have resolved this problem perfectly. But his hesitations as to whether one should make a "better understood, more intelligent" choice, or whether one should eliminate "abnormal cases," do reflect a problem encountered by any statistician who is familiar with his terrain. This "previous knowledge" is often acquired through direct contact, through familiarity (though not always, as appears in the case of "variables controlled" by general statistics). He is then close to Dumont's idea of "culture," or Tönnies's "community." The observer always has an intuition of some general knowledge of whole situation, of a person, a social group, or even of a nation.

At the close of their analytical work, researchers sometimes say "we must put back together what we've pulled apart." Once again, we have here the problem of generalization: to reconstruct the whole in its unity. To return to the whole on the basis of one of its parts: this fantasy constantly troubles scholars. And in the most recent developments in sampling methods, techniques of simulation, made possible by powerful computers, allow this reconstruction to be simulated by generating a large number of subsamplings based on the sample, and by studying the distribution of statistics thus produced. This method has been termed the *bootstrap,* for it suggests the dream of pulling oneself up by one's own efforts. This dream of reconstructing the universe on the basis of one of its parts is not unlike the "complex and laborious procedure" suggested by Leibniz for passing from a singular to a universal culture. What has changed is the cognitive and political mechanism.

8

Classifying and Encoding

The aim of statistical work is to make a priori separate things hold together, thus lending reality and consistency to larger, more complex objects. Purged of the unlimited abundance of the tangible manifestations of individual cases, these objects can then find a place in other constructs, be they cognitive or political. Our goal is to follow the establishment of the formalisms and institutions that have made massive use of these objects socially and technically possible. While mathematical tools and statistical administrations have, especially in the 1980s, been the subject of historical surveys often mentioned above, this is less true of the conventions of equivalence, encoding, and classification that precede statistical objectification. Taxonomical questions derive from intellectual and theoretical traditions that differ greatly, and that do not communicate much among themselves. I shall recall some of them here before mentioning a few significant examples of historical work on these classifications, concerning natural species, branches of industry, poverty, unemployment, social categories, and causes of death. Not only are the objects of this research different, but their perspectives differ too. Taxonomy is, in a way, the obscure side of both scientific and political work. But the study of taxonomy cannot be reduced to the unveiling of hidden relationships between these two dimensions of knowledge and action—as is sometimes the case in a sociological criticism of science, which swings directly from a purely internalist position in terms of the progress of knowledge, toward an opposite, externalist position in terms of power relationships and social control. The question, rather, is to study in detail the nature of the bonds that make the whole of things and people hold together—which in fact precludes the opposition between analyses described as technical or social.

Statistics and Classification

Among the intellectual disciplines that include theoretical or empirical reflections on taxonomy, we find statistics (from the German school and Quetelet until Benzécri); the philosophy of sciences (Foucault); natural history; linguistics; sociology and anthropology (Durkheim and Mauss); cognitive psychology (Piaget and Rosch); law; and even medicine. Economics, in contrast, at least in its theoretical aspects, is a science in which taxonomic work was scarcely conceived as such. More generally, the relative silence of certain human sciences in regard to problems of classification and, even more so, of encoding (that is, the decision to attribute a case to a particular class) is linked to the division between pure and applied science. Taxonomy and especially coding are perceived as technical and practical problems, often solved from one day to the next by practitioners and not by theoreticians. This is also why the more original and fertile reflections on these questions can be produced in fields such as law or medicine, where dealing with singular cases is a major component. For most of the human sciences, however, the passage from singularity to generality and the construction of consistent classes of equivalence are essential theoretical and practical problems, giving rise to very varied cognitive and social tools. Statistics is one such tool, but is far from being the only one: a major goal of this book is to help to shed light on the way in which these various resources have been historically constructed.

From its origins, German statistics of the eighteenth century presented itself as a vast nomenclature intended to provide classification for a general description of the state, using words and not numbers (Chapter 1). In the nineteenth century, treatises on statistics were often still plans for classification, for information that was henceforth mainly quantitative, generally resulting, like types of subproducts, from administrative practices (Armatte, 1991). As such, statistics was a new language that helped unify the state and transform its role (Chapters 5 and 6). But at the same time, with the hygienist movement (in France) and public health movement (in England) a "moral statistics" was developed, influenced by the tradition of political arithmeticians, and partly distinct from the government statistics resulting from the German school. The combination, symbolized by Quetelet, of traditions previously foreign to each other led to the establishing of a "science of society" (or "sociology"), which relied on the records of administrative statistics (the general registry, causes of deaths, and justice) in order to discern the laws that governed marriage, suicide, or crime (Chapter 3).

From then on this double, administrative or "moral" dimension of statistics was the origin of the two directions taken by the taxonomic activities of later statisticians. On the one hand, as tributaries of administrative records, they worked on constituting and defining categories and on encoding singular cases, in a perspective that shows some affinity with that of legal or administrative specialists (the word "category" itself derives from the Greek term *kategoria*, connected with judgment rendered in the public arena). But on the other hand, as interpreters of their own productions they tried to infer, on the basis of ever more complex mathematical constructs, the existence of underlying categories revealed by statistical regularities or by particular forms of distribution, notably the law of errors or the binomial law (the future normal law). Later on, the factor analysis of psychometricists—detecting a "general intelligence"—or the methods of classification resulting from data analysis (Benzécri, 1973) would be situated in this same general trend.

These two perspectives of research on taxonomy were partly foreign to each other. They were often completely unaware of the other's existence. The first, inspired more by nominalism, held category to be a convention based on a practice duly codified by stable and routinized procedures—for example, the general registry, medical bulletins on causes of death, the activity of the police and tribunals, and, later on, the techniques of conducting surveys through more or less standardized questionnaires. The second, in contrast, was more realistic, issuing from Quetelet's initial alchemy of transforming subjective averages into objective ones by means of his "average man." It held category as being revealed by statistics. For Adolphe Bertillon, the bimodality of the distribution of the heights of conscripts from the Doubs *proved* that this population was descended from two distinct ethnic origins. Psychometrics and its rotations of factorial axes, or the mathematical methods of ascending or descending classification, are situated in this perspective. It is not my intention here to contrast these two points of view, denouncing one on the basis of the other; rather, I shall try to consider them together, by reconstructing their geneses and their respective autonomization, and by comparing the taxonomic reflections of statisticians to those of other thinkers.

The Taxonomy of Living Beings

In his book *Words and Things,* Michel Foucault (1970) presented a personal version of the birth of taxonomy in the "classical era" (the seventeenth and eighteenth centuries) in the sciences of language (general grammar), life (natural history), and of exchange (wealth). During that period, according to him:

> The sciences always carry within themselves the project, however remote it may be, of an exhaustive ordering of the world; they are always directed, too, towards the discovery of simple elements and their progressive combination; and at their centre they form a table on which knowledge is displayed in a system contemporary with itself. The centre of knowledge, in the seventeenth and eighteenth centuries, is the *table.* (Foucault, 1970, pp. 74–75)

This scientific project corresponded to the goal of German statistics, with its interlocking and coordinated nomenclatures, then to the goal of nineteenth-century administrative statistics as expounded by Moreau de Jonnès. In his archeology of taxonomy, Foucault never alludes to the tradition of statistics. And yet his analysis of the contrast between Linné, for whom "all of nature could enter into a taxonomy," and Buffon, for whom it was "too varied and rich to be fitted into so rigid a framework," introduces debates that recur throughout the history of statistical classifications: encoding as a sacrifice of inessential perceptions; the choice of pertinent variables; how to construct classes of equivalence; and last, the historicity of discontinuities.

The eighteenth-century naturalist observed plants and animals far more than was previously the case; but at the same time he "voluntarily *restricted* the field of his experience." He excluded hearing, taste, smell, and even touch in favor of sight, from which he eliminated even color. Residues of these exclusions, his objects were seemingly filtered, reduced to sizes, lines, forms, and relative positions. Thus evidence, senses other than sight, and colors were all sacrificed. But the compensation for this loss was the possibility of revealing (Linné) or constructing (Buffon) the order of nature. Both of them chose the same four variables to characterize the structure of species, a common elementary encoding then used in two different ways to set in place the cartography of life. Among these variables, two could result from countings or measurements: *numbers* (pistil and stamens) and *sizes*

(sizes of the different parts of the plant). In contrast the two others—the *forms* and *relative dispositions*—had to be described by other procedures: identification with geometric forms, possible resemblance to parts of the human body. This ensemble of "variables" allowed a species to be designated, or assigned a *proper noun,* by situating it in the space of variables thus constructed.

This manner of proceeding distinguishes eighteenth-century naturalists from both their sixteenth-century predecessors and their twentieth-century successors. Previously, the identity of a species had been ascertained by a unique distinguishing feature, an emblem incommensurate with those of other species: one bird hunted at night, lived on water, and fed from living flesh. Each species was singular by the very nature of the characteristics that singularized it. But the eighteenth-century naturalist, in contrast, created a context of equivalence and comparability within which spaces were positioned in relation to one another according to *variables.* These spaces could then be regrouped into families, endowed with common nouns and based on these propinquities and differences. These differences were described in a more external (using sight) and analytical (enumerating the elements) manner. Beginning with Cuvier, the zoologists of the nineteenth century retained this general way of comparing things, but henceforth applied it in a more synthetic manner to organic unities with internal systems of dependency (skeleton, respiration, circulation). The analytical classifications of the eighteenth century, based on networks of differences observed in botany especially, were thus situated between the sixteenth-century idea of the unique distinguishing feature and the theories of animal organisms of the nineteenth century (Foucault, 1966, p. 157).

This rearrangement into families of species, creating common nouns based on proper nouns, relied on codified descriptions by means of the four types of variables. But even within this already limited descriptive universe, the techniques enabling families to be classified on the basis of networks of differences could be founded on opposing principles. Linné's *System* was antithetical to Buffon's *Method.* Any statistician concerned with the creation and practical use of a nomenclature has encountered a basic opposition of the type symbolized by the controversy between Linné and Buffon (even if its modalities have changed by virtue of the availability of more elaborate formalisms). Of all the features available, Linné chose certain among them, *characteristics,* and created his classification on the basis of those criteria, excluding the other traits. The pertinence of such a selection, which is a priori arbitrary, can only be apparent a posteriori; but for Linné this choice represented a necessity resulting from the fact that the "genera" (families of species) were *real,* and determined the pertinent characteristics: "You must realize it is not the characteristic that constitutes the genus, but the genus that constitutes the characteristic; that the characteristic flows from the genus, and not the genus from the characteristic" (quoted by Foucault, 1966, p. 159). There were thus valid natural criteria to be discovered by procedures that systematically applied the same analytical grid to the entire space under study. Valid criteria were real, natural, and universal. They formed a *system.*

For Buffon, on the other hand, it seemed implausible that the pertinent criteria would always be the same. It was therefore necessary to consider all the available distinctive traits a priori. But these were very numerous, and his *Method* could not be applied from the outset to all the species simultaneously envisaged. It could only be applied to the large, "obvious" families, constituted a priori. From that point on, one took some other species and compared it with another. The similar and dissimilar characteristics were then distinguished and only the dissimilar ones retained. A third species was then compared in its turn with the first two, and the process was repeated indefinitely, in such a way that the distinctive characteristics were mentioned once and only once. This made it possible to regroup categories, gradually defining the table of kinships. This method emphasized local logics, particular to each zone of the space of living creatures, without supposing a priori that a small number of criteria was pertinent for this entire space. In this perspective, moreover, there was continual slippage from one species to another. Any demarcation was in a sense unreal and any generality was nominal. Thus, according to Buffon:

> Our general ideas are relative to a continual scale of objects, of which we can only clearly see the middle while the opposite ends recede, ever more eluding our considerations . . . The more we increase the number of divisions of natural productions, the closer we shall draw to the truth, since only individuals really exist in nature, while genera, orders, and categories exist only in our imagination. (Buffon, quoted by Foucault, 1966, p. 159)

Buffon's perspective was thus more nominalist, whereas Linné's was realist. Moreover, Buffon suggests re-creating objects through their *typicalness,* regrouping them around a "clearly visible middle," whereas the opposite ends "flee and escape our consideration." This method is antithetical to Linné's *criterial* technique, which applied general characteristics presumed to be universally effective. These two ways of classifying things were analyzed later by cognitive psychologists (Rosch and Lloyd, 1978) and used, for example, to describe the creation and use of socio-occupational nomenclatures (Boltanski and Thévenot, 1983; Desrosières and Thévenot, 1988).

Adanson, who was close to Buffon, summarized the difference between *method* (Buffon) and *system* (Linné) by emphasizing the absolute or variable nature of the principles of classification employed, pointing out that the method must always be ready to correct itself:

> Method is some sort of arrangement of facts or objects connected by conventions or resemblances, expressed by an idea that is general and applicable to all these objects, without, however, viewing this basic idea or principle as absolute, invariable, nor so general that it cannot brook an exception . . . Method differs from system only in the idea that the author attaches to his principles, seeing them as variables in method, and as absolutes in system. (Adanson, quoted by Foucault, 1966, p. 156)

Theoretical taxonomists are spontaneously drawn to Linné's way of proceeding, and mistrust Buffon's: what is the good of a method when its principles fluctuate with the various difficulties encountered? And yet any statistician who, not simply content to construct a logical and coherent grid, also tries to use it to encode a pile of questionnaires has felt that, in several cases, he can manage only by means of assimilation, by virtue of propinquity with cases already he has previously dealt with, in accordance with a logic not provided for in the nomenclature. These local practices are often engineered by agents toiling away in workshops of coding and keyboarding, in accordance with a division of labor in which the leaders are inspired by the precepts of Linné, whereas the actual executants are, without knowing it, more likely to apply the method of Buffon.

The two approaches described thus give rise to questions about the nature and origin of the breaks or boundaries between classes. Linné's perspective, based on the combination of a small number of criteria, defines theoretical places in a potential space. These places may be filled to varying degrees, but one does not a priori know why. Buffon's method, on the other hand, leads to a multidimensional continuum, within which the cuts are nominal: "Categories exist only in our imagination." Neither Linné nor Buffon can thus account for the existence of roses, carrots, dogs, or lions. These discontinuities could only result, for eighteenth-century naturalists, from a historicity of nature, from a chain of contingent events, of avatars and inclemencies, independent of the inner logic of the living world as described by both Linné and Buffon. This type of contingent historical explanation has also appeared, for example, since the 1970s in a scenario proposed by certain anthropologists in order to describe the appearance of discontinuity between apes and man (walking in an upright position), a scenario based on modifications in the relief, climate, and vegetation in a certain region of East Africa several million years ago. I can now compare this approach with one that describes the emergence of social groups in particular historical circumstances: the debate between realist and nominalist positions, though present in both cases, assumes different forms, the analysis of which lends originality to the study of the nomenclatures used to construct social and economic statistics, and no longer merely to classify vegetable or animal species.

The Durkheimian Tradition: Socio-logical Classifications

Through these eighteenth-century debates, naturalists have given us the idea of system and of criterion (Linné); the criticism of systematicity; and the intuition of distinctly local forms of logic (Buffon). This discussion gave taxonomical reflection a density that purely logical speculation could not engender alone. It unfolded in the nineteenth century through debates on the evolution of species, fixism, and transformism, Cuvier, Lamarck, and Darwin. This trend finally provided food for statistical thinking, with Galton's and Pearson's efforts to transpose Darwinian analyses of living beings to the human race, thus developing classifications with a biological foundation to hereditist and eugenicist ends. This contributed to a physical anthropology, centered on the study of the human body and its variations (Chapter 4).

But anthropology was also the source of an entirely different tradition of analyzing and interpreting classifications, having to do with social classifications and their affinities with the most elementary logical acts. Durkheimian sociology particularly emphasized the close links between social groups and logical groups: it even held that the structure of the former ordered the mechanisms of the latter. This point of view was strongly upheld by Durkheim and Mauss in their ground-breaking text of 1903: *De quelques formes primitives de classification: contribution à l'étude des représentations collectives* (On several primitive forms of classification: a contribution to the study of collective representation). This reflection was very different from that of the naturalists. It aimed not at ordering elementary observations of the world but, rather, at describing and interpreting the classifications used by primitive societies. This ethno-taxonomy led scholars to view classifications *from the outside,* as objects already constituted, and no longer to construct and use them, as did Linné and Buffon. It drew attention to the links between indigenous classifications and scientific classifications by emphasizing the social basis for classifications used in primitive societies:

> Not only do the division of things into regions and the division of society into clans correspond exactly, but they are also inextricably intertwined and confused . . . The first logical categories were social categories; the first classifications of things were classes of men into which these things had been integrated. It was because men were grouped and imagined themselves in the form of groups that they ideally grouped other beings, and the two modes of grouping began merging together to such a point that they became indistinct. Phratria were the first genera; clans, the first species. Things were supposed to be an integral part of society, and their place in society was what determined their place in nature. (Durkheim and Mauss, 1903)

This hypothesis had the great advantage of considering classifications as objects of study in themselves, and not simply as grids, or tools, through which intermediary the world was discussed. But it was used by Durkheim and Mauss to establish scrupulous correspondences between the social and symbolic classifications used by primitive societies, and not for the mainly statistical tools used for describing urban and industrial societies. A structuralist perspective of this kind was applied in modern French sociology (Bourdieu, 1979), based on the statistical techniques for the factor analysis of correspondences, techniques that made it possible to construct and exhibit multidimensional spaces combining varied social practices. The social class then served as a guiding thread and an invariable in interpreting the regularity of the structural oppositions described with the help of these schemas, to the extent that the classifications were registered in a stable topology (the field), defined precisely on the basis of these oppositions.

Although Durkheim and Mauss devoted major efforts to establishing affinities among social, symbolic, and logical classifications, they did in contrast explicitly exclude "technological classifications" and "distinctions closely involved in practice," since according to them symbolic classifications were not linked to actions. This untoward cleavage between society and science on the one hand and technique on the other limited their analysis. It also made it impossible to encompass at a single glance the set of operations that made it possible to order things and people and give them consistency for thought and action, which it seems hard to distinguish between:

> These systems, like science, have an entirely speculative goal. Their objective is not to facilitate action but to make people understand, to make intelligible the relationships that exist between them . . . These classifications are intended to bind ideas together, to unify knowledge; as such, they are scientific labors and constitute a first philosophy of nature.(87) The Australian does not divide the world among the totems of his tribe with a view to regulating his own behavior or even to justify his practice; he does so because, the idea of totem being cardinal for him, it is necessary to situate all his other knowledge in relation to that idea.
> (Note 87) In this, they are clearly distinguished from what might be termed technological classifications. It is probable that, in every era, mankind has more or less clearly classified the things he eats according to the procedures used to obtain them: for example, animals that live in water, or in the air, or on land. But initially, the groups thus constituted were not connected to each other and systematized. They were divisions, distinctions between ideas, not tables of classification. Moreover, it is obvious that these distinctions were tightly bound up with the practice of which they express only a few aspects. That is why we have not spoken of it in this book, in which we are trying to cast light on the origins of the logical procedure that is the basis of scientific classifications. (Durkheim and Mauss, 1903)

In thus carefully distinguishing a sphere in which the social, symbolical, and cognitive aspects are "inextricably intertwined and confused" from the world in which one "regulates one's behavior" and "justifies one's practice," Durkheim and Mauss miss the opportunity to apply their theory to the categories that inform action. Among these categories, for example, taxonomies that underlie administrative action—especially in productions of a statistical kind—are in fact tools for classification and encoding, linked to action and decision making, and cannot be separated from the social network into which they are introduced. Thus when in 1897 Durkheim used the statistics of suicide to ground his macrosociological theory, he no longer asked himself about the possible social variations of *declarations* of suicide—something which, given the religious stigma attached to suicide, could have been pertinent (Besnard, 1976; Merllié, 1987). These two parts of his work are disconnected. At one and the same time, but separately, he created an objective sociology based on statistical regularities, and a sociology of "collective representations" that did not apply to the tools of knowledge and to the representations underlying statistical objectification. In this perspective, it was not simply a matter of raising a criticism of statistical expedients resulting from a possible subdeclaration of suicides, but more generally of analyzing the very construction of the object as revealed through its administrative encoding. Indeed, the rigid dichotomy traced by Durkheim and Mauss between, on the one hand, schemes underlying pure knowledge and, on the other, distinctions pejoratively described as "nonsystematized" and "tightly bound up with practice," ultimately reinforces a realist position, in which an object exists independently of its construction. This construction is thereby reduced to an operation of "measurement," inferior to the process of conceptualization.

Thus the theoretical and practical examination of the definitions, nomenclatures, and encodings used by the social sciences has long been rendered almost unimaginable by this division of labor, and by the distinction between an ideal of scientific knowledge aiming at truth, and daily action dependent on categories deemed impure and approximative, if not biased. This sharing of tasks offers advantages both to scientists and to those who take action and make decisions, if only by making possible the existence of a space of expertise, useful to both groups. It can only be outstripped by a detour, which reinserts the work of production and the use of so-called administrative knowledge (by the state, businesses, or other social actors) into the larger universe of the production of any branch of knowledge. This production is then seen as a formatting and stabilization of categories sufficiently consistent to be transported or handed down directly while still maintaining their identity in the eyes of a certain number of people. Attention is thus drawn particularly to the social alchemy of *qualification,* which transforms a case, with its complexity and opaqueness, into an element of a class of equivalence, capable of being designated by a common noun and integrated as such into larger mechanisms.

The Circularity of Knowledge and of Action

The encoding stage—an often unassuming moment hidden in routinized chains of production—is more apparent when it is itself an aspect of a decision laden with further consequences. This is noticeable in three domains, in which statistics have long been produced and utilized: law, medicine, and education.

Judges apply the law and the penal code, but this application is only possible after a preliminary investigation and public proceedings during which the nature of the acts committed is debated through various arguments. Judgment involves an interpretation of the rules, themselves the result of law and the jurisprudence accumulated in previous cases (Serverin, 1985). The law constitutes an ancient and inexhaustible reserve of resources for identifying, discussing, and describing acts, and for reclassifying them in categories that derive from a general process, itself defined in a more or less simple way. It is a process that lends itself well to statistical registration: the *Compte général de l'administration de la justice criminelle* (General accounting of the administration of criminal justice), established in France since 1827, is one of the oldest administrative statistics (M. Perrot, 1974).

Doctors and public health officials describe diseases and causes of death, for clinical (diagnostic) purposes, with the goal of prevention (epidemiology)—or even in order to manage health systems (hospital budgets) and social protection (compensation for acts). The introduction, beginning in the 1830s, of "numerical methods" in medicine, either when epidemics break out or else to study the effectiveness of treatments, has given rise to lively controversy, centered on the singularity of illnesses and ill people and on the conventions of equivalence applied (Chapter 3).

Last, schools are institutions that tend to qualify the individuals who attend them. They provide durable tests that enable these individuals to be divided into categories with specific qualities, capable of comparison with other categories, found in employment. The definitions of these categories are often based on scholarly qualifications, diplomas, grades, and length of study.

In each of these three fields, taxonomy is associated with both the construction and the stabilization of a social order; with the production of a common language allowing individual acts to be coordinated; and last, with a specific and transmissible knowledge employing this language in descriptive and explanatory systems (especially statistics) capable of orienting and triggering action. According to this perspective, the interactions between knowledge and action can be presented in a circular manner by including two categories often used (and sometimes discussed): "data" and "information." "Data" (literally "givens") appear as the result of an organized action (whence the ambiguity of the word). "Information" is the result of the formatting and structuring of these data through nomenclatures. "Knowledge" and "learning" result from the reasoned accumulation of previous information. The categories of classification first ensure the equivalence of singular cases, and then the temporal permanence of these equivalences allows the action to be triggered again. Thus the circle "action-data-information" can be closed, not only logically but historically, with nomenclatures acting as repositories of accumulated knowledge.

To apply this schema more precisely, especially to statistical productions, we must analyze the degrees of generality and the fields of validity of the investments of forms constituted by these branches of knowledge structured by more or less stable taxonomies. In particular, data recorded directly by statistical institutions (surveys, censuses) are distinguished from those that result, apparently as subproducts, from activities of management having aims other than to produce information. These two categories of knowledge are often opposed in reference to a project of scientific knowledge, the second being considered less satisfactory and more makeshift in the absence of the former, the cost of which would be too high. But the comparative history of bureaus of statistics shows that, even in the first case, the themes of the surveys, the questions asked, and the nomenclatures used are marked by the forms of public action predominant in a given country and era. The dynamic periods in the life of these bureaus correspond precisely to the times when they manage to link their investigations closely to crucial current questions (Chapters 5 and 6), simultaneously producing the categories of action and the means to evaluate them: one thinks of England during the 1840s with Farr and the public health movement; Prussia during the 1860s and 1870s with Engel and the Verein für Sozialpolitik; the United States during the 1930s with the Depression and unemployment; France during the 1950s and 1960s with planning and growth.

The mutual interactions of knowledge and action are, in the case of surveys conducted directly by statistical bureaus, more global and less immediate than in cases where the data are subproducts of administrative action. Until the 1930s, moreover, these data have always constituted the main material of official statistics, with the important exception of population censuses—even if these censuses were originally, primarily, administrative counting operations, for military and financial purposes. But in every case, the ability to constitute—or not—obligatory points of passage for the other actors depended on the possibility of firmly connecting the successive links of this circle, for this language and these forms of knowledge and action. This is shown, on the contrary, by periods of reflux in these links: in Germany, after 1880, in France, between 1920 and 1940; in the United States, Great Britain, and, to a lesser extent, France, since 1980.

Within the perspective outlined here, and on the basis of works dealing with the history of taxonomies and statistical variables, I shall discuss the classifications of industrial sectors; definitions of the wage-earning classes and of unemployment; socio-occupational nomenclatures; and those of diseases and causes of deaths. The way in which statisticians have perceived and identified objects, describing and treating them in categories, assembling and distributing them in tables, not to mention the misunderstandings and criticisms they have met with—all this informs us about the transformations in society, in a way that is quite different from that of long series based on theoretically stable procedures, indexes of prices, production or external trade. Often, the historians who use these statistical series deplore changes in the means of recording and classifying objects, since these interrupt their continuity. Sometimes they seek recipes for reconstructing, through approximate estimates, fictitious series that would result from more constant procedures for constructing and classifying data. They thus run the risk of applying anachronistic schemes of perception, breaking the circle of knowledge and action specific to a country or an era (the same is true of international comparisons). Such methods of adjusting series (conceivable only in the proximity of taxonomic discontinuity) will not be proposed here; rather, I shall examine certain of these discontinuities, together with possible historical interpretations.

Industrial Activity: Unstable Associations

The link between forms of administrative or economical action and the knowledge that accounts for them is mutual, because this knowledge results from descriptions carried out on the occasion of such action (lists, records, account books, market-price lists), and because these inscriptions are themselves necessary for stabilizing these actions. The existence of reasonably permanent, solid classes of equivalence is linked to the specific constraints of the *general treatment* of separate operations, and to the standardization entailed by these constraints. In the preceding chapters I have emphasized the general treatment resulting from the activity of the state and its administration, to the extent that the historical (and etymological) bond between statistics and state was a close one, beginning with the relationship between the registry (the obligatory recording of births, marriages, and deaths) and the calculation of the "population trends" in the eighteenth century. But other forms of general treatment came from transformations in the economic world, first with the expansion of commercial dealings at the end of the Middle Ages (banks, currency, prices valid for expanded markets) and then, in the nineteenth century, with the standardization of industry and the appearance of large firms.

Branches of quantitative knowledge (for example, accounting) had thus existed, for use in commercial trading, prior to and independently of those that served the state bureaus. But around 1750 the complex relations between these two worlds of knowledge gave rise to *political economics* (J. C. Perrot, 1992), which discussed ways in which the state could act in the economic sphere, enhancing both its own revenues and the power of the kingdom. Mercantilism and physiocracy were more or less global responses to these questions. In this perspective the monarchic state, from Colbert onward, tried to set up an inventory of the country's economic activities. This inaugurated a long series of statistical surveys, mainly of industry. But the forms of state production and economic action evolved so much that these attempts to describe industry—carried out intermittently in the course of more than three centuries—employed completely different categories as time went by. These categories have been presented by Guibert, Laganier, and Volle (1971). They were among the first to think of examining the changing nomenclatures as objects significant in themselves, and no longer as mere tedious events obstructing the construction of long series of variables connected by economic models.

The field of productive activities was more favorable to such reflective excursions than that of the population. Indeed, the administrative records of births, marriages, and deaths—or even the conscription of men in well-mapped territories—were stable enough to result in simple, robust taxonomies (age, sex, family status, parish, department). In contrast, production and commerce were not recorded in such state records, and their relations with the state were a subject of constantly renewed controversies. For this reason the forms of action taken by the state, and thus of its scrutiny, were constantly evolving, successively addressing different aspects of this activity in the most varied cognitive registers. The criteria for dividing industry would change several times, and despite an apparent stability even words experienced a radical change in content.

At the end of the eighteenth century, shortly before the Revolution, the physiocrats' political and intellectual plan tried to promote a new concept of relations between the state and the freely developing economic activities that fructified nature, the source of all wealth. But freedom of trade, which favored general prosperity and enhanced fiscal revenues, did not preclude the state's having a desire to know about these activities. In 1669 Colbert had already stipulated that "the situation of the kingdom's fabrics be noted in numerical terms"—referring essentially to textiles. In 1788 Tolosan, the general intendant of trade, "drew up a table of the main industries in France, with an accompanying evaluation of the products manufactured by each of them" (SGF, 1847). The nomenclature that Tolosan proposed provided a framework for industrial statistics until 1847. The major categories referred to the *origin* of the raw materials employed: mineral products, vegetable products, animal products. This led to the division of textiles into two distinct groups: cotton, linen, and hemp were counted among the vegetable products, wool and silk among the animal products. "Weaving tapestries" and "furnishing" appeared among the latter. The term "furnishing" did not refer to furniture (made of wood), but essentially to carpets (made of wool), which justified their being classified with animal products. This nomenclature, a product of the physiocrats' economic descriptions, still inspired the appraisal of the industry published by Chaptal in 1812.

In the nineteenth century, surveys of industry were not regular occurrences, although attempts to survey it were made in 1841 and 1861. Later, from 1896 to 1936, the five-yearly population census was used for providing indirect information about economic activities. During the 1830s and 1840s industrial techniques were rapidly transformed, and new machines appeared. The problem was then to identify fresh objects that had not yet been defined or indexed. These entities issued from the old world of "arts and trades" (craftsmanship) and gave rise to "manufactures." The survey of 1841 therefore aimed less at counting than at locating and describing the as yet small number of industrial establishments. The creation of statistics began with the drawing up of a list (a *state:* the polysemy of the word reflects a basic activity of the state and of its statistics), of an index, a catalogue—material forms of the logical category of "classes of equivalence":

> One must carefully examine the general table of patents in every department, and extract from it a list of the manufacturers and factory

> owners whose establishments exceed the category of arts and trades and belong to the manufacturing industry, either because of their nature, their scale, or the value of their products . . . In each *arrondissement,* in accordance with ideas furnished by documents, one must proceed to a detailed survey, the goal of which is to establish by means of numbers the yearly industrial production provided by each factory, firm, or plant. One must only take into account, however, establishments that keep at least ten or so workers busy, excluding those that employ a lesser number, as having generally to fit within the category of arts and trades, which will be explored later. One must collect statistical data relative to the industrial establishments, either by asking the owners or directors or, for want of such information, by officially proceeding to assessments based on public repute or any other means of investigation. To this end one must consult all the educated men who can provide, confirm, verify, or correct the necessary information. (SGF, 1847, quoted by Guibert, Laganier, and Volle, 1971)

In its innovative phase industry rebels against statistics because, by definition, innovation distinguishes, differentiates, and combines resources in an unexpected way. Faced with these "anomalies" the statistician does not know what to do, for they are unlike anything experienced in agricultural statistics, in which "products are easily related to similar expressions":

> In certain manufactures, only one kind of raw material may be used, from which ten different kinds of manufactured products are obtained, whereas in others, in contrast, only a single product is manufactured from ten raw or variously prepared materials. These anomalies present great difficulties to the execution of statistical tables which, essentially subject to the analogy of types, the symmetry of grouped figures, and the similarity of their spacing, cannot lend themselves to these enormous disproportions. Nothing like this had been found in agricultural statistics, the products of the earth being easily related to similar expressions; and neither did this drawback occur in the old attempts at industrial statistics, given that its practitioners had constantly held aloof from obstacles, dwelling only on the surface of things. (SGF, 1847)

Rather than statistics, the survey of 1841 thus presented a series of monographs on various establishments, mainly emphasizing techniques and machines. Though frustrating for historians of macroeconomics, this survey proves valuable for historians of sciences and techniques, since they discern therein the conjunction of scientific discoveries and their translation into industry. This "statistics," in a sense still close to that of the eighteenth century, addressed performances rather than averages and aggregates. While maintaining the general framework established by Tolosan, it did not emphasize large categories. The statistics of the mining industry—for which Le Play, the creator of monographs on workers' budgets (Chapter 7), was responsible in 1848—was also inspired by this minutely descriptive concept, rather than by national additions.

In 1861 a fresh industrial survey presented activities as no longer classified in terms of products, but in terms of their *destinations.* Industrialization was now well under way, and competition with the other European countries, especially England, was intense. Employers had organized themselves into professional unions (unions of employers) according to families of products, in order to defend their positions: protectionist, when selling competing products, but free traders when dominant or when purchasing their raw materials. The Franco-English trading treaty of 1860 inaugurated a period of free trade, and discussions were henceforth structured by the way professional groups were classified, by branches of products (a firm could belong to several unions if it manufactured several products). This system of branch unions endured, even growing stronger during the extreme shortage of the 1940s, and still exists today. Economic activity was then identified according to its products.

Beginning in the 1870s, the configuration changed again. The economic policy of the Third Republic shifted back to protectionism, and the administration no longer felt sufficiently confident in regard to the firms to organize direct surveys of their activities. Information on industry and other economic sectors was produced in a circuitous way, thanks to the integration of the SGF into the new Office du Travail in 1891 (Chapter 5). Thereafter emphasis was placed on employment, unemployment, and the various occupations, which this office had the job of studying. The population census of 1896 included questions on the economic activity and the address of the firm where each person questioned worked, as well as on his occupation. The systematic use of these twenty or so million individual bulletins on active persons was centralized in Paris thanks to the new Hollerith mechanographical machines with their punched cards. It allowed a statistics of industrial and commercial interests to be drawn up, apportioning the active population according to the activity and size of firms and the sex and type of employment of the people questioned. But the rearrangement of activities gave rise to a fresh problem, since information no longer dealt with products, their provenance or destination, but with the trades of the people employed. This prompted a rearrangement of activities and establishments

> so as to separate as little as possible the occupations that are most often grouped together or juxtaposed in practice . . . Neighboring industries resemble one another through the analogy of industrial procedures, an analogy that generally determines the association of several individuals within the same establishment or industrial center. (SGF, 1896, quoted by Guibert, Laganier, and Volle, 1971)

The criterion for rearrangement, therefore, was now the *technique of production,* as reflected by a certain combination of trades and by the "analogy of procedures." The use of the word *"profession"* (occupation) to designate both individual activity and the group of firms employing a family of techniques and branches of knowledge intimated one of the possible orientations of the taxonomy of industrial activities. It remained predominant from 1896 to 1936, in a remarkably stable series of population censuses that presented this statistics of industrial establishments. After 1940 the criteria of products and even of raw materials reappeared at the same time as "branch surveys" carried out by professional unions. Subsequent nomenclatures combined these diverse approaches, relying mainly on a "criterion of association" that tended to group within the same aggregate activities frequently associated within the same firm. The application of this empirical criterion led variously to rearrangements based on products or techniques, depending on whether the firm's identity was defined by the market of an end product made by different techniques or, on the contrary, by specific industrial equipment and knowledge leading to different products (Volle, 1982).

From Poverty to Unemployment: The Birth of a Variable

The shift in the criterion for classifying activities, during the 1890s, from a market-driven logic to an industrial and professional logic can be linked to a change in the way the law and the state codified economic life. Around this period, in fact, the relations between employers and wage earners began to be regulated by specific labor legislation, and no longer simply by the Civil Code, which treated the work contract—inasmuch as it involved the "renting of services"—as a contract of exchange between two individuals. Previously, the state's interventions in business had mainly concerned customs tariffs, and controversies between free traders and protectionists largely determined the statistical formatting of economic activity. From now on, however, questions raised by work and the wage-earning classes were the driving force of original institutions with a statistical vocation. In most industrial countries "labor offices" were created at this point; their role was both to prepare new laws and create the language and statistical tools informing this legislative and administrative practice (Chapters 5 and 6). Thus, for example, the law of 1898 regulating the compensation of work-related accidents, based on the idea of occupational risk established on a statistical basis, was one of the first texts explicitly to define firms as distinct legal entities, different from individual employers (Ewald, 1986). One of the most striking aspects of this development was the social, institutional, and statistical construction of a new object that would gradually replace the idea of *poverty: unemployment* became unimaginable without a statute of wage earners that linked and subordinated employees to the firm.

The histories of social classifications developed and used in Great Britain and France respectively during the nineteenth and twentieth centuries are profoundly different. However, both converge on this point, in this period during the 1890s when new social relationships were being debated—as was a new language for dealing with them, which included the idea of unemployment. The historical concatenations were not the same. Industrialization had begun to take root in Britain in the eighteenth century. People left the countryside and towns grew rapidly in size throughout the nineteenth century, with dramatic results. The rift between classes was viewed and analyzed in terms of poverty, the causes of poverty, and remedies that could lessen its effects. The economic crisis of 1875 was devastating. Riots took place even in the heart of middle-class London. The typically British debates on social reform and eugenicism can only be understood in this context. In France, on the other hand, industrialization and urbanization developed far more slowly. For a long time the old occupational and social fabric, often woven around trade solidarities, continued to influence the social taxonomies. The new laws prepared during the 1890s resulted more from the political compromise effected by the Third Republic in the wake of the Commune (a civil uprising in Paris in 1871) than from direct economic upheavals—as was the case in Great Britain. But despite these subtle differences in historical context between the two countries, the idea of unemployment took shape around this time, henceforth defined as loss of wage-earning employment, which could be taken over and compensated as a risk by specialized institutions.[1]

Before this major turning point, the idea of poverty had been a subject of significant debate in Great Britain from the perspective of constructing statistical tools. Two ways of construing and assessing this poverty were then compared, one (used by Yule in 1895) based on the activities of relief institutions that had existed since 1835, the other through the organization of special surveys, the most famous being those of Booth. In each of these cases the circle of description and action was present, but in a different way. In the first case it was incorporated into the preceding, previously created categories of institutions, almost disappearing in the process. In the second case, in contrast, it structured a complex taxonomic construction that was still being created, the aim of which was to propose measures to eliminate the unfortunate consequences of poverty.

The Poor Law of 1835 was the source of a dual system of relief. On the one hand, the workhouse linked the indoor relief granted able-bodied men to obligatory work and to particularly dissuasive living conditions in a workhouse (the principle of "least eligibility"). On the other hand, assistance was administered in the recipient's home, or at least outside the workhouse (whence the name "outdoor relief") to women, old men, or to the sick. The cost and effectiveness of these systems were debated for decades. Yule intervened in this debate in 1895, seeing it as an opportunity to transfer the new statistical tools of Karl Pearson from the domain of biometrics into social and economic questions. In relying on the calculations of correlation and regression through least squares, he sought to prove that increasing indoor relief did not reduce poverty (Chapter 4). But the data used in assessing both local poverty (indicator of results) and the means theoretically used to relieve it had all derived from the management of local bureaus (Poor Law Unions) administering the relief. This circularity seems to have concerned neither Yule nor those to whom he presented his study. In this case, the identification between poverty and the various branches of public assistance was sufficiently entrenched for the object to be assimilated into the very action that was supposed to fight it. There was nothing exceptional about this assimilation: it can be observed in regard to illness (doctors and hospitals); delinquency (police and justice); or, more recently, unemployment. Already closed, the circle was no longer visible.

Booth's surveys, begun during the 1880s, played a further role (Booth, 1889). They contributed to the criticism of these ready-made taxonomies, precisely because these old forms of action were no longer appropriate. By drawing attention to the characteristics of family incomes (level and above all, regularity of income) they greatly enlarged the field of objects considered in identifying *several* different forms of poverty. These latter derived from different treatments, according to a complex socioeconomic reasoning, because they mixed individual and moral references with other, more macrosocial ones (Hennock, 1976, 1987; Topalov, 1991). The proposed nomenclature was highly detailed (eight hierarchical categories, designated by the letters A to H, virtually dividing up the entire social urban space), because it was intended to demolish the equivalence often presumed to exist among three distinct ensembles: the *dangerous classes;* the *poor;* and *workers* in general. The hierarchy was established on the basis not only of income levels but also of their *regularity.* This prepared the transition from the old idea of poverty to the as yet nonexistent idea of unemployment—the temporary loss of a wage-earning position that guaranteed a regular income. The families surveyed were assigned to these categories (encoding) by "school-board visitors," social workers appointed by the city of London to check that children were attending school. The surveyors took into account a set of indications observed in lodgings. Thus the population was divided into eight categories in terms of percentages, initially in the East End of London, which was thought to be the poorest, then in the entire city (using a less detailed nomenclature).

Category A, which was the lowest, constituted the small minority of individuals who were deemed completely infamous, whose sources of income were either dishonest or unknown: layabouts, criminals, louts, and alcoholics. They generated no wealth and were behind the worst disorders. Fortunately they were not very numerous: 1.2 percent in the East End, and 0.9 percent for the entire city. Close to the preceding and visibly more numerous, however, the persons in category B were from very poor families, with "casual" incomes, in a state of chronic destitution. They constituted 11.2 percent of the population of the East End, and 7.5 percent of the population of London. This ensemble (*A* + *B*) of the "very poor" presented the most serious problems, and seemed largely unable to benefit profitably from locally administered public assistance.

This was not the case with the groups of "poor," themselves subdivided into two, depending on the regularity of their incomes. The constituents of Group C had only intermittent incomes, and were usually victims of competitiveness subject to the vicissitudes of seasonal unemployment. They would have worked regularly, if they could, and they were the ones who were supposed to receive relief, since they were close to achieving stability. They made up 8.3 percent of families in the East End of London. Group D was close to the previous category, but the small though regular incomes of its members were not enough for them to emerge from poverty. They formed 14.4 percent of the population of the East End. The total (*C* + *D*) constituted the "poor," distinguished both from the "very poor" (*A* + *B*) and from those with more comfortable situations (categories *E* through *H*).

These latter categories were arranged into two groups: (*E* + *F*), the "comfortable working class," and (*G* + *H*), the "lower and upper middle classes." Among them, those of borderline category *E* workers with "regular standard incomes" who were above the "poverty line" were by far the most numerous: 42.3 percent in the East End. Group *F* (13.6 percent in the East End) represented the working-class elite. The sum (*E* + *F*) of workers with comfortable or very comfortable incomes constituted 55.9 percent of the East End and 51.9 percent of the entire city. The *G*s formed the lower middle class (employees, liberal professions of the second rank). Last, the *H*s were the upper middle class, those who "could maintain servants." Groups (*G* + *H*) formed 8.9 percent of the population of the East End, but 17.8 percent of the whole of London.

The detailed taxonomy thus defined, and the accompanying percentages, both for the East End and for the whole of London, were important because they supported a precise line of action, and each of the successive divisions had its importance and supported an argument. The boundary between *A* and (*B*, *C*, *D*) isolated a "truly dangerous" (the "louts") but not numerous group: scarcely 1 percent of the population. This enabled the more alarmist arguments to be refuted. Then came the gap between (*A* + *B*) and (*C* + *D*), distinguishing the "very poor"—almost incurable in London itself—from the simply "poor" who, if assured of regular and sufficient income, could join the vast cohort of *E*s, the stable working classes. If, on closer examination, the boundary between *B*s (with "occasional" income) and *C*s (with "intermittent" income) seems rather vague, this is precisely because the only difference between them was the wager that they could or could not make just enough progress to rejoin the *E*s, if only they were given help. Last, the boundary between the two major groups (*A* through *D*) and (*E* through *F*) constituted the "poverty line." The advantage of this latter was that it cut through the identification between "poor" and "workers": those workers who were above the poverty line formed more than half the population of London.

The fact that in the same districts of London, the "very poor"—who could not be helped, unless they moved out ($A + B$)—lived side by side with the "poor," who could conceivably be helped ($C + D$), offered certain disadvantages: not only did the "very poor" set a bad example, but this cohabitation gave rise to unfortunate rivalry in regard to possible jobs and salaries. One solution was thus to remove the "very poor" from London altogether, relieving the East End and making the labor market more favorable to the "poor." This proposal by Booth, which according to him was suggested by the results of his study, later provided the inspiration for surveys conducted in other English cities by Rowntree and Bowley. The aim of these studies was to compare poverty levels in the various cities, thus opening the way for national, rather than local, policies based on the social sciences rather than charity (Chapter 7).

A Continuous, One-dimensional, Hierarchical Social Space

In successfully carrying out an operation that combined taxonomic construction, field surveys, numerical evaluation, and the definition of an action, Booth sought to put into motion objective things, marked and joined together according to a model that was simultaneously moral, psychological, sociological, economic, cognitive, and political. While its hierarchical and one-dimensional categorization was seldom adopted and used directly, especially by governments, its general drift strongly influenced the empirical social sciences in Britain until the present time. Combining the percentages of Booth's eight categories with the supplementary hypothesis (absent in Booth) that this classification reflected a continuous variable distributed according to a normal law, "civic worth," or "ability," Galton was able to standardize this scale and transform it into a measurable quantity. Thus naturalized, Booth's categories were immediately distributed onto the successive intervals of a continuous numerical table, in which every individual theoretically had a place. As in the case of height, this individual attribute was distributed according to a normal probability law. It could be measured by aptitude tests, and was later construed as "general intelligence factor" *g* by Spearman (1903). Its heredity was studied in the same way as that of physical attributes. Thus two taxonomic perspectives were combined. Although Booth, like most of his contemporaries, was intellectually close to the eugenicists, his classification was introduced into a construct that was more economic and social. Galton's project, on the other hand, which rested on a biological description of individuals, was intended to resolve the problem of poverty by limiting reproduction in the more inept classes.

The particular context of the political and scientific controversies in Great Britain between 1890 and 1914 gave rise to a model of social classification that differed from the French model. It represented social space as a continuous one-dimensional scale, summarizing the situation of individuals in a single quantifiable indicator. Categories were like successive gradations along this scale. The rather political and judicial logic of classes of equivalence, which regrouped persons of equal rights, or cases that merited the same general treatment, was opposed by another logic deriving from the life sciences, in which qualities were inherent to individuals, distributed in a continuous manner, and resulted from a measurement. Coding always signified a reduction—in other words, a sacrifice—but in several different ways. A singular case could be reduced either to a discrete category (that is, uniform within it, but discontinuous in relation to others, the ensemble not necessarily being ordered), or to a position on a continuous scale (or possibly on several scales).

The difference between these two elementary mathematical formalisms has sometimes been interpreted as an opposition between a holistic sociology that presupposed the existence of social groups essentially distinct from individuals and another, individualist, more Anglo-American sociology that refused to grant categories such a degree of exteriority. But this interpretation conceals more than it reveals. On the one hand formalism, whether continuous or discrete, was used to shore up descriptions of people corresponding to distinct principles: innate genius, certified scholarly competence, wealth, lineage, creativity . . . (for the differences among these qualities, see Boltanski and Thévenot, 1991). On the other hand, the categories existed once individuals, relying on some or other descriptive principle, established and strengthened links (for example, judicial ones) and produced common objects, making hold a collective thing which, in as much as it was not demolished, could be called a class. But the reality and the consistency of these categories remained, of course, a subject of controversy, the analysis of which constituted a subject of research. From this point of view the formalisms analyzed here, in the form of discrete classes or of continuous spaces with one or several dimensions, played a part in these debates. Emphasis was then placed on the connection among these descriptive schemes, descriptive principles, and the categories of political action based on them, the whole constituting a more or less consistent system each part of which supports the others.

The debates—on poverty and its remedies, on unemployment, on the functioning of the labor market and the means to regulate it—that shook Great Britain between 1890 and 1914 are rich in examples of all this, in that several scientific and political tools were present in a nascent state, and were discussed and competed with one another. These tools had not yet been included in black boxes long since closed, with contents everyone forgot. In British controversies on social questions of this period two major types of models were opposed, of description, explanation, and action. The problem they dealt with was that of poverty and wretchedness, and more generally of what would later be termed social inequalities.

For the "environmentalists," heirs of the public health reformers, the causes of poverty were to be found in the unrestrained urbanization and in the uprooting of populations then piled together in wretched slums where disease, alcoholism, and prostitution ran rampant. The urban environment, habitat, hygiene, and the inevitable accompanying degeneration of moral values explained the accumulation of situations of distress. Descriptions were based on local surveys, on information gathered by the Poor Law Unions. These municipal bureaus were charged not only with relief but also with the registry, with population censuses, and with local statistics on health and epidemics (Chapter 5). The officials of the General Register Office (GRO) who coordinated them from London were close to this environmentalist current, which based its action on local demographic and medical statistics generated by the census, the registry, and the verification of causes of death (Szreter, 1984). For them the idea of a "social milieu" was linked to geographical environment, urban district or rural parish, to the agglomeration of factories and unhealthy dwellings rather than to the more general idea of "social class," defined by equivalent conditions throughout the land, especially by occupational circumstance. In contrast, this idea of social class is found among their opponents, both hereditorists and eugenicists, in a general construct that may seem strange a century later, for nowadays a social group defined on the basis of occupation is thought best to express the idea of "milieu," as opposed to possible individual attributes of hereditary origin. The comparison of these two contexts of controversy, separated by a few decades, shows to what extent scientific and political combinations can have varied, if not opposite, configurations.

For the eugenicists, biological and hereditary attributes were what explained the inequalities in men's situations. Civic value and aptitude were expressed through a person's occupational and social value. For this reason, the ordered scale of Booth's categories could serve as a basis on which to standardize this measure. The heredity of this social scale, as borne out by statistics, is today interpreted in terms of the economic and cultural influence of the original milieu. But around 1900, it was read more as a proof of innate and biological character. In the nascent empirical social sciences, the fact of belonging to an occupation appeared an indicator of aptitude inherent to an individual, his biological lineage, and his race. This interpretive grid exerted a strong influence in research in the human sciences until the 1950s or later, mainly in the interpretation of matrices termed—depending on the point of view adopted—social heredity or social mobility (Thévenot, 1990). This political trend was opposed to measures involving social assistance and support for the poorest, in that these measures increased and strengthened the most inept segments of the population. It even went so far as to recommend that certain handicapped or ill-adapted persons be sterilized—a measure that actually became legal in a few countries, even outside Nazi Germany (Sutter, 1950).

The controversy between the two currents of thought is a complex one, and its terms underwent several changes. It is mainly revealed through differences in the use of statistical tools and nomenclatures. Before 1905 or thereabouts social reformers, allied to the statisticians of the GRO, used mainly geographical data, relating (for a detailed breakdown of the country) demographic, economic, social, and medical statistics. Through these spatial correlations they showed the links among the various components of poverty. Statistics that covered the territory in detail and involved a sizable administrative infrastructure (that of bureaus of assistance and of the life registration) allowed these comparisons, which were well suited to local policies still administered by parishes or counties. The eugenicists, in contrast, used statistical argument in quite a different way. They tried, in their laboratories, to formulate general laws of heredity with the help of new mathematical methods. They drew conclusions that were not specific to particular localities, but were valid for the English nation as a whole. They formed political alliances at the highest level and their statements were enhanced by the prestige of modern science, whereas their opponents projected a more obsolete image: social relief was still linked to charity, churches, and the conservative tradition of resisting science and progress.

But the debate changed its configuration between 1900 and 1910. A new generation of social reformers, often with ties to the Labour party, henceforth expressed the problems of poverty in terms of regulating the labor market and passing laws to provide social protection, rather than relying on local charity. An example of this evolution is the debate over the "sweating system," a then frequent form of organization of production. Work was subcontracted by the employers to intermediaries (subcontractors) who then recruited the necessary workmen themselves. This system was often denounced, and the subcontractors were considered shameless exploiters. Parliamentary commissions studied the question, and concluded that subcontractors could not be held responsible for flaws that were inherent to labor legislation itself. As in France by other means, laws were thus passed providing a more precise definition of the wage-earning work force and the duties of employers, and defining unemployment as a break in the bond between wage-earning workers and their bosses—a bond previously obscured by subcontracting. Thus the statistics of the active population, wage-earning labor, and unemployment assumed a consistency quite different from nineteenth-century statistics.

The evolution of the categories of social debate—henceforth centered on policies more national than local—also had repercussions for the categories of statistical description developed by the environmentalists of the GRO. Their eugenicist opponents emphasized analyses in terms of social classes, dealing for example with the birth rate and infant mortality, measured over the entire country. Around 1910 this mode of discussion, hinging on nationwide statistics, was more in keeping with the public debate in general than it had been in 1890. This nationalization of the space of description and of action prompted the statisticians of the GRO to construct in their turn a nomenclature of social classes, used for the census of 1911 (Szreter, 1984). Like Booth's nomenclature, it was hierarchical. It comprised five classes, arranged around the boundary line between the middle and working class. In the middle classes a distinction was drawn between *professionals* (I) and so-called *intermediary* groups (II). Among the manual workers, there were *skilled* workers (III); *semi-skilled* workers (IV); and *unskilled* workers (V).

This scale of five classes was dominated by the typically British (subsequently, typically American) group of people who pursued a *profession,* in the English sense of an activity involving both a specific training at a university level and a common culture based on the awareness of its usefulness for the collectivity. The hierarchy governing this categorization was based on a quality closer to the eugenicists' idea of aptitude and civic worth, rather than on attributes of wealth and power, which could underlie an economic or sociological classification. The one-dimensional, "professional" (in the English sense) structure of this nomenclature left its mark on the majority of social classifications subsequently used in the English-speaking countries—until today. Later, a group of *managers* was introduced into category II, but until quite recently (1990) employers were placed after the *professionals*. This has been partly forgotten now; yet through this set of features (one-dimensionalness, hierarchy, continuity, implicit reference to an individual's social value) this taxonomy retains a trace of the scientific and political constructs of the eugenicists of the turn of the century and of surrounding debates. This distinguishes it from the French and German classifications, generated in entirely different political and cultural worlds.

From Crafts to Skilled Jobs

The French socio-occupational nomenclature is marked less by a particular social philosophy, like the English one, than by diverse stages in the history of the organization of labor during the past two centuries. The succession and partial conservation of these strata from the past partly explain the apparent multiplicity of the criteria employed. Moreover, this very heterogeneity has made it possible to conceive and describe a social field that has several dimensions, is more complex than the Anglo-American scale of aptitude, and is closer to Buffon's method than to Linné's system. We may summarize the long history of this nomenclature in three phases, which explain the current characteristics of each. The first phase was still marked by the structure of trades, in the old sense of the word. The second phase, after the 1850s, witnessed the emergence of the distinction between salaried and nonsalaried workers. The third was characterized, after the 1930s, by a hierarchy of the wage earners' force, codified by conventional grids linked to the training system (Desrosières, 1977; Desrosières and Thévenot, 1988).

Despite the abolition of corporations in 1791, the social organization and vocabulary of the *métiers* (both crafts and trades) remained prevalent in the nineteenth century (Sewell, 1983). In 1800 Chaptal's questionnaire was sent to the prefects (Bourguet, 1988; Chapter 1 above); like Tolosan's questionnaire on industrial activities, it bore the mark of physiocratic thought, distinguishing persons according to the source of their incomes: the land, the state, "industrial and mechanical work," and all the others, "unskilled labor." The enormous third category included all those who—masters or journeymen, doctors or men of law—had the common characteristic of practicing a trade based on skill acquired through apprenticeship, a trade from which their specific income or position derived. Everybody else—unskilled workers, servants, or beggars—formed a "separate class." The separation between masters and guildsmen, or journeymen (which later became the distinction between non-wage-earning and wage-earning workers), was not yet pertinent. Nor was the aggregate of a "working class," which only began to be conceived as an entity after the uprisings of 1832 and 1834, or the junction of journeymen and manual laborers, distant ancestors of "skilled" and "unskilled" workers respectively. In contrast, the group of "those employed by the state" survived in France, despite occasional eclipses.

The corporatist organization by crafts and trades shaped a view of the social world that formed a constant background against which occupational taxonomies were situated. Based on the family transmission of knowledge and patrimonies, the distinction between master and journeyman long embraced the father-son model, and was only gradually transformed into the employer-employee relationship of twentieth-century labor legislation. Engendered in the nineteenth century by the analysis of capitalist relationships, theories of social classes—of which Marxism was a systematic development—ignored this model of familial transmission, since they were partly constructed to oppose it. This model did, however, survive in several ways: either through the apologia that certain Christian conservatives, disciples of Le Play, later turned it into (Kalaora and Savoye, 1987); or, on the contrary, in the sociological criticism of the 1960s and 1970s, which criticized the injustice of economic and cultural reproduction as effected by means of the family. This form of social bond is still important in the underlying characteristics of current nomenclature. During the nineteenth century it essentially took the form of a list of "professions," in the French sense of the old artisanal and commercial trades. In the censuses of 1866 and 1872, for example, it served to evaluate the "number of individuals that each profession provides with a living, either directly or indirectly." The professions were listed horizontally in tables, while vertical columns listed "individuals who really practiced professions; their family (relations living off the work or fortune of these individuals); and servants (in their personal service)." And yet in 1872 a third distinction appeared, crossed with the preceding ones, between "heads or employers, clerks or employees, workers, day laborers." Three points of view were combined. The first two were linked to the familial trade structure. The third was close to the opposition between master and journeyman. But the distinction between "workers" and "journeymen and hard laborers" still appears in 1872 and 1876: the construction of a working class that included the "unskilled workers" of the future was not yet apparent.

The importance of this trade structure was also shown by the fact that, until the 1940s, the two nomenclatures known today as *individual activity* and *group (or sector) activity* remained the same, although one in principle classified people and the other, businesses: the identification of baker-bakery or doctor-medicine was typical of the world of trades. However, the internal divisions of a firm between employers and workers, then between manual workers and nonmanual employees (that is, between blue- and white-collar workers), had appeared previously, but constituted another, transversal cut of "positions *within* the occupation." The distinction between individual and collective activities was not consistent with the logic of trades. This led to a mixture of taxonomies, until the clear definition of salaried labor, structured into categories codified by rules of law and collective conventions, imposed a decided mark. This happened in two stages: the emergence of an autonomous labor legislation, at the end of the nineteenth century, and the extension of hierarchical grids of skilled jobs defined in terms of training, between 1936 and 1950.

Previously, the distinction between employers and workers was not always clear. As in Great Britain, many small producers worked as subcontractors. They were then simultaneously dependents of people giving orders, and employers of journeymen. The silk workers of Lyons formed a very different social group from the rich "manufacturers" who provided them with the material and sold the finished products. In the building, the workers were given a task and then recruited other workers. The census of 1872 shows, within the actual "occupational situation" of the employers, a category of "worker leaders attached to arts and trades," distinct from workers and properly so-called journeymen. These intermediary situations were still reflected by the presence, in the continuous series of censuses between 1896 and 1936, of an important category of *isolés*. This category, distinct both from that of employers and of workers, grouped very small market (nonsalaried) producers—agricultural, craft-related, or commercial—and people who worked at home, receiving the raw materials and addressing the task. Those commenting on the censuses themselves expressed doubts as to whether these "isolated ones" (23 percent of the active population in 1896, and still 14 percent in 1936) were closer to the workers or the employers. But gradually these intermediary states, in between wage-earning and non-wage-earning workers, became less numerous, as labor legislation and fiscal administration drew clear distinctions between these forms of employment and income.

Thus, in transverse relation to a list of occupations presented in horizontal rows, the censuses taken between 1896 and 1936 gave, in vertical columns, five "positions within the occupation," ancestors of the present socio-occupational categories: the heads of an establishment; the employees (white collar); the manual workers (blue collar); the "unemployed workers"; and the *isolés.* The last major evolution took place between 1936 and 1950, and involved the definition of *levels* of salaried employment. These were ordered according to the durations and types of training received, and inscribed in collective conventions of branches, or in the "Parodi categories" of 1946 (named after the then minister of labor) for the business sector—or even in the general statute of civil service, for administrations. This new criterion introduced in part a one-dimensional hierarchy, absent from the previous taxonomies, which distinguished manual workers and employees among those receiving salaries. Among the manual workers, the difference between craft workers and the others had previously existed only indirectly: in the occupations listed in rows, there was first a detailed list of denominations of crafts and trades—sometimes archaic ones—then subsequently, in a single row, appeared "manual laborers, day laborers, hard laborers." The employees, for their part, still included engineers, technicians, and accountants. No specific grouping corresponded to the category of *"cadres"* or *executives,* which was recognized—socially, in terms of unions, and finally in terms of statistics—between 1936 and 1950, and which today forms an essential part of French social taxonomy (Boltanski, 1982).

Collective conventions and grading scales developed after this period were intended to define standard categories of employment, relating training as guaranteed by nationally valid diplomas to jobs, salaries, rules of promotion, welfare systems, and means for electing personnel representatives. In the branches of major industry (beginning with metallurgy), the specific vocabulary used in these texts would partly replace the old vocabulary used in trades. But in the statements of occupation recorded in the census these two vocabularies coexist: one finds OS2s, P3s, blacksmiths, and joiners (Kramarz, 1991). The large categories of collective conventions—and especially the three electoral colleges of delegates to business committees—served as a basis, around 1950, for the construction of the socio-occupational categories by INSEE (Porte, 1961). These colleges were the manual workers (themselves subdivided by collective conventions into "manoeuvres," or unskilled workers; "ouvriers spécialisés," or semiskilled workers; and "ouvriers qualifiés," or skilled workers); the "employees-technicians-low-level-executives"; and last, the "cadres," or actual executives. This electoral procedure and these colleges helped clarify boundaries that previously were often rather vague. They were also an element in the creation of nationwide equivalences, both between regions and between sectors, between positions previously described in specific vocabularies that were local and incommensurable. But this standardization did not extend to the entire socio-occupational space, imposing a sole classificational criterion on it. Other qualifying principles, emerging from a long history, were mixed with the principle of job qualification by means of training and national diplomas.

Four Traces Left by the French Revolution

The structure of the nomenclature reflects the history of the original manner in which, in nineteenth- and twentieth-century France, social links were woven and consolidated on the basis of professional solidarities and antagonisms. More precisely, a number of characteristics of the indivisible whole formed by the French social structure and its terminological representation derived from particular features that dated back to the Revolution of 1789. Thus the identity and consistency of four contemporary social groups—farmers, clerks, workers, and "cadres" (executives)—can be related respectively to the sharing of agricultural land; the establishment of a unified state; the influence that a specific civic language had on the workers' movement; and last, the creation of engineering schools linked to the state and distinct from the university.

The division of land in various legal forms, such as rented farms, sharecropping, or farming by the owner, allowed small-scale farming—more widespread than in other European countries—and a crowded rural life to be maintained. The population of active persons living in very small communes reached its apogee around 1850 and only decreased gradually thereafter, whereas in England during the same period the countrysides had been emptied, often by force. Associated with this rural life was an economy in which craftsmen, small industry, and small trade featured prominently. The rapid industrialization and unfettered urbanization that characterized early nineteenth-century England—and to a lesser extent pre-1914 Germany—were less apparent at the time in France. This was responsible for the images of "moderation," the "happy medium," and "rational progress" claimed by the Third Republic and its radical socialism, images subsequently denounced during the period of growth that occurred between 1950 and 1975. The explicit presence of groups of farmers and employers, as distinct from the salaried workers, separated the French socio-occupational nomenclature from its Anglo-Saxon homologues, which had no such distinction. It was the sign of a historical permanence, affirmed and asserted by specific representative organizations.

National unification and the establishment of a centralized state had the effect of enabling a civil service with a heightened civic sense to be created, which was separate from the local clientele networks: the prefectoral body, teachers, fiscal administration, the army, the postal service, state engineers, judges, and statisticians. The national character of this civil service—created through recruiting, training, and a geographical sweep of assignments—had important consequences for its criteria for assessing situations and day-to-day decision making. The existence of such bodies, endowed with strong cultural homogeneity, helped make it possible to establish a statistical system that endowed social debates with elements of a common language, of which statistical classifications were one component.

The particular civic language resulting from the Revolution thus helped fashion the characteristics of the French workers' movement, with its emphasis on equality, the importance of state power, and revolutionary upheaval. A peculiarity of the French social corps was that other social groups were often conceived and organized according to a trade union model inspired by that of the workers' movement—even when the ostensible purpose was to distinguish them from it. This involved transferring a system of values and comparable modes of organization and representation from the trade union movement onto the other groups. With this hypothesis in mind we may read the history of the trade unions of executives, educators, and even farmers, employers, and the liberal professions. It was precisely these forms of organization and representation that shaped a vision of French society in terms of "socio-occupational groups," traces of which are found in both social and political life, and in the tables of official statistics. The widespread use of this taxonomy, both in common parlance and in specialized works, distinguished France from other countries such as Great Britain, the United States, or Germany, which had different political traditions. In particular this explains the distance between this composite representation and the continual, one-dimensional scale of the Anglo-Saxons: the groups existed with distinct identities, for all that the historical work of social construction, both trade unionist and political, took place and still produced effects.

Last, the creation of state engineering schools, begun under the Ancien Régime with the École des Ponts et Chaussées (School of Bridges and

Roads), was partly responsible for a specific characteristic of the French state: the fact that socially recognized technical skills were often internal to its mechanisms, whereas in the Anglo-Saxon countries the same skills tended to be external to the state, appearing, for example, in the form of *professionals,* a term untranslatable in French. This relative importance of the engineering profession in France, even in the private sector, partly explains the appearance, in the late 1930s, of a social group of executives formed around the kernel of engineers: *cadres,* a term not easily translated into English. The classifications reflect these different ways of qualifying people. In France, salaried executives were distinguished from employers and PDGs, whereas the English and Americans lumped salaried *executives* and employers under the title of *managers.* In France, the significant demarcation line was between salaried and non-wage-earning workers. In the English-speaking countries, it was between academic skill and power—even if, naturally, a number of *professionals* (especially engineers) worked as employees for firms. An interesting borderline case, again in France, was that of the *liberal professions,* a cross between the two taxonomic principles of specific competence and type of income. They were a sort of French equivalent of the Anglo-Saxon model of *professionals,* since until the 1930s or thereabouts the term "liberal profession" could still encompass wage earners such as educators in the public sector. This is shown by the examples given on the reverse of the census bulletin between 1896 and 1946.

The multidimensional character of French social classification, a product of the strata in its history, became apparent when the factor analysis of correspondences (Benzécri, 1973) was applied to the results of surveys about social categories: from 1970 on, this allowed representations to be made of the social space, the axes of which corresponded to the various taxonomic practices. By this expedient the two perspectives of research on classification—often said to have turned their backs on each other—were able to meet. The analysis of words used to state an occupation, and their rearrangement into classes, derives from the first perspective, whereas the distribution of points representing individual cases may lead to typological constructions, which corresponds to the second perspective. In these analyses of data, the most explanatory factor (in terms of variance analysis) is a combination of the levels of income and training, close to the single British scale, but it is often interpreted differently. Moreover, a second factor, *transverse to the preceding one,* opposes non-wage-earners to wage earners, and wage earners in the public sector (especially educators) to those in firms. The social, religious, cultural, and electoral behavior and opinions of the various groups are often distinguished better according to this second factor than according to the traditional social scale of the first factor (Bourdieu, 1979). Furthermore, these statistical methods allow the particularity of individual cases to be preserved in representation, since these cases can appear as particular in the charts that best summarize a space of data comprising many dimensions. Even the names people use to cast light on their occupations can be reproduced in these graphics. Like a geological cut showing past strata, these charts show different linguistic spaces, corresponding to various moments in the history of the social taxonomies. Thus in one zone of the factorial chart we find joiners, bakers, train drivers, and manual laborers, whereas the opposing zone groups together machine operators, P3s, and OS2s (Desrosières and Gollac, 1982).

In France the combination of taxonomic principles dating back to different moments can be seen in the socio-occupational terminology, an instrument of synthesis much used in the empirical social sciences. The history of Germany, however, which is richer in radical discontinuities, has not allowed the old criteria to be conjoined with the new ones to nearly the same extent. Yet these criteria, products of a very dense social history, survive like eroded bluffs, which the German statisticians—sociologists and labor economists—cannot succeed in integrating into the Anglo-American kind of social taxonomy largely adopted in empirical works. These old objects, historical curiosities, as it were, of German sociology, are the *beamte* (civil servants), the *arbeiter* (manual workers), and the *angestellte* (employees). Each category has corresponding rules, forms of welfare system and retirement pay, and representative organizations. In exchange for strict duties of loyalty and obedience to the state the *beamte*—products of the civil service of eighteenth-century Prussia—have high guarantees of employment, confirmed during the 1950s after the establishment of the Federal Republic. The *arbeiter* are manual laborers, historically represented by powerful, specific unions and by the Social Democratic party. Last, the *angestellte* comprise the nonmanual wage earners working for firms: this corresponds, in France, not only to the "employés," but also to the intermediary occupations and *"cadres."* The clear definition of this group dates back to the 1880s, when Bismarck instituted the first laws ensuring a welfare system (Kocka, 1989). Nonmanual workers wanted to be distinguished from manual workers and their trade unions and political organizations. They thus formed themselves into a group, adopting the old model of the *beamte* (clerks). Their loyalty to the firm, they felt, had to be distinguished from the revolutionary and assertive spirit of the workers. The presence, in the census questionnaires, of these various titles—which German statisticians and sociologists seem to consider anachronisms—is a hangover from this period.

One or Several Urns: Taxonomy and Probability

In 1893, at the International Congress of Statistics held in Chicago, the Frenchman Jacques Bertillon presented two plans for international classifications, the aim of which was to harmonize the definitions of statistical coding in two different domains: *occupations* and *diseases*. The first was justified not only by the problems connected with labor legislation but also by demographic questions. Death rates varied according to the professions of the deceased: certain causes of death occurred more frequently in certain occupations. Attention had been drawn to the excessive death rate of workers in the printing industry. Cases of death through saturnism (lead poisoning) had been pointed out in France. And yet the statistical tables of causes of death according to occupation did not confirm this result: the number of printers who died of saturnism was insignificant, whereas phthisis occurred twice as often among printers as among ordinary men.[2]

The second plan for international classification presented in Chicago in 1893 was entitled *Three Plans for the Nomenclature of Illnesses (Causes of Death—Causes of Inability to Work)*. In the discussion of death according to profession, Jacques Bertillon (a doctor and statistician, grandson of the botanist and demographer Achille Guillard, and son of the doctor and statistician Adolphe Bertillon) thus established a link between two quite different taxonomies, dealing with occupations and diseases. More precisely, the occupation appeared here as a *risk factor,* that is, as the definition of a subwhole within which a random event (disease, inability to work, death) was more likely than for "most men." This point of view joined that of the optimization of a classification in relation to certain descriptive criteria: the best division was the one that increased differences between classes and reduced internal differences. But it specified this in cases where the analytical criterion was seen as a risk, a probabilizable event, thus joining a tradition of statistical thinking already pioneered by Laplace, Poisson, and Cournot (Chapter 3)—even though Bertillon, a man of action and an administrator, did not often refer to their philosophical speculations. Because it was directly associated with medical practices, whether clinical (therapeutic) or public health–oriented (prevention), this approach drew attention to the conventions used in creating classes of equivalence constituted by risk factors. Thus not only occupation, age, or sex could be seen as risk factors—inasmuch as they were categories differentiating the probability of death—but also disease itself, in that it did not always lead to death. Information that was *useful for action,* provided by the identification of a cause of death, was precisely what influenced the delicate problem of choosing, from among all the events preceding death, the one that would be deemed *the* cause of death. The coding of death certificates was a procedure standardized by the instructions of the World Health Organization (WHO), the result of a long history and complex controversies.

The International Classification of Diseases (ICD) and causes of death then adopted by the ISI under the name of the "Bertillon classification" was still in use until the 1940s, though subject to regular revisions every ten years. These revisions were first the responsibility of the French (through the ISI) until the 1940s, then of the WHO in Geneva after 1955. Even before the adoption of the Bertillon plan in 1893 this terminology had been debated forty years earlier, at the first International Congress of Statistics held in Brussels in 1853 at Quetelet's prompting. Two conflicting points of view had been presented, concerning whether to give priority to the "etiological principle" (seeking the initial cause) or to the "topographical principle" (noting the symptoms and their localization). The first, defended by the Englishman William Farr (founder of the GRO) was naturally of greater interest to the epidemiologist; but the second, upheld by the Genevan Marc d'Espine, was more easily applied by the doctor filling out the death certificate. The two points of view were not entirely opposed, since each was very conscious of the two constraints: usefulness, and ease of application. It was a matter of hierarchizing the criteria: "First Farr applied an etiological principle, then a topographical one, whereas d'Espine subordinated etiological division to topographical division, and then topographical division to division according to an evolutive mode" (Fagot-Largeault). In 1853 Achille Guillard (the grandfather) presented a resolution in Brussels asking Farr and d'Espine to reach an agreement, but they proved unable to do so. Nonetheless, the problem remained in the Bertillon family. First Adolphe and then his son Jacques worked on it, in their capacity as officials in the Paris bureau of statistics. The solution proposed in 1893 was ambiguous. Although claiming to be in the tradition of Farr (perhaps to placate the English), Bertillon in fact rejected etiological classification (according to initial causes): if one referred to it—he opined—one ran the risk of relying on uncertain theories based on provisional hypotheses, resulting in statistical returns which "in a few years would be unusable or give rise to laughter." He constantly repeated this argument—as, for example, in 1900:

> An etiological classification is no doubt more satisfactory to the scholarly mind; but at the present time, at least, it seems impossible to adopt it, for before very long it would certainly cease to be applicable. (Bertillon, 1900)

So rigid was his point of view that in 1920, at the Paris meeting of the International Commission charged with the decennial revision, twenty-five years after Pasteur's death in 1895, Bertillon still considered bacteriology a mere passing fashion, like all the others:

> It is all the more necessary to take the anatomical seat of diseases as a framework for classification, in that this is the sole classification that does not vary. Fifty years ago diseases were divided into fevers, inflammatory diseases, trophic, diethical, dietetic diseases . . . divisions long since obsolete. Today, it is bacteriology that forces itself upon the pathologist; we can already glimpse an era when an excess or lack of internal secretions will be what captures our attention. Shall we need to change our terminology each time? (Bertillon, 1920)

After hailing Bertillon at length for his thirty years of assiduity (he died in 1922), the same commission nonetheless adopted a proposal to classify general diseases in accordance with an etiological principle, arguing that "even if the causal agent has not yet been identified in every case, certain diseases are so obviously infectious in nature that it is not erroneous to connect them with those of which the infectious agent is known." Once the principle of coding according to initial cause was, at least in theory, retained, the question of the methods of recording and encoding this cause arose. This was debated during the 1920s and 1930s. Several questions were raised. How should the death certificate be formulated? Who should select the actual cause of death, among the various causes listed? By what procedure? In addition, what steps should be taken to ensure that all this was identical in every country?

Michel Huber, the director of the SGF, made a report on these points in 1927 with a view to the fourth revision of the international classification. How many questions should the certifying doctor be asked? And in what order (chronological, or starting in the present)? How far back into the deceased person's past should the questions go? If in chronological order, this might "cause the doctor to certify old facts he has not himself observed . . . In asking him what, in his view, was the principal cause, one is asking him to state the cause that must be taken as a basis for statistics." This debate turned on the question: Who should bear the cost of encoding? Who assumed the dual responsibility, cognitive and economical, of reducing the diversity and uncertainty, and of constructing categories on which to base action? In 1948 an international model of death certificate was adopted, comprising four questions: the immediate cause; the intermediary cause; the initial cause; and other pathological states (concomitant causes). The French model only included three of them, omitting the "intermediary cause." Theoretically, but for explicit instructions specific to certain particular cases, the "initial cause" was what served as a basis for the encoding. This choice was justified because it was the "most useful" information from the point of view of public health: "To prevent death, the important thing is to interrupt the chain of events or to begin treatment at a certain stage." (*International Classification of Diseases*, 1948)

Thus, of the long chain of events that preceded death, one event had to be chosen. The expression "initial cause" is deceptive: the sick person smoked, worked in a mine, liked riding a motor bike—was this an initial cause? Moreover, one could not put the *cause* of death in the same category as *death itself* (the heart stops beating). The criterion of choice seemed to be as follows: among the events, one retained the one that significantly and clearly increased the probability of death, without, however, this probability being equal to 1 (certain death). What mattered was the *variation* of probability, clearly different from both 0 and from 1. The analysis of causes specified the urns in which random drawings were carried out, with a view to some future intervention: "The convention designating the medical causes of death is that . . . cause is what gives us a foothold if we have as our objective to struggle against disease or death" (Kreweras). The taxonomic convention was clearly linked to a system of conditional probabilities that guided action, in relation to an average death rate—the probability of death in the absence of any information. Insurers are familiar with this tension between the two opposing positions, involving either one or several urns: either one demands the same insurance premium for everyone or, on the contrary, one specifies the premiums in relation to a host of different imaginable probabilities.

On the occasion of the sixth revision of the ICD, in 1948, it was decided, after lively debates, to eliminate old age from the list of causes of deaths. This choice is deeply connected with the foregoing circumstances. During old age it becomes more and more difficult to distinguish causes clearly—that is, to calculate conditional probabilities attached to well-identified states, on which specific actions may hinge. But old age also often confronts the certifying doctor a posteriori with a chain of events from which it is difficult to choose. The difficulty results from having to add together, in statistical tables, the causes of death for old people, persons of mature age, and young people, because the action cannot—at least from certain points of view—have the same significance in every case. Statistical equivalence here triggers an insoluble debate between distinct and incompatible moral principles, even though each is coherent and legitimate (Fagot-Largeault, 1991). According to one (the deontological principle) each person has a unique value that is incommensurate with any other. One cannot balance the life of an old man with that of a young man. According to the other (teleological) principle, a common good exists that is superior to individuals, justifying arbitration on the part of the collectivity, especially in allotting limited economic resources to potentially unlimited public health actions.

Making a Story Hold

Because medicine is continually stretched between treatment in the singular and treatment in general, its long history—and the history of its methods of observing and generalizing—highlight numerous moments of statistical practice: the selection of pertinent traits; the constitution of categories; modelization with a view to taking action. Many nineteenth-century statisticians (Farr, the Bertillon family) were doctors. The controversies concerning Dr. Louis's numerical method, or the hygienists' use of averages, hinged precisely on methods of identifying and naming cases (diagnosis), and of intervening on the basis of knowledge previously accumulated in taxonomies (Chapter 3). The negotiated construction of an international classification of diseases was a stage in the formation of a common body of knowledge. Anne Fagot-Largeault's detailed analysis (1989) closely connects the two stages of classification (the debate over etiological and topographical forms) and encoding (the form and treatment of the death certificate). It can be adopted in the perspective—so typical of statistical procedure—of the systematization of these two stages: one, of the actual nomenclature (Linné's goal) and two, of the encoding (by more or less automatic algorithms).

In so-called traditional medical practice, in which "medicine is an art," the doctor recognizes and names a disease on the basis of his own experience and that of his predecessors (D'Amador): "This is a typhoid case." He combines the symptoms he observes and the situation he witnesses in a single word, which lends consistency to the story and makes it hold, assimilating it with other, identical case histories of previously known consistency. This basic act of recognition and designation ("this is a . . .") calls on previous knowledge, giving it new life by reactivating a category—just as a path only survives if it is regularly taken. But the taxonomic principles behind knowledge thus accumulated through experience are often local and heterogeneous, and doubly so. The criteria of observation and generalization varied depending on the type of disease (Linné's style of criticism), and on the doctors and medical schools. Nineteenth-century debates—notably in regard to the classification of sick people—reveal an attempt both to systematize classification and to develop a language common to the entire medical world. Thus the network that made particular stories hold grew broader, and entailed increasingly solid references. But this development came at a price: part of the knowledge resulting from the practical skill and intuition of the doctor-artist, applied during individual consultations, may be considered lost: sacrificed in order for the case to be inserted into the network of general, solid categories, of the instruments of measurement and analysis with which medicine has equipped itself (Dodier, 1993). This tension is not particular to medicine. The practice of law, of tribunals, of judicial expertise, also seeks to make stories and situations hold, identifying and naming them in order to record them in more general forms (as, for example, "professional misconduct" [Chateauraynaud, 1991]).

But in emphasizing what makes a configuration stable, particularly by resorting to standardized tools and systematic taxonomies, one can no longer contrast—as has sometimes been done (notably by Durkheim)—universal and scientific knowledge contained in a scholarly project to other forms of knowledge said to be indigenous, local, partial, nonsystematic, or action-oriented. Quite the reverse: inasmuch as they are used by turns in daily battles whose aim is to make obvious and indisputable a particular understanding of the world, the diverse modes of knowledge can only be treated symmetrically, without any one of them being endowed with an a priori privilege. This methodological choice has nothing to do with a denunciation of the illusion of science (or of statistics), in the name of other knowledge unjustly abased and unrecognized. It tries only to shed light on complex situations in which scientific or statistical resources are called on to rival or to complement others, and thereby tries to understand what makes them hold, and what may constitute proof. How is a consensus established, or not; or an agreement as to how to give account of a situation? What is the role of statistical formalisms in the toolbox of instruments of proof and convincing arguments?

The systematization and automatization of the procedures offer great advantages. This is true not only from an economic point of view, in terms of costs—for example, of encoding—but also from the standpoint of seeking agreement, of objectifying a meaning common to the various actors, by detaching these processes from personal intervention and programming them into machines. The establishment of agreement is then shifted to the negotiated construction of the actual machinery. Even then, however, controversy regarding these mechanisms can always be triggered anew. The diagnostic algorithms produced by expert medical systems are a case in point. The question of their validity and their performance is still undecided, as is shown by this quotation from Anne Fagot-Largeault; indeed, there could be no better conclusion to a chapter on taxonomy and encoding:

> How well do these algorithms of imputation perform? To judge this, one must know by what standard to evaluate these performances. At first sight, the best algorithm is the one that most often approaches the truth; in other words, that assigns in every case the degree of causality that best corresponds to reality. But we do not know the truth, since that is what we are looking for. The best algorithm is therefore the one whose imputations coincide with those of the best expert. But who is the best expert? Fine: the best algorithm is the one that judges like a consensus among experts. And if there is no consensus among experts? Then the best algorithm is the one that creates consensus, by combining the viewpoints of the experts who agree to submit to it. Good; but what if all the algorithms do that, each in its own way? Then the best algorithm is the one whose judgments differ least from those of the other algorithms. Unless no better algorithm can be found? (Fagot-Largeault, 1989, p. 360)

9

Modeling and Adjusting

The history of economic analysis sometimes tends to confuse three developments: its mathematization, its quantification, and its recourse to the language of probabilities. All three would seem to signify a reconciliation between economics and the so-called hard sciences (notably physics); also, a break with political philosophy and literary disciplines, from which political economics was born. Not only are these three tools different, however, but they have long been deemed incompatible. The difficulties attendant upon their union, during the successive episodes of the construction of econometrics between 1900 and 1950, show the enormity of the work required to combine forms issuing from traditions that were profoundly foreign to one another. The accompanying figure summarizes a few of these affiliations in genealogical form, pinpointing the moments when previously distinct—even contrary—contributions were combined. Around 1900, four could be identified (only the second and the third have been described in the preceding chapters): mathematical economics; descriptive historicist statistics; statistics as a technique for analysis (then developed by biometrics); and probabilistic mathematics as a language allowing the consistency of the induction to be tested.

Each of these four lineages could be the topic of a different historical narrative, construing distinct facts and emphasizing episodes that have no connection among themselves. For example, the history of the linear regression model follows the path (shown by framed areas in the figure) that describes the formulation of least squares, the normal law, averages, regression, multiple correlation, and maximum likelihood. The works of Stigler (1986) and Gigerenzer et al. (1989) enable us to reconstruct it, while the book by Morgan (1990) shows the use that model was put to by the first econometricians. We can also draw attention to meeting points, when the translations needed to connect a priori heteroclitic things had not yet been

A Genealogy of Econometrics before 1940

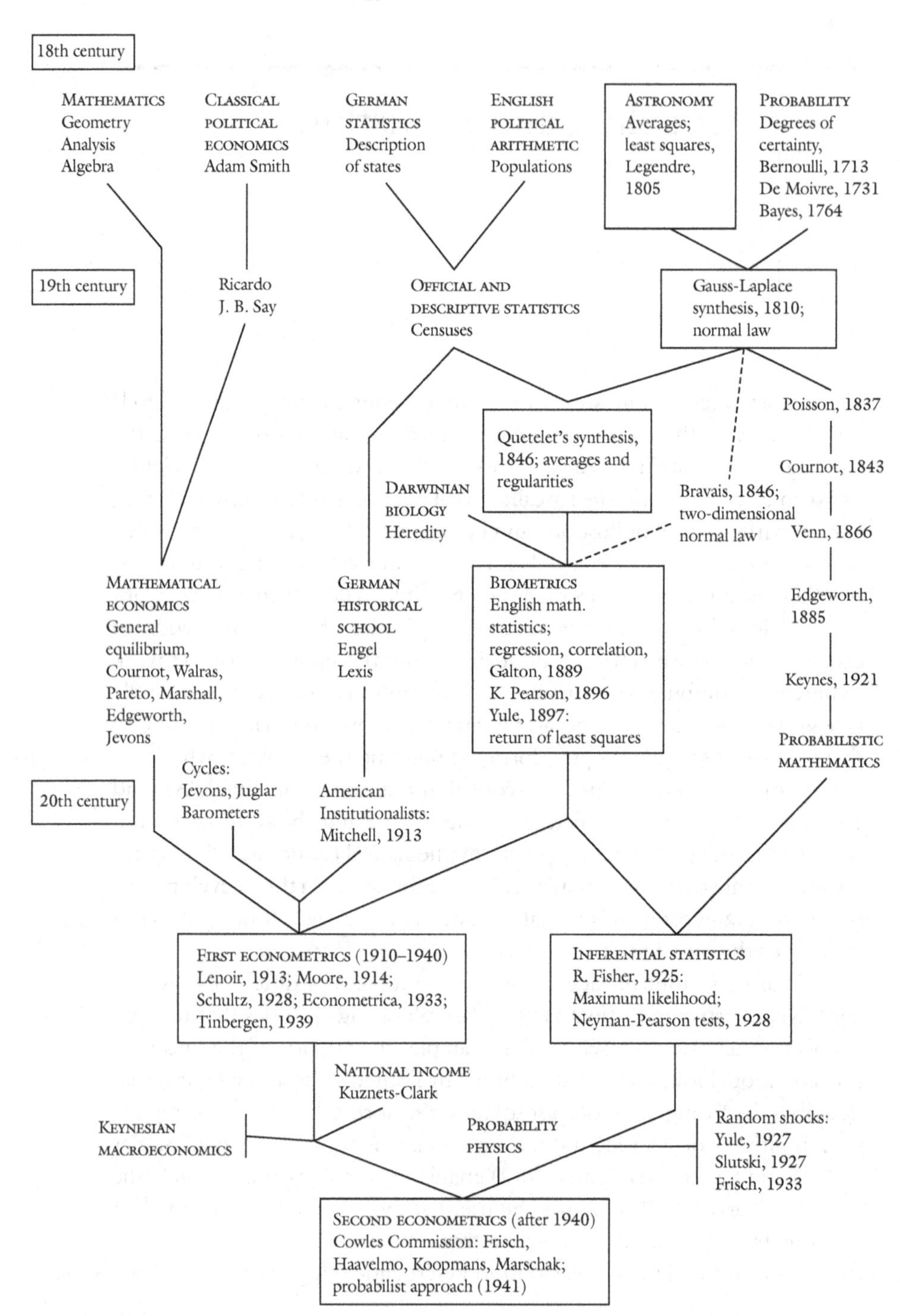

routinized and confined in standard boxes. The birth of econometrics is rich in such moments.

Early in this century, the first attempts to estimate the laws of supply and demand by Marcel Lenoir (1913) and Henry Moore (1914) raised a new question, later referred to as a problem of identification: what relationship should be established between theoretical laws and observed data? During the 1920s the Polish mathematician Neyman (familiar with the works of Borel and Lebesgue) met Egon Pearson, son of Karl Pearson; their encounter marked the rediscovery, after an entire century, of statistics and probability. Finally, during the 1940s, the Cowles Commission adopted the tools of Neyman and Pearson, thus completing the synthesis between the four traditions. This synthesis gave rise to new debates, as evidenced by the controversy between Vining and Koopmans (1949), but it subsequently emerged as the main solution to the previously raised contradictions between the languages of economic science, of statistics, and of probability. The present chapter describes a few of these oppositions and methods used in dealing with them. It outlines two contrasting tendencies. One proceeds more from data to theory: constructing indexes, analyzing cycles, assessing the national income. The other proceeds from theory and tries to connect data to it, either to estimate the parameters of a theory while taking it for granted or, conversely, in order to verify it. Thus the relationship between theory and data can be engaged in three completely different ways: in manufacturing new laws from data; in measuring the parameters of a law presumed to be true; or in accepting or rejecting a theory by means of statistical tests based on probabilistic models.[1]

Economic Theory and Statistical Description

The nascent mathematical theory of the nineteenth century aimed at reducing the inexhaustible complexity of the production and exchange of goods to a small number of simple hypotheses, comparable to physical laws. Its second goal was to reconstruct, while allowing itself to be guided by the energetic hand of mathematical deduction, an edifice whose principal merits were that it possessed both logical consistency and the possibility of being amended and enriched indefinitely, by increasing the initial hypotheses. But this way of reducing complexity was perceived, at least until the 1920s, as largely incompatible with this other form of reduction described in the preceding chapters, a reduction achieved by recording, encoding, and statistical aggregation—of which nothing seemed to guarantee that they were constructed in the conditions suggested by the basic hypotheses of mathematical economics. The difficulty of connecting these two modes of reduction can be expressed in various terms: the impossibility of applying a truly experimental method; the a priori unlimited number of explanatory factors; the variability among individuals; and errors in measurement. Much of this opposition was linked to doubts as to the possibility of transposing into the social sciences formalisms born of sciences such as physics, which presume a "homogeneity of nature"—that is, the spatial and temporal permanence of general laws reducible to simple expressions. The historicity and diversity of situations and cultures were invoked by the German tradition of descriptive statistics in order to combat the hypothetico-deductive mathematical economics of the Austrians, of Walras and Marshall. Linked to the officials in statistical bureaus of German states within the scientific and militant association of the Verein für Sozialpolitik, the economists of the historical school accumulated statistical compilations but did not allow themselves to deduce general laws from them.

For their part, mathematical economists had reservations about statistics which they suspected of jumbling together facts resulting from complex, unknown, and above all uncontrollable interactions (Ménard, 1977). Before 1900, the gap between these two categories of economists, statisticians and mathematicians, was very wide. The former rejected not only deductive constructions based on general laws postulated a priori, but the more radical among them rejected even the possibility of inducing laws based on empirical data. With few exceptions (Lexis, for example) they ignored the developments of nascent mathematical statistics, even criticizing Quetelet's *average man* as a meaningless abstraction. But the counterpart of this rejection of generalization and abstraction was a tendency to accumulate empirical and scholarly research, and to organize monographic or statistical surveys. This tradition of observation and description influenced French sociologists (Simiand, Halbwachs), and also the institutionalist American economists who had studied in Germany during the 1880s. One finds a trace of this in Mitchell, who founded the National Bureau of Economic Research (NBER) in 1920, or among the economists advising the Census Bureau during the 1930s. Thus at the turn of the century an opposition existed between an economic science that was mathematicized and hypothetico-deductive, and another science that was historical, descriptive, generally inductive, and sometimes described as "literary." The use of statistics was more an activity of economists practicing the second kind of science than an activity of partisans of the first kind. One of the first economists to plunge into what later became econometrics was the American Henry Moore (1869–1958). Moore did so while polemicizing against economic theories formalized according to a priori hypotheses; he thus attached himself, in his own way, to the tradition that favored the examination and analysis of empirical data.

But Moore stands out from the German economist statisticians in that, in his study of business cycles published in 1914, he largely employed the tools of multiple correlation born of English biometrics. The significant moment in the spread of these tools throughout the world of economic and social statistics came at the congress of the International Statistical Institute (ISI) held in Paris in 1909: in particular Yule, Bowley, and Lucien March—director of the SGF—made speeches on these topics. The shift in the meaning of the word "statistics," from the administrative or moral statistics of the nineteenth century toward the mathematical statistics of the twentieth century, then became clear. Correlation and regression enabled previously separate objects to hold together. They constructed a new type of spaces of equivalence and compatibility, of an order superior to those of taxonomy described in the previous chapter. In the groping attempts to explore this unknown continent certain experimental connections were tried that did not become stabilized later on. Thus Hooker (1901) analyzed the correlation between fluctuations in the marriage rate and those in the business cycle. At the SGF Henry Bunle (1884–1986) carried out a similar study in 1911, applying techniques inspired by those of the English, which Lucien March had expounded in 1905 and 1910 in the *Journal de la société statistique de Paris* (Desrosières, 1985).

In contrast, other means of establishing equivalence, tested thanks to these new methods, led to the first developments in econometrics. On the one hand, a comparison of fluctuations in the diverse indicators of economic cycles anticipated the construction of dynamic macroeconomic models. On the other hand, attempts to evaluate the curves of supply and demand on the basis of data concerning prices and quantities exchanged heralded the resolution of models with simultaneous equations. Thus, in 1913, at Lucien March's prompting another member of the SGF, Marcel Lenoir (1881–1927), published a remarkable thesis that passed largely unnoticed at the time: *Études sur la formation et le mouvement des prix* (Studies on the formation and movement of prices). The structure of this book, which is divided into two distinct parts, is significant. The first part, "Formation of Prices: A Theoretical Study," presented the mathematical theory of indifference curves and the determination of quantities and stable prices through the intersection of the curves of supply and demand. It was clearly situated in the context of the nascent mathematical economics. The second part, however, "Movement in Prices: Statistical Studies," made use of a large number of statistical series that were economic, monetary, and financial. Its aim was to "explain" (in the sense of multiple regression) the variations in prices of various goods through variations in production and other variables. This work had almost no impact in France. Lenoir was not an academic. He died prematurely in 1927 in Hanoi, where he had created a statistical bureau for Indochina. He was one of the very first statisticians (*the* first in France, for a very long time) to connect the three traditions of mathematical economics, descriptive statistics of administrative origin, and mathematical statistics.

At around the same time, Moore had also analyzed the cycles and relationships between prices and quantities, basing his comments on empirical observations. But whenever these seemed to contradict the commonly accepted theory he rejected it, instead of discussing the method that allowed identification of an observed regularity. Thus, having recorded a positive correlation between the quantities and prices of pig iron, he thought he had found an increasing demand curve, contrary to theoretical hypotheses. The later criticism of this interpretation led to the analysis of *simultaneous* variations in supply and demand curves; Lenoir for his part had given a good theoretical description of them, but had not proposed a statistical method to resolve such a system of structural relationships. During the 1920s, some American economists (but only a few Europeans, and no Frenchmen) developed and discussed Moore's works, especially in regard to agriculture (Fox, 1989). But numerous criticisms remained, concerning the possibility of rediscovering the laws of economic theory based on empirical data: disrespect for the rules of *ceteris paribus* (all things being equal); omitted variables; errors in measurement. The most basic, that of the "nonhomogeneity of nature," referred back to the earlier historicist point of view.

Beginning in the 1930s, probabilistic language offered a framework for thinking out several of the obstacles that hitherto prevented empirical observations and theoretical formalisms from holding together. If, a posteriori, it appears that this language, which was more than two centuries old, has provided effective tools for treating these difficulties, the connection occurred tardily. The probabilistic mode of thinking led an independent life, both in relation to economic theory and, more surprisingly, in relation to statistics, after the Gauss-Laplace synthesis was taken up in diluted form by Quetelet. The causes of this resistance, offered by economics on the one hand and statistics on the other, were not superimposed; nor did they coincide with the mutual mistrust of economists and statisticians (whereas, frequently, these various criticisms are confused, and thus further linked to the rejection of mathematicization, which is another matter). It is important to make a careful distinction between these arguments.

Degree of Belief or Long-Term Frequency

The relative eclipse, in the nineteenth century, of the use of probabilistic formulations, both by economists and by every kind of statistician, has already been mentioned in Chapter 7, in regard to representative sampling. Whereas, in the eighteenth century, Laplace had used it to evaluate the *probable error* in a measurement of the French population based on a sample, this method was rejected by Quetelet and his disciples, following Keverberg's critique of 1827: nothing guaranteed sufficient uniformity throughout the land, and therefore nothing ensured the uniquity and constancy of the probabilistic urn from which the sample would be drawn. Quetelet associated statistics with ideas of exhaustiveness and rigorous accuracy, which were necessary for them to be accepted by a broad public and, especially, by the administration. He rejected sampling procedures that were reminiscent of the hazardous and acrobatic calculations made by political arithmeticians in the English manner. The establishment of a solid administrative infrastructure, covering the whole country, seemed capable of closing the door on these methods, which were deemed poor expedients.

But Quetelet's point of view also reflects the shift that occurred, between the 1820s and 1840s, in the way the language of probability was understood and used. From its very origins this language had been stretched between two interpretations (Shafer, 1990). The first, termed subjective, dominated during the eighteenth century, notably in the work of Bayes and Laplace. It envisaged probability as a state of mind, a "reason to believe," an estimate of the degree of confidence to be placed in an uncertain statement, concerning either the past (for example, the guilt of a defendant) or the future. It could therefore involve a unique or rare event. The second, in contrast, termed objective or frequentist, viewed probability as a state of nature, observable only through the multiple repetition of the same event, identified with the drawing of balls from an urn of constant but unknown composition. Of course, ever since Bernoulli (1713: the law of large numbers) and De Moivre (1738: the normal law as the limit of binomial draws), it has been possible formally to connect these two interpretations, at least in cases such as tossing coins or games of dice. Similarly, two centuries later, the axiomatized formulations of Kolmogorov would seem to have brought about the definitive disappearance of this opposition. Even so, from the standpoint of their being used as argumentative tools—especially to support choices and decisions—these two interpretations have retained considerable autonomy, as is shown by the return, since the 1930s, of subjective probabilities and Bayesian statistics.

Quetelet's rhetoric was centered on ideas of the average man and statistical regularity, as observed when data bearing on a large number of cases had been gathered by the bureaus of official statistics that developed as a result of his influence during the 1830s and 1840s. It helped to make the frequentist vision prevalent, relegating the subjective perspective to mere speculation, founded on sand. In particular, the idea of specifying the "a priori probability" needed, in Bayesian reasoning, in evaluating the probability of a cause given an observed effect, was rejected as arbitrary and unjustified. But the success that Quetelet seemingly guaranteed through probabilistic language in its frequentist form in fact caused it almost to disappear from the statisticians' world for almost a century, until Gosset (Student), Fisher, and Neyman. The routinization of the use of the Gaussian law as the limit of a binomial law to a large extent excluded questions posed in terms of "reasons to believe," or of the degree of stability that can be granted a statement. These questions, which constituted the philosophical backbone and specificity of probabilistic language, disappeared from the horizon of the statisticians who produced and utilized figures; henceforth, they were treated by philosophers and logicians. While famous economists (Cournot, Edgeworth, and Keynes) were interested in probability, this part of their work was almost entirely disconnected from the handling of statistical data and their constructions in economic theory. On the contrary: these seemingly paradoxical cases show the difficulty of the work and the scope of the controversies necessary, between 1920 and 1940, for these three languages to be combined into the single language of econometrics.

However, the frequentist perspective associated by Quetelet with the production and success of ever more abundant statistical data did, by the circuitous path of physics and the kinetic theory of gases, help define new objects, which were termed "statistics." The fact that the microscopic parameters of the position and speed of particles are unknown did not prevent these gases from being described in a deterministic way at a macroscopic level (Maxwell, then Boltzmann). This new formulation of the question of determinism—according to which order could emerge from chaos—was close to the one Quetelet promoted in the social sciences. It was probably even influenced by it, through the intermediary of the astronomer Herschel (Porter, 1986). At this time (mid-nineteenth century), uncertainty about microscopic parameters was still formulated in terms of the impossibility of observing them directly. Beginning in the 1920s, however, it was expressed according to a far more radical form of probabilism, as a rejection of the very existence of simultaneously determined (though unknown) measurements, for the position and speed of a particle (Heisenberg's relationships). From then on the frequentist, probabilistic point of view was no longer simply a convenient choice of method linked to the imperfection of the instruments of observation: it also claimed to reflect the actual nature of the phenomena described. Thus, far from the statistics of statisticians, the probabilistic mode of thought made headway among physicists, arriving at formulations during the 1930s that subsequently influenced the founders of a new econometrics, based precisely on this mode of thought. Tinbergen and Koopmans had degrees in physics and were prepared to reimport into the social sciences the probabilistic schemes (naturally, more sophisticated) that these sciences had given nineteenth-century physics (Mirowski, 1989a).

As for the statisticians emerging from biometrics and using Karl Pearson's formulations of regression and multiple correlation (Yule, March, Lenoir), they did not, at least before the 1920s, view these schemes as necessary. Pearson's philosophy of science, which excluded the idea of causality—or reduced it to empirically observed correlations (Chapter 4)—did not directly prepare for an exploration of the question of fitting empirical data to a theoretical model external to these data, for a possible probabilistic expression of this adjustment. The regressions then calculated by Yule and Lenoir did not include explicitly described residual disturbances nor, a fortiori, hypotheses as to their distributions of probability. It is significant that, within the Pearson school, probabilistic formulations only appeared at the moment when the hypotheses underlying the frequentist perspective (that is, the "large numbers") were manifestly unrealist, with Gosset (alias Student) and Fisher, because both men worked on problems involving a limited number of observations.

Gosset was paid by a large brewery. He refined techniques for industrial quality control based on a small number of samples. He therefore had to determine the variances and laws of distribution of estimates of parameters calculated from observations that obviously did not satisfy the supposed "law of large numbers." Similarly Fisher, when working in a center for agricultural experimentation, could only proceed to a limited number of controlled tests. He mitigated this limitation by artificially creating randomness, itself controlled, for variables other than those whose effect he was trying to measure. This technique of "randomization" thus introduced probabilistic chance into the very heart of experimental proceeding. Unlike Karl Pearson, Gosset and Fisher were thus led to use distinct mathematical notations to designate, on the one hand, the theoretical parameter of a probability curve (a mean, a variance, a correlation) and, on the other hand, the estimate of this parameter, calculated on the basis of observations so insufficient in number that it was possible to neglect the gap between these two values, theoretical and estimated.

This innovation about mathematical notation marked a decisive turning point that made possible an inferential statistics based on probabilistic models. This form of statistics developed in two directions. The *estimation* of parameters, which took into account an ensemble of recorded data, presupposed that the model was true. It did not involve the idea of repeated sampling, and could be based either on a Bayesian formulation, or else on a function of likelihood, of which a maximum was sought. The information produced by the model was combined with data, but nothing indicated whether the models and the data were in agreement. In contrast, *hypothesis tests* allowed this agreement to be tested, and permitted possible modification of the model: this was the inventive part of inferential statistics. In wondering whether a set of events could plausibly have occurred if a model was true, one compared these events—explicitly or otherwise—to those that would have occurred if the model were true, and made a judgment about the gap between these two sets, in a typically frequentist perspective.[2] Thus estimation went hand in hand with a subjective point of view (if not a Bayesian one), whereas hypothesis tests relied on the frequentist idea (Box, 1980).

The duality of viewpoints on probability could also be related to situations in which this language constituted a resource for supporting a choice or a decision. Frequentism was associated with treatment in general, by a collective decider, of problems for which the creation of equivalences and aggregates was politically and socially plausible. The urn had to be constructed and the colors of the balls could no longer be a subject of debate. Taxonomic questions were resolved and hidden away in tightly closed black boxes. Statistical treatments, in the sense of the law of large numbers, were possible. They supported decisions made by the state that optimized a collective good, or ones made by an insurance company. In the latter case, taxonomic conventions could again be called into question if it was desired to highlight a subcategory facing a more particular risk, thus modifying the system of premiums; also, if it was pointed out that the persons least affected by accidents were relinquishing their insurances, thus increasing costs for those who remained.

Subjective probabilities, in contrast, were brought to bear in choices that did not involve an idea of repetition. This might, of course, involve an individual decision; but it could also involve a choice made by the state: Should war be declared? Should some controversial treaty be ratified, on due consideration of the subjective estimate of the risks involved in the two possible decisions? This involved an at least approximate evaluation of the a priori probability of an unknown event, either past or future. Widely used in the eighteenth century, then rejected in the nineteenth, this measurement of a "degree of belief" was again taken seriously and formalized, during the 1920s and 1930s, by the Englishmen Keynes and Ramsey; by the Italian actuary De Finetti; and by the Bayesians who were inspired by Savage (1954). The problems of judicial decision making (whether or not to condemn a defendant) or medical decisions (diagnosis and treatment) derived from this proceeding. Encoding, as described in the preceding chapter, was the turning point between the two points of view. It could be conceived in relation to its effect on statistical addition (a "fuzzy encoding" was spoken of) or from the point of view of the individual and the decision that concerned him.

Randomness and Regularity: Frisch and the Rocking Horse

The originality of probabilistic language lay in the fact that it was not simply a field well axiomatized by mathematics, but also a flexible resource for arguments and decisions, capable of being put to very varied uses, in constructions functionally very different from one another. This linguistic crossroads was involved in several ways in the development of econometrics, during the 1930s and 1940s: it provided means for treating random shocks, errors of measurement, omitted variables, and the irreducible variability of economic situations. But in order for such questions to be mastered with the help of probabilistic models, the problem of relating the empirical data to economic theory had to be conceived of as capable of being formalized, and no longer simply as an argument in polemics concerning either the realism of the theory or the pertinence and significance of the data. The tools allowing the tension between theories and data to be interpreted in probabilistic terms were conceived and constructed during these years. They were combined into a unified technology of formalisms originating in different fields. The first economists had calculated partial correlations and multiple regressions, but they had not worked on the gap between the data and the model, by formulating a residual of linear regressions. Then the analysis of economic cycles was transformed by the idea—surprising, a priori—that random shocks could sustain regular oscillations. Finally, the inferential statistics of Ronald Fisher (1890–1962), Jerzy Neyman (1894–1981), and Egon Pearson (1895–1980) constructed tools that made it possible to measure and test the relationships between data and models. Theories of estimation and hypothesis tests provided a standardized language for expressing proof, which was necessary for the grounding of scientific statements, at least in the experimental sciences (Gigerenzer et al., 1989).

The sense common to the varied uses of probabilistic reasoning has been summarized by Hacking (1990) in a significant phrase as "Taming chance." A typical episode in this history of the rationalization of chance was the formalizing of the theory of "random shocks," which related series of unforeseeable events, arising "by chance," to the maintenance of relatively regular cycles (Morgan, 1990, chap. 3). This understanding of random shocks, combining the intuitive perceptions of Yule (1927), Slutsky (1927), and Frisch (1933), completely transformed the question of the recurrence of economic cycles, itself posed recurrently since the mid-nineteenth century. Until then there had been two opposing concepts of these cycles. Some thinkers tried to make links appear with periodic phenomena quite external to economics: sunspots (Jevons, 1884), or phases of the planet Venus (Moore, 1914); transmission, they opined, could result from meteorological cycles, themselves connected to astronomy. Others gave up trying to discern the general laws of cycles (mainly from their periodicity), and treated each of them as a unique event (Juglar, 1862; Mitchell, 1913). These two interpretations could be linked to the two families of sciences capable of presenting models and metaphors for economics: mechanics, astronomy, and physics, on the one hand, and the biological sciences on the other.

This opposition between an external explanation and a rejection of explanation was presently overtaken by the idea of the self-reproduction of more or less regular cycles, based on irregular and unforeseeable external impulses. For this, it was enough simply to suppose that the state observed at time t was a linear function of states observed at previous periods: $t - 1$, $t - 2, \ldots t - k$. Thus, by relying on previously elaborated mathematical theories on difference equations (for series) or differential equations (for continuous functions), Yule and Frisch showed that, under certain circumstances, systems of equations linearly relating states described at lagged intervals offered solutions comprising regular, damped oscillations. In this case, the only effect of the random shocks was to restart the oscillations, the frequency of which was determined by characteristics internal to the system (an idea close to that of *proper frequency* in the theory of physical resonance). Thus, by distinguishing the mechanisms of (random) impulse from those of (periodic) propagation, it was possible to account for the regularity of cycles, without invoking external regularities, sunspots, or the phases of Venus.

The initial idea of this ingenious formalism came, in an indirect way, from a simple but surprising observation made by a Russian statistician, Slutsky, and published in 1927 (in Russian, with a summary in English) in the Institute of Moscow's review of research on economic cycles. Studying the effect of replacing a time series by its *moving average*—calculated, for example, over a period of ten months—he observed that this common calculation, intended to smooth out the series so as to make their long-term tendencies appear, *itself engendered* cycles, of a ten-month periodicity. Thus, calculating the moving averages of the series of drawings of the Moscow lottery, he established a curve that strangely resembled the fluctuations of the London Stock Exchange. This troubling discovery made a vivid impression on some analysts of economic cycles, to whom it suddenly suggested a possible explanation that no one had thought of until then. This case derived precisely from the model just mentioned: the moving average was a linear operator applied to a sequence of ten observations. Its solution presented a stable period component. Slutsky's discovery indicated a way to analyze the propagation of stable cycles based on random shocks.

The same year (1927) Yule analyzed the manner in which random errors could disturb and mask harmonic series that had been artificially created, for example by a pendulum; in so doing, he observed that shocks applied erratically to it continually modified the amplitude and phases of the oscillations, but not their frequency. He imagined a child randomly bombarding the pendulum with peas. Chance disturbances thus restarted the motion, but did not alter the frequency linked to physical characteristics proper to the pendulum. A common feature of these two simple demonstrations—Slutsky's and Yule's—was that they proceeded by means of *simulation,* beginning with a series of completely random events (drawing lotteries, throwing projectiles), making visible regularities on the basis of this pure chance. During the same period (the 1920s) Fisher used the methods known as "controlled randomization" for agricultural experimentation. In the field of sampling surveys, random choice definitively prevailed over "purposive selection" (Chapter 7). Last, industrial manufacture began systematically to use randomly drawn lots in controlling series of standard products (Bayart, 1992). Thus not only could random phenomena be tamed by hypotheses concerning their distributions, but they could even be used *positively,* to engender experimental conditions likely to bring out and consolidate scientific facts. This constituted a decisive transformation in the use of probabilistic reasoning. Previously associated (even in its frequentist version) with the idea of incomplete knowledge, with uncertainty and unpredictability, it was henceforth thought capable of generating recognized facts by itself. This reversal of perspective occurred in parallel with the development of a radically probabilistic physics, even in regard to elementary particles.

This new concept of randomness and its effects was taken up by Ragnar Frisch (1895–1973) and Jan Tinbergen (b. 1903), breaking with the old ways of interpreting business cycles. These interpretations were torn between, on the one hand, the quest for rigorous regularity linked to extra-economic causes (Jevons, Moore) and on the other hand, the monographic analysis of each cycle taken separately (Mitchell). The starting point of this story was the method proposed by Frisch (1928 and 1931) for analyzing time series. Techniques for decomposing these series into cyclical, superimposed components with constant frequencies had already been used for a long time, notably by Moore (the periodogram method). But the fundamentally deterministic nature of this method of adjustment could lead to interpretations deemed absurd (as in Moore's case) or, in any case, having no particular justification. Moreover, it did not allow the observed irregularities to be accounted for, especially in regard to periodicity. Starting with the idea (close to Mitchell's) that each cycle had local properties, Frisch developed a procedure to bring out superimposed oscillations without, however, prejudging the constancy of their frequencies and amplitude.

The crux of his method was to eliminate, one after another, increasingly long oscillations of approximate frequencies by fitting, at every stage, a curve that passed through the *points of inflection* of the previous one. This procedure gave rise to increasingly smooth curves that eventually culminated in a long-term tendency. It could thus make seasonal cycles visible, then short-term cycles (three to four years), medium-term cycles (eight to nine years), and so on. Analytically, the method amounts to calculating first differences $(X_t - X_{t-1})$, then second differences $[(X_t - X_{t-1}) - (X_{t-1} - X_{t-2})]$, then increasingly complex linear combinations of variables phased in time. Thus, a system of linear operators enabled a time series to be broken down into a set of harmonic series variable in time (and no longer strictly periodic, as with Moore's periodogram). Comparing this method with the result obtained by Slutsky, Frisch (1931) examined the effect of such linear operators combined with purely random shocks. He observed irregular periodic oscillations resembling the series that had indeed been observed. This was the origin of the "rocking horse" model, which he presented in 1933 in an article aptly entitled "Propagation Problems and Impulsion Problems in Dynamic Economics."

Frisch ascribed the rocking horse metaphor to Wicksell, and also the idea of distinguishing two separate problems: one, the propagation of oscillations through linear operators intrinsic to the system; and two, the initial triggering and retriggering of oscillations by means of external shocks:

> Wicksell seems to be the first who has been definitively aware of the two types of problems in economic cycle analysis—the propagation problem and impulse problem—and also the first who has formulated explicitly the theory that the source of energy which maintains the economic cycles are erratic shocks. He conceived more or less definitely of the economic system as being pushed along irregularly, jerkingly . . . These irregular jerks may cause more or less regular cyclical movements. He illustrates it by one of those perfectly simple yet profound illustrations: "If you hit a wooden rocking-horse with a club, the movement of the horse will be very different to that of the cane." (Frisch, 1933, p. 198, quoted by Morgan, 1990, p. 92)

A comparable metaphor—waves causing a ship to pitch and roll at sea—had already been imagined by Jevons to describe the cycles. The essential idea was that "more or less regular fluctuations could be induced by an irregularly acting cause; this did not involve any synchronism between the initial cause and the oscillations of the system." This manner of perception disqualified the research into correlations between astronomical and economical phenomena pursued previously. It also introduced the idea that, once a shock had been administered, its effects were propagated in an increasingly damped form, and that this propagation could be modeled by linear operations linking lagged variables. Thus, by processes of random simulation rather than by the actual analysis of real series, Frisch conceived a formulation that inspired the first dynamic macroeconometric models of Tinbergen (1937, 1939), applying lagged variables that had been observed first in the Netherlands, then in the United States. Frisch's article (1933) had a considerable impact. Samuelson (1947) declared that it had brought about a revolution in economics comparable to the one produced in physics when classical mechanics gave way to quantum mechanics. The comparison is indicative of the influence exerted by the new physics, which had abandoned the determinism of the nineteenth century in favor of probabilistic constructions, not only from a macroscopic and frequentist point of view, but even from the standpoint of elementary mechanisms (quanta). Frisch was aware that his contribution lay in relating two distinct ideas, the combination of which provided the model's originality:

> Thus by connecting the two ideas: (1) the continuous solution of a determinate dynamic system and (2) the discontinuous shocks intervening and supplying the energy that may maintain the swings—we get a theoretical set-up which seems to furnish a rational interpretation of the movements which we have been accustomed to see in our statistical time data. The solution of the determinate dynamic system only furnishes a part of the explanation: it determines the *weight system* to be used in the cumulation of the erratic shocks. The other, and equally important part of the explanation lies in the elucidation of the general laws governing the effect produced by linear operations performed on erratic shocks. (Frisch, 1933, pp. 202–203, quoted by Morgan, p. 97)

Remedies for the Crisis: Tinbergen's Model

In clearly distinguishing, and then articulating, a deterministic dynamics internal to a system and the effect of random external shocks on such a system, Frisch opened the way for efficient models that were directed toward economic policy and no longer purely speculative, as were those of the theoretical economists. During the critical period of the 1930s the hitherto unknown idea emerged that governments could act upon the general trend of an economy—especially the recurrence of cyclical crises and the chain of misfortunes they involved. If the cycles were not completely deterministic but could be disturbed by random shocks, two different forms of action were conceivable, corresponding to the two aspects of Frisch's model. On the one hand, the government could intervene promptly by administering a deliberate shock, the effects of which had been analyzed in advance by means of the model. On the other hand, the relationships internal to the dynamic system could also be modified by so-called structural reforms, with long-term effects. Tinbergen conceived and developed his first econometric models with a view to constructing such a tool, resulting from the conjunction of theoretical formalisms, the available statistical data, and the urgent need for an economic policy different from those in force before the crisis. This situation led him to a pragmatic way of proceeding, a kind of theoretical-empirical tinkering, removed both from the formal edifices of pure economics and from the descriptive statistics born of the historicist tradition as represented by Mitchell or by the technique of Harvard's "economic barometers" (Armatte, 1992).

Invited in 1936 by the Economic Association of the Netherlands to study policies capable of combating the Depression, Tinbergen gave a positive response: he had, he said, "given up physics," his original specialty, "in favor of economics, thinking this science more useful to society." His main idea was to simplify the overall economy, reducing it into a small number of variables and relationships, forming a model that could be subjected to experimental tests—like a scale model of an airplane in a wind tunnel. But the arbitrary nature of the simplification could only be challenged in the actual process of its application, which proceeded by a repeated iteration between hypotheses born of theories, and statistical estimates. Tinbergen viewed this practice as an "art," an art of "cooking," which should not be concealed from the public:

> I must stress the necessity for simplification. Mathematical treatment is a powerful tool; it is, however, only applicable if the number of elements in the system is not too large . . . the whole community has to be schematised to a "model" before anything fruitful can be done. This process of schematisation is, of course, more or less arbitrary. It could, of course, be done in a way other than has here been attempted. In a sense this is the "art" of economic research . . . The

> description of the simplified model of Dutch business life used for the consideration of business cycle causes and business cycle policy commences with an enumeration of the variables introduced. The equations assumed to exist between these variables will be considered in the second place. This order does not correspond exactly to the procedure followed in the construction of the model. One cannot know *a priori* which variables are necessary and what can be neglected in the explanation of the central phenomena that are under consideration. It is only during the actual work, and especially after the statistical verification of the hypotheses, that this can be discovered. As a matter of fact, the two stages mentioned are really taken up at the same time; it is for the sake of clearness that the exposition is given in two stages. A glance at the "kitchen" will nevertheless be occasionally made in order to avoid creating an impression of magic. (Tinbergen, 1937, pp. 8–9, quoted by Morgan, 1990, p. 103)

The simultaneously artistic and craftsman-like aspect of Tinbergen's work, the importance of the glance into the kitchen and of the requisite familiarity with statistical series and their peculiarities, are enhanced by the absence of means for rapid automatic calculations. One cannot allow oneself the luxury of trying countless regressions. Tinbergen used a graphic method of "plate stacking": statistical series, traced one over another, caused the relationships to appear visually, facilitating the selection of pertinent multiple regressions and reducing the burden of calculation. But this selection was not purely empirical, it was also guided by the suggestions of the available economic theories. Estimates of equations were made separately by means of ordinary least squares (the idea of a bias of simultaneity did not yet exist); but sometimes parameters already estimated in another equation were integrated into a new one. Differences between observed variables and those reconstituted by the model were carried over onto the graphs and made explicit in the forms of residuals in the actual equations. "Statistical verification" did not rely on probabilistic tests, but on this examination of the deviation between the data and the model.

An essential feature of this way of modeling was that the estimated relationships linked variables that were *lagged,* the choice of these lags being an essential ingredient in the culinary art that Tinbergen developed. In fact, once a system of 22 equations had been decided on—thought to summarize the whole of the Dutch economy—it was analyzed with a view to making a possible cyclic structure of its solution appear. To this end, Tinbergen combined the relations linearly in order to eliminate progressively all the variables except one. The final result was a linear function linking a variable Z_t to its previous values of Z_{t-1} and Z_{t-2}. This difference equation, analogous to the one analyzed in Frisch's "rocking horse" model, led, in certain conditions, to damped periodic oscillations. These constituted the backdrop of the economic fluctuations, on which random shocks or deliberate actions of economic policy were superimposed. Taking into account and combining the coefficients of the 22 equations, the central equation thus obtained was supposed to synthesize the whole of the structure of the economy, but it did not have an obvious economic significance. This technique has been described as "night train analysis," for it seems to transport the reader into darkness, passing from economically interpretable relationships to a cyclical structure emerging from the difference equation linking Z_t, Z_{t-1}, and Z_{t-2}.

The cyclical model can be disturbed either by external events or by governmental decisions. In the former case, Tinbergen examines the effects of an imported international cycle. In the second, he distinguishes two types of decisions: those that bear only on external shocks (short-term policies), and those more profound decisions that affect the coefficients of the central system of relationships (structural policies). He thus studied, from the standpoint of macroeconomic regulation (especially with a view to reducing unemployment), six possible governmental interventions: a program of public works; protectionist measures; a policy of industrial rationalization; the lowering of monopoly prices; reductions in salaries; and finally, devaluation. Similarly, compensatory measures of an imported cycle were studied through manipulations of rates of exchange or public investments. Through this modeling by means of a system of equations, Tinbergen introduced powerful constraints into debates on the connections between economic theories and their application to contra-cyclic policies. Both theoretical hypotheses and government measures had to be formulated precisely, if precise conclusions were desired. In addition (although Tinbergen did not clearly explain this), economic theory, statistical data, and the instruments of economic policy would henceforth have to be expressed in a common language, thus allowing free exchange between the scientific centers producing the theory, statistical offices, and the official committees preparing the policies. After the war, national accounts played an essential role in coordinating, if not unifying, this triangle of science, information, and action.

The Dutch model of 1936 did not have much impact outside Holland.

But shortly afterwards the League of Nations asked Tinbergen to test a broad spectrum of theories of cycles that had been gathered and published by Haberler (1937), also at the request of the League of Nations. Published in 1939, the two successive reports present first, a detailed discussion of the method, and second, the results of the model, as applied to the United States. They had considerable impact and gave rise to many comments, both favorable and critical. They constituted the original point of departure for the tradition of macroeconomic models, which reached its zenith during the 1960s and 1970s. The first report tested the "verbal theories" (that is, theories not formalized by equations) gathered by Haberler, from the standpoint of their ability to be translated into relationships that could possibly be checked and estimated by statistical procedures. Or to put it more exactly, it was not so much a matter of actually checking a theory—something which, according to Tinbergen, statisticians could not do—but of seeking its limits of validity, and the cases in which it is not applicable. Thus, in a Popperian perspective that linked the truth of scientific facts to their nonfalsification—in other words, to their being solidly inscribed in a network of other facts—Tinbergen defined the tasks of the economist and the statistician, and the respective role of correlation, that tested the relationships, and of regression, that measured their force:

> The theories which he submits to examination are handed over to him by the economist, and with the economist the responsibility for them must remain; for no statistical test can prove a theory to be correct. It can, indeed, prove that theory to be incorrect, or at least incomplete, by showing that it does not cover a particular set of facts:—but, even if one theory appears to be in accordance with the facts, it is still possible that there is another theory, also in accordance with the facts, which is the "true" one, as may be shown by new facts or further theoretical investigations. Thus the sense in which the statistician can provide "verification" of a theory is a limited one.
>
> On the other hand, the role of the statistician is not confined to "verification" . . . the direct causal relations of which we are in search are generally relations not between two series only—one cause and one effect—but between one dependent series and several causes. And what we want to discover is, not merely what causes are operative, but also *with what strength each of them operates:* otherwise it is impossible to find out the nature of the combined effect of causes working in opposite directions. (Tinbergen, 1939, p. 12, quoted by Morgan, 1990, p. 109)

The solidity of the general construction is guaranteed and tested in several ways. The model can be applied to other countries or to other times. Structural coefficients calculated for several partial periods are compared. Tinbergen also tested the normality and time autocorrelation of residuals, to verify that they could be assimilated into drawings in a normally distributed space, and that the so-called random shocks were indeed random. He further calculated the variance of the estimates and coefficients to be certain of their significance. But in other cases the verification—or rather the nonfalsification, in the sense suggested above—was made in terms of economic likelihood, and of the model's ability to account for the historical evolutions observed. Thus a marginal propensity to consume was rejected if it was above 1, or again if it was lower for low-income households than for those with high incomes.

Engineers and Logicians

Tinbergen's voluminous reports to the League of Nations gave rise to lively controversies. The most famous was the one triggered by Keynes, in a review of the first report, published in 1939 (at that time he was unfamiliar with the second report, which anticipated certain of his criticisms and responded in advance). If, on the one hand, it appeared that Keynes was unaware of the most recent developments regarding the dynamic models of cycles (notably Frisch's model concerning random shocks), he nonetheless raised basic questions about statistical induction and the econometric method.[3] The two positions were characteristic of two ways of conceiving the relationships between theory, observation, and action, born of different intellectual and professional traditions. Tinbergen's position was analogous to that of an engineer who conceives and tests prototypes using mock-ups, or scaled-down models, then actual-size models. His epistemology can be linked to that of Karl Pearson and his *Grammar of Science.* "True causes" are unknowable, or are simply hypotheses that can always be proved false. A model is good if it makes things hold together in a stable manner, and allows alternative economic policies to be tested and their effects to be measured.

The Keynesian perspective, however, which was linked to its history, was entirely different. It emerged from an academic tradition of political economics (mainly through his father) and philosophy of knowledge, as debated at Cambridge in the early part of the century. Keynes's *Treatise on Probability* (1921) aimed at founding a logic of induction. His concept of probability was antifrequentist, and close to the idea of "degree of belief."

For him, probability was a logical relationship between a statement and the information that supported it. His study derived from the philosophy of sciences, and more precisely from logic. He had already expressed his idea of probability and statistical induction in his youth, on the occasion of a controversy that pitted him against Karl Pearson. Pearson had directed a study intended to demonstrate that the alcoholism of parents did not influence the physical and intellectual aptitudes of their children.[4] In a series of articles published in 1910 and 1911 in the *Journal of the Royal Statistical Society,* Keynes criticized Pearson's procedures of statistical induction: Pearson, he complained, extrapolated on the basis of small samplings. Above all he did not make sure, in comparing two samplings, that causes other than alcoholism might not in fact explain the results recorded (Bateman, 1990). The theoretical arguments he opposed to Pearson were developed and systematized in the *Treatise* of 1921, and taken up again in his discussion with Tinbergen in 1939. The tools of statistical inference were transformed during the 1930s (notably by Karl Pearson); but Keynes pursued—in regard to the methods of correlation and regression henceforth used by econometricians, and in particular by Tinbergen—the question on the complete enumeration of causes that he had already raised in 1910:

> The method of multiple correlation analysis essentially depends on the economist having furnished not merely a list of the significant causes, which is correct so far as it goes, but a *complete* list. For example, suppose three factors are taken into account, it is not enough that these should be in fact *verae causae;* there must be no other significant factor. If there is a further factor, not taken account of, then the method is not able to discover the relative quantitative importance of the first three. If so, this means that the method is only applicable where the economist is able to provide beforehand a correct and indubitably complete analysis of the significant factors. (Keynes, 1939, p. 308, quoted by Patinkin, 1976, p. 1095)

Quoting this passage, and Keynes's critique of Tinbergen, Patinkin adds: "What could be a better description of specification bias?" Then, seeking to show that Keynes had in his way intuitively grasped the main problems of the nascent discipline of econometrics, Patinkin suggests that he had also anticipated the idea of the simultaneous equation bias through his critique of time lags—a central feature of Tinbergen's model:

> Professor Tinbergen is concerned with "sequence analysis"; he is dealing with non-simultaneous events and time lags. What happens if the phenomenon under investigation itself reacts on the factors by which we are explaining it? For example, when he investigates the fluctuations of investment, Professor Tinbergen makes them depend on the fluctuations of profit. But what happens if the fluctuations of profit partly depend (as, indeed, they clearly do) on the fluctuations of investment? (Keynes, 1939, pp. 309–310, quoted by Patinkin, 1976, p. 1095)

Among the causes at work that the economist cannot include in the required complete list, Keynes explicitly mentions nonmeasurable factors, such as psychological, political, or social variables, speculations, or states of confidence. But Keynes would only have to change his words slightly, replacing "nonmeasurable factor" with "nonmeasured factor," to emerge from the category realism that permeates his discourse, then envisaging the conventions of designation, of creating equivalence and encoding objects as integral parts of the process of knowledge. The *Treatise on Probability* closely links the inductive argument not only with the number of cases in which a proposition is verified but also with the idea of *analogy*, positive or negative; the likelihood of a proposition is only increased by a new proposal to the extent that this latter *differs* from the preceding ones, for factors other than the one studied. This idea of *negative analogy* allowed possible causes to be eliminated from the list. But this purely logical idea that Keynes had of analogy presupposed that the construction of the equivalence underlying it was outside the field to be studied: this was a philosopher's point of view, and not a sociological one.

Keynes's methodological realism is equally apparent in the other criticism he formulated, some thirty years later, of the thought processes of Pearson and Tinbergen. The central idea of this critique was the theme of the nonhomogeneity of nature, in space and in time. The probabilistic urns were of variable composition; nothing guaranteed the stability of the structural coefficients calculated by Tinbergen. Keynes refused to see the conventional and practical side of Tinbergen's model, judged by its resistance to tests and its ability to test variants of economic policy, and not according to whether it expressed true reality. This criticism led him to emphasize the intrinsic variability of nature, and therefore to reject conventions of equivalence that Tinbergen must adopt to make his model work. But when in his *Treatise on Probability* Keynes analyzed the logical conditions of a legitimate induction, from fictitious examples dealing with the color of birds, he observed that his reasoning was only valid on condition one supposed that the varieties of birds and color were *finite in number*—which implied a taxonomic reduction. He justified this essential hypothesis of "varieties finite in number" by the felicitous results it had provided in the past, but this "was still a hypothesis, and not something that could be proved" (quoted by Bateman, 1990). He was thus very close to taking taxonomic conventions seriously. But the logician's viewpoint he adopted to treat induction prevented him from seeing that equivalence was not something to be proved (logically), but something to be constructed, sometimes at great expense; something that then had to resist tests that were designed to deconstruct it.

On the Proper Use of Adjustment

The debates about the models of Frisch and Tinbergen gave new content to the old, recurrent debate on the relative role of theory and of observations, of deduction and induction, in economic science. The development of inferential statistics, estimating parameters and testing models, allowed close links to be forged between two worlds—the world of theory and the world of description—previously unaware of each other. The range of possible discourses relating to this question began to be diversified during the 1930s. At that time the means of joining the two worlds could be summarized in terms of three principles: theory was primary and could not be falsified; nothing of a general nature could be stated and one could only describe the multiple; provisional models could be tested. Moreover, deviations between observations and generalizations (whether they were described as theories, laws, regularities, or models) could be interpreted according to various rhetorics. These two problems, distinct though connected, were at the heart of the question of the realism of statistical constructions. For want of a good answer, such as would have been provided by the judgment of history, we can describe the paths by which answers have been formulated. In particular, we can emerge from the confrontation between theories and observations by introducing the question of the use, and inscription of the formalizations in networks of measures, both meterological and political. The way in which Tinbergen constructed, used, and justified his model provides a good illustration of this perspective. From this point of view the 1930s and 1940s were crucial, for at this juncture new connections were created, providing the basis, even today, for the language of an economic rationalism that is efficient and no longer purely speculative as was previously the case.

Until then, the links between economic theory and statistical data had been tenuous because the two ways of proceeding were opposed. One of them, modeled after the physical sciences, supposed a priori that general principles of maximization and optimization directed individual behavior, and deduced from this a determinist (at least in theory) representation of economic life. The other, inspired first by Quetelet, then by Karl Pearson and Yule, perceived the observed regularities and correlations as the only "laws" or "causes" a scholar could really speak of. In the former case, one could, at best, measure the parameters of a theoretical model presumed true a priori. In the second, laws could only emerge from the abundance of data. A third attitude was possible, which required a theory to be proved, subjected to criticism, and confirmed or rejected in view of the observations. This attitude was rare in the nineteenth century: it was the position of Lexis criticizing the regularities displayed by Quetelet (Chapter 3). In all three cases, the rhetorics allowing the language of the laws to be combined with those of statistical tabulations were different.

For those who postulated general theoretical principles, the idea of rediscovering them in a heap of statistical records seemed for a long time almost utopian and hopeless. This was approximately what Cournot and Edgeworth said. Both engaged in deep reflections on the importance of probability as a basis and support for knowledge; but their work in economic theory did not make use of these tools. Keynes himself devoted much energy to analyzing the role of probability in induction, but remained reticent and skeptical about Tinbergen's attempts at a global modelization of the economy.[5] More profoundly, the model of physics, based on the homogeneity of nature, objects, and the forces that bind them, was inadequate to assume the variability of societies. The theme of the absence of a controlled experimental situation and of the nonrespect of conditions *ceteris partibus* reappears constantly in the recurrent debate on whether it was possible to apply to economics the empirical techniques that served the natural sciences so well. The successive procedures conceived by statisticians all appear, at some point, as an answer to the problem of the impossibility of applying the experimental method to the social sciences: Quetelet's regularities and subjective methods; Lexis's test of the uniqueness of the probabilistic urn; Yule's partial correlation and multiple regression; Fisher's analysis of variance and technique of randomization. But these repeated efforts were not enough to seduce the economists, at least until the 1920s. Lenoir's book of 1913 juxtaposed—but did not mix—the deductive way of proceeding and estimation by means of regressions of the fluctuations of supply and demand (but his intent to compare them is nonetheless clear). Last, Keynes, criticizing Tinbergen in 1939, clearly thought that an econometric model could not be the source of new economic ideas: empirical work did not give rise to discovery, but could only illustrate theories or estimate their parameters. If the model contradicted theory the data were false, but not the theory.

At the other end of the epistemological spectrum, one can discern a purely descriptive approach—the most radical of all—and an inductive approach that sought to bring out regularities and laws on the basis of observations. The former was typical of the German historical school, and of some American institutionalists. It was characterized by the idea that situations were incommensurate among themselves, and that no general ahistorical regularity could be detected without artifice. The analysis of cycles was paradoxical ground in which to apply such a concept: first Juglar (1819–1905), then Mitchell (1878–1948) attempted to describe and analyze cycles by maintaining the idea that each of them was unique in its kind and derived from different explanations. This naturally limited the use of these studies to purposes of forecasting. The second, more inductive approach, accepted the existence of underlying regularities, of a macroscopic order emerging from microscopic chaos. Karl Pearson's *Grammar of Science* presented this way of seeing the world, which is sometimes described as instrumentalist in its refusal to concern itself with the ultimate reality of things and of causal relationships. This antirealism also had its sympathizers on the side of the natural sciences, for example in Mach, whom Pearson admired. In this perspective, the deviation between the observations and the "law" extracted from them was reduced to the residuals (not "explained" in this particular type of causality) of regression. These residuals could be interpreted either as errors in measurement, or as the reflection of "omitted variables," or as a variability of unknown origin (which amounts to almost the same thing). No probabilistic model was necessary since the regularities did not have a logical status distinct from that of observations, whereas later on inferential statistics carefully distinguished between the theoretical random space (the urn) and the sample of observed data (the balls drawn from it).

A third possible attitude in relation to the interaction between theories and observations was to test the consistency of the models in the light of the data—or better still, to give oneself functions of the cost of the consequences of accepting or rejecting a hypothesis, as did Jerzy Neyman and Egon Pearson. In this case, the question of realism escaped from the traditional alternative that opposed the underlying philosophical positions of the first two attitudes. It anticipated the rhetorical splitting characteristic of modern uses of arguments based on empirical observations. The linguistic operator of this splitting is the expression "everything takes place as if," which allows the parallel formulation of two kinds of phrases, one realist, the other instrumentalist (Lawson, 1989). The cost functions associated with the acceptance or rejection of the model enable the truth of an assertion to be linked, at least virtually, with a larger network of statements; they also allow the judgment of consistency to be applied to this network, and not simply to isolated statements. The extension of this network and the list of beings and statements involved are then very variable. The controversy changes radically in nature, in henceforth relating to this list. The ideas of "paradigm" or "style of reasoning" refer back to a closely related concept: the test hinges on the coherence, solidity, and fruitfulness of a group of statements emerging from a system of records, definitions, and encoding, connected by what they say and what they do (Crombie, 1981; Hacking, 1991). This concept allows the questions raised in Chapter 8, on taxonomy and encoding, and those mentioned in this chapter, on modelization and adjustment between theories and data, all to be brought together within the same propositions, although these themes are almost always treated in separate contexts, by separate persons.

Beginning in the 1930s the language of probability enabled the network of records and a priori modelizations to hold in an entirely new way compared with the previous Manichean opposition between the importance of "theory" and "observations." The notion of *likelihood* was presented by Fisher as an alternative to Laplace's old idea of "inverse probability," allowing scholars to move beyond the opposition between subjective and objective probabilities. The central idea was to retain, within a precisely defined and circumscribed family of probability laws, the one that made a set of observations most "likely" (in other words, the law for which the probability of occurrence of these observations was the greatest). In this perspective, economic theory interposed to define and circumscribe the family of laws within which this method of "maximum likelihood" was applied. This synthesis was presented in its most general form by Haavelmo (1944). It also influenced the work of the Cowles Commission, whose program of the 1940s provided the foundation of modern econometrics (Epstein, 1987). It allowed, at least in principle, the majority of objections previously raised against the marriage of economic theory and descriptive statistics to be integrated into the same formalism—a marriage which the new church of inferential statistics thereafter welcomed and blessed.

In this perspective, the various forms of possible deviation between the observed values of a variable "to be explained" and its value as calculated (or "reconstructed") by the model thus optimized can be interpreted in various ways, by distinguishing the *errors in specification* and the *errors in measurement*. The former have to do with the translation of the theory in terms of virtual causal relationships (in general, linear). The selection of "explanatory" variables and the type of linkage retained among these variables constitute the specification of the model, leading to the delimitation of a family of probability distributions. The deviations in adjustment (or residuals) can be interpreted as resulting from certain relationships or certain variables. In particular, a *simultaneity bias* can result from the fact that several relationships unite several variables at the same time, and the omission of one of them distorts the definition of the space of possible distributions, within which the most likely one is sought. This particular case is illustrated by the history of the attempts made, between 1920 and 1940, to estimate laws of supply and demand on the basis of data concerning the prices and quantities exchanged of certain products (Christ, 1985). The fact that the two theoretical curves moved simultaneously when affected by other variables (for example, revenue in the case of demand, and cost of production in the case of supply) was the source of a puzzle that finally led to the formulation of the "simultaneous equations" model.

But the residuals (deviations between variables observed and reconstructed by the model) could also be interpreted in terms of *irreducible variability*, or of *errors of measurement*. The former case was rather similar to that of the omitted variables—so much so that people gave up imagining and explaining "factors" that justified this residual variability. The list of pertinent facts susceptible to inclusion in the network scholars were trying to build could not be extended indefinitely, for reasons involving the economy of the actual formatting. The goal of a modelization was to reduce complexity by investing in the selection and standardization of the facts described, the expected benefit of which was that it allowed this partial modelization to be connected with a larger ensemble of representations and actions. This investment involved a sacrifice, in this case a residual variability, comparable to the internal variability lost by a taxonomist in constructing an equivalence class. But, of course, it is always possible to proceed from this unexplained variability toward the formulation of new explicative variables, thus reducing the residual variance, just as the taxonomist can always provide details of his classifications.

Last, residuals can be seen as resulting from errors in measurement. In this case, the work of recording and encoding described in the preceding chapter is presumed to be enclosed and isolated in a black box, distinct from the network the model is intended to account for. The very choice of the expression "error in measurement" implies a realistic epistemology, according to which objects preexist—at least in theory—the work of identification, definition, and delimitation analyzed above. Or at least this sequence of acts and the facts they involve are presumed exterior to the list of objects and utterances retained as pertinent. They are not even saved by the rhetorical operative phrase "everything takes place as if." But a position denouncing this delimitation and the construction of this list is not tenable, except when it recommends another one, presumed to be more complete. It would be better to take note of the fact that the realistic definition of pertinent objects has profound justifications, both economic and cognitive, in terms of the finality of the model and its inscription in networks of description and decision. The hypothesis that objects exist prior to their construction allows them to be used as conventions of reference, as elements lending order to chaos, serving as "objective" reference points—that is, common to intrinsically different subjects (individuals). From this point of view, a society in which men could manage without a realist position of this kind is utterly unthinkable. The conceptual divide between the interior of the black box (the recording procedures described in Chapter 8) and its exterior (relating the outputs of closed boxes, with intent to develop boxes of a superior order) is reflected, in the language of econometrics, by the distinction between, on the one hand, errors in measurement and, on the other hand, residual variability and omitted variables. It may appear arbitrary, but it is deeply inscribed in institutions, in routines of descriptions, and in implicit epistemologies, without which no rhetoric connecting science and action would be conceivable. Nevertheless, during periods of crisis and innovation these routines of encoding and taxonomy are destroyed and other modes of action instituted, endowed with other indicators. That is precisely what happened during the 1930s, a decade especially rich in innovations of this type. Among them we may include Frisch and Haavelmo's formulation of a rhetoric encompassing the two previously opposed ideas of basic law and observed regularity, ideas henceforth united through the idea of the autonomy of a relation.

Autonomy and Realism of Structures

Tinbergen's macrodynamic model, described above, was explicitly inspired by Frisch's model (1933), especially in the use of one final difference equation, with a single lagged variable. This "night train" equation was supposed to sum up all the structural dynamics of the model. Yet Frisch (1938) criticized Tinbergen: the relations that Tinbergen estimated on the basis of observations were not "structural" but "confluent," in the sense that they resulted from a more or less complex combination of basic relationships, impossible to observe directly, but which reflected the essence of the phenomena. This idea of confluence had been conceived by Frisch in 1934. After constructing, in 1933, his "rocking horse" model on the basis of fictitious parameters characterizing the "structure" of its central kernel, he sought means for estimating these parameters. But in working on a system with several equations, he stumbled on the question of "multicollinearity": since certain relationships could be combined linearly to describe others, nothing guaranteed the stability of the adjustments performed on the original equations or on the combined equations.[6] Now, the observable relations, the coefficients of which can be estimated, are in general such linear combinations of essential but unobservable "structural" relations. In this terminology, the idea of structure occupies the place of "reality," which logically precedes the observations—just as, for Quetelet, the average man, though invisible, constituted a more profound reality than did his contingent but visible manifestations. But henceforth this reality of a superior order was explicitly connected to a construct developed elsewhere: that of economic theory, which was not the case for Quetelet. However, although they were observable, the confluent relations offered a major drawback: they depended on one another. More accurately, a change in the parameters of one modified the parameters of the others. They were scarcely "autonomous," whereas structural relations were more so. Frisch expressed this idea clearly in his commentary on Tinbergen's model:

> [The autonomous features of a system are] the most autonomous in the sense that they could be *maintained unaltered while other features of the structure were changed*. The investigation must use not only empirical but also abstract methods. So we are led to constructing a sort of *super-structure*, which helps us to pick out these particular equations in the main structure to which we can attribute a high

> degree of autonomy in the above sense. The higher this degree of autonomy, the more *fundamental* is the equation, the deeper the insight which it gives into the way in which the system functions, in short the nearer it comes to being a *real explanation*. Such relations are the essence of "theory." (Frisch, 1938, quoted by Aldrich, 1989, p. 24)

Thus the economist, who tries to invent a new way of expressing deductive economics with empirical observations, grants the former a privileged status in identifying pertinent relations. Ten years later, however, Frisch himself would express some pessimism as to the possibility of attaining autonomous relations, which he even envisaged seeking through direct surveys of the behavior of individuals:

> It is very seldom indeed that we have a clear case where the statistical data can actually determine numerically an autonomous structural equation. In most cases we only get a covariational equation with a low degree of autonomy . . . We must look for some other means of getting information about the numerical character of our structural equations. The only way possible seems to utilize to a much larger extent than we [have] done so far the interview method i.e., we must ask persons or groups what they would do under such and such circumstances. (Frisch, 1948, quoted by Aldrich, 1989, p. 25)

Frisch is fairly close here to Quetelet and his average man. The combination of the two approaches, one theoretical and hypothetic-deductive, the other empirical and statistico-inductive, is still devoted to epistemological controversies that have neither head nor tail (in the sense that one does not know in which direction, theoretical or empirical, to proceed), unless one connects this repetitive face to face to the third principle, constituted by the use of these constructions and their inscription in a network of actions based on measurements. Haavelmo set off in this direction in his program-text, published in 1944, on "The Probability Approach in Econometrics." Here he gave three reasons for seeking autonomous relations: they were more stable, intelligible (interpretable), and above all useful in economic policy. To illustrate his reasoning he referred to the way a car functions:

> If we conduct speed tests of a car, by driving on a flat, dry road, we establish a precise functional relationship between pressure on the accelerator (or the distance between the pedal and the floor) and the corresponding speed of the car. We need only to know this relation-

> ship to reach a given speed . . . [But if] a man wanted to understand how [automobiles] work, we should not advise him to spend time and effort in measuring a relationship like that. Why? Because (1) such a relation leaves the whole inner mechanism of a car in complete mystery, and (2), such a relation might break down at any time, as soon as there is some disorder or change in any working part of the car . . . We say that such a relation has very little *autonomy,* because its existence depends upon the simultaneous fulfilment of a great many other relations, some of which are of a transitory nature. On the other hand, the general laws of thermodynamics, the dynamics of friction, etc., etc., are highly autonomous relations with respect to the automobile mechanism, because these relations describe the functioning of some parts of the mechanism *irrespective* of what happens in some other parts. (Haavelmo, 1944, pp. 27–28, quoted by Aldrich, 1989, p. 27)

This concept of autonomy is directly linked to the possibility of using the model to test the consequences of decisions in economic policy, by analytically separating the effects of diverse actions:

> If then the system . . . has worked in the past, [the economist] may be interested in knowing it as an aid in judging the effect of his intended future planning, because he thinks that certain elements of the old system will remain invariant. For example, he might think that consumers will continue to respond in the same way to income, no matter from what sources their income originates. Even if his future planning would change the investment relation . . ., the consumption relation . . . might remain true. (Haavelmo, 1943, p. 10, quoted by Aldrich, 1989, p. 28)

The notion of autonomy reflects the idea that one can, at least as an intellectual hypothesis, isolate an elementary economic mechanism by fixing the values of the other factors. Haavelmo speaks of "free hypothetical variations," which he opposes to "variations restricted by a system of simultaneous relations." This idea was not, despite appearances, simply a repetition of the old condition of *ceteris paribus.* It is more subtle, in that it complements the idea of simultaneity. To the extent that several structural (and thus autonomous) relationships act independently of one another in a stochastic manner (that is, a probabilistic one), the distribution of probability of observed data is limited by this system of simultaneous relations. The conditional relations linking the variables *in the space limited in this way* are different from autonomous relations, which describe hypothetical relations that fail to account for the entire system. Estimates (by the method of maximum likelihood) must be made in this restricted subspace, otherwise they would feed off a bias of simultaneity. The two ideas of autonomy and simultaneity, characteristic of Haavelmo's construct, are thus inseparable. In the most favorable cases, a system of simultaneous equations can be resolved by matrix algebra, allowing the formulation of equations reduced by a linear combination of the autonomous relations, then allowing unbiased estimates of the structural parameters to be deduced from them. But these limited equations—"confluent"—in the sense used by Frisch, have the drawback of not being autonomous. Their parameters depend on the entire system, and they cannot be used directly to test economic policies. The tests involve making the whole model work, which can be more or less complex according to the number of relations and variables conventionally retained as pertinent.

Thus in the early 1940s a synthesis of previously disconnected cognitive and political elements was developed: old economic theories (the law of supply and demand) or new ones (Keynesian macroeconomics); statistical records henceforth considered as respectable partners; estimates of national income; inferential statistics based on probability; linear algebra and the techniques for diagonalizing matrices; and last, a socioeconomic situation that made it possible to imagine governmental intervention designed to regulate the global economic balance. This previously unknown mix was to make possible the success of macroeconomic models inspired by Tinbergen's model (De Marchi, 1991), subsequently supported by the work of the members of America's Cowles Commission: Haavelmo, Klein, Koopmans, Marschak, and Wald. The latter thus often refer to two cases in which the estimate of a structural model enabled the government to take an optimal decision: modification of tax rates or price control (Epstein, 1987). Inspired by Keynesian economic arguments, they emphasized the incentive effects prompted by possible decisions through reliance on their structural models. But, in this sense, the word "structure" has quite a different nuance from the one given it by other economists influenced by the institutionalist current, such as those of the National Bureau of Economic Research. When these economists speak of structure, they call to mind procedures (legal, regulatory, or habitual) that direct the functioning of markets: antitrust laws, trade-union laws, wage conventions, collective negotiations, unemployment compensation. For the econometrician, in contrast, the word structure refers to deep-seated economic mechanisms underlying the observed statistical regularities, and theoretically expressible by autonomous relations. This difference in perspective highlights the controversy that in 1949 opposed, through Vining and Koopmans, the tendencies of the two groups.

Their own concept of economic action did not prevent the members of the Cowles Commission from making occasional suggestions which, though framed in a Keynesian perspective of regulation through public demand, could nonetheless prove surprising. Thus Epstein relates how in 1945, Marschak, Klein, and the atomic scientist Teller all intervened at the request of a group of scientists alarmed at the development of nuclear weapons. The text they coauthored suggests that, in order to minimize the effects of a nuclear attack on the United States, the entire population should be relocated in new cities spread out along large axes, whence the name of "ribbon cities." They estimated that spending twenty billion dollars a year for fifteen years would make this project viable. It would, however, involve "giving up many habits that the populations of large cities have acquired in the course of the past century." Was this a serious project or an intellectual prank? In any case, it certainly reflects a period in wartime when a massive, coordinated intervention on the part of the state could be conceived and justified by means of macroeconomic arguments, supported by a new type of descriptive system: national accounts.

Three Ways of Calculating the National Income

Among the ingredients that allowed the first econometric models to be constructed and applied, national accounts occupied a central place. It was a consistent and exhaustive system for defining, classifying, and aggregating economic actions with a view to describing them and relating them through these global models. Through the scope of the investment in formulating and circulating statistical records that it involved, it may be compared to other major scientific ventures guaranteed only by public will and financing, ventures that have increased in number since the 1930s, especially in regard to military needs: nuclear research; research into aeronautics, space, and electronics. The comparison is not fortuitous, for it was during or immediately after the two world wars (in the United States for the first; in Great Britain and in France for the second) that public bodies made significant efforts to calculate the national income and its breakdown according to different points of view. In these periods of intense activity, when the individual firms were mobilized and coordinated according to a unified design, the task of inventorying, classifying, and adding up the resources and their uses in the form of national aggregates was consistent with the deliberate concentration of means of different origins around major scientific projects. But, aside from these exceptional wartime periods, it was also in contingencies of crisis or particular political mobilizations that this construction of coordinated networks of recordings and nomenclatures of economic actions was undertaken: in the United States during the 1930s, France during the 1950s and 1960s, and in eastern Europe during the 1990s (in this case it had already been carried out, though in a different way, for centralized planning).

Although the first attempts to calculate these national aggregates date back to the late nineteenth century (not to mention the assessments of William Petty and Gregory King in the seventeenth century), their significance changed completely between 1900 and 1950. The very word "national income," long used to designate this form of accounting, is significant. The initial perspective was focused on the formation of incomes: the research of Arthur Bowley (1919) concentrated on incomes that originated in different sectors of activity, and how they were distributed according to the different social classes. Then, during the 1920s and 1930s, the analysis of cycles prompted Wesley Mitchell and Simon Kuznets (at the NBER) and Colin Clark (in Great Britain) to construct time series, making apparent the *uses* of the goods produced (investment, final consumption, or intermediate consumption by the firms), rather than simply the incomes they generated. Finally, the perspective opened up by Keynes in his *General Theory* (1936) and the problems raised by the financial cost of war led to a generalization of the breakdown of the national product into the three components ($C + I + G$) of its final use: consumption, investment, and government expenditure (Patinkin, 1976).

These various stages corresponded to different ways of conceiving the role and scope of public economic action, and therefore of formatting the bookkeeping descriptions, based on sources that also differed greatly since they were linked to actions with distinct goals. Already we find here the three perspectives used in calculating the national income (or, later on, the gross national product): sectoral origin (agriculture, industry, services); the sharing of the factors in production (salaries, profits, taxes); and finally, the use it was put to, expressed through final demand (consumption, investment, public expenditure).[7]

The first two viewpoints were typical of Bowley's research (1919) or of the very first publications of the NBER, which was founded in 1920.

These latter concerned the long-term growth of the economy—reflected by the growth in national income—and also its distribution according to various divisions (industrial and agricultural workers; capital and work; size of the firms). According to the preface of the first work published by the NBER (1921), the goal was to determine whether "the national income was capable of providing a decent standard of living for everyone, and whether this income grew as quickly as the population, and whether its division among individuals developed more or less unfairly." Bowley worked within the perspective of the surveys on poverty conducted by nineteenth-century social reformers (Booth, Rowntree: see Chapters 7 and 8). His work of 1919 aimed at "describing the total and the sources of the aggregated revenues of the British people, and the proportions of this aggregate that went to the various economic classes." He based his findings mainly on the statistics produced by the administration of income taxes, which comprised few exemptions. These studies did not contain assessments of sequences of successive years, and did not refer to the question of cycles. In a later work (1927), Bowley even alluded to the "industrial instability" that deprived his assessments of the national income of a "reasonable degree of permanence"—showing clearly that his goal was not to analyze economic fluctuations, as did his successors at the NBER, especially Mitchell and Kuznets.

This shift in focus in questions associated with the assessment of the national income, from the division of revenues toward an explanation of business cycles, modified their architecture. Indeed first Mitchell (1923), then Kuznets (1934) saw in the "volatility of investment" the principal explanation for the analysis of cycles, which became a crucial question after 1929. Thus Kuznets explained, in the 1934 bulletin of the NBER, that the annual estimate of the net formation of capital was not simply a measurement of the growth in wealth; it also provided information on the "cyclical variations in the production of equipment goods, which are far removed from those for goods consumed over a short period of time." The differences in the rhythms of fluctuations of these two categories of goods caused the national income to be broken down according to this criterion of end use, by distinguishing investment from end consumption. These categories were present in the theories of cycles that preceded Keynes's *General Theory* (1936), but the publication of this work lent strong momentum to this type of analysis. In particular, it caused the component G (government expenditure) to be isolated in the breakdown ($C + I + G$) of the national income—something that had not been done previously. This third perspective of the analysis of the national income, centered on the breakdown of the eventual uses of production, was well adapted to Keynesian theories of cycles, but unlike the other two perspectives, it could not rely on statistical records already in existence. The long-standing censuses of firms conducted in the United States allowed the assessment of values added by branches, which addition led to a gross product, in the case of analysis according to sectoral perspective. The management of income taxes yielded a statistical subproduct that could be used in the perspective of "production factors" (at least in the case of Great Britain, though not in the United States, where this tax did not yet exist). In contrast, the perspective of "end use" involved establishing new surveys, dealing with household consumption, investments made by firms, and public expenditure.

The development and use of national accounts led to a major extension of the tasks and role of official statistics. This trend began in the United States during the 1930s, in Great Britain during the 1940s, and in France during the 1950s and 1960s. Previously, the first research into national income had been carried out by academics (Bowley, Clark) or by nongovernmental research bureaus, such as the NBER. From 1932 on, however, Kuznets organized a collaborative effort between the NBER and the Department of Commerce, on which depended administrative statistical services that were radically transformed during the 1930s, especially with the first uses of the sampling method (Chapters 6 and 7). In Europe, in contrast, political and administrative officials did not begin to learn and speak the language of macroeconomics and national accounts until during the war (in Great Britain), or after it (in France). Thus Richard Stone, the founding father of national accounts in Britain, commented in 1976 on the prewar situation in these terms:

> In Britain, non-academic interest in the national income was not aroused before the second world war. The concept of demand management did not exist despite the economic troubles we went through in the 1920's and 30's. Politicians and civil servants simply did not think in these terms and the suggestion that Bowley and Stamp and, above all, Colin Clark were doing work of potential importance for practical policy purposes would have been met with incomprehension and, no doubt, with derision had it been comprehended. (Stone, letter to Patinkin, quoted by Patinkin, 1976, p. 1115)

But if this new language was to be understood, adopted, and used naturally, it was not enough for the mere idea to forge a path. It also had to be able to rely on a dense network of records and stabilized tools. These not only made it credible, they paradoxically led it into oblivion through its inscription in data banks, short-term economic indicators, and arguments in daily use. During the 1930s these networks did not yet exist, and figures did not inspire confidence, as is shown by Keynes's skepticism in regard both to Colin Clark's assessments and to Tinbergen's models. According to contemporary accounts, Keynes trusted more in his intuition than in figures provided by statisticians. If a figure did not suit him he modified it, and if, by chance, one satisfied him, he was amazed: "Heavens, you've found the right figure!" (anecdote of Tinbergen, quoted by Patinkin).

Testing Theory or Describing Diversity?

In 1930, at the instigation of Irving Fisher and Ragnar Frisch, the Econometric Society was founded. Thus the word econometrics made its appearance, modeled on biometrics and psychometrics. Its avowed goal was to treat economic questions by integrating both statistical method and mathematical reasoning, and in 1931 it numbered ninety-one members, including Bowley, Schumpeter, Keynes, Tinbergen, Sraffa, and the French Darmois, Divisia, and Roy. It brought together people from quite different intellectual traditions before a common acceptance of the word "econometrics" gradually emerged, during the 1940s and 1950s, mainly (though not solely) around the program of the Cowles Commission (Malinvaud, 1988). This commission was created in 1932 by a businessman and investment advisor, Alfred Cowles: just when the crisis was at its height, he offered financial support to the Econometric Society so that it would study the causes of cycles and crises, and the means to forestall them. The intellectual and professional traditions brought together by Fisher and Frisch were expressed in the journal *Econometrica,* founded in 1933. Simplifying slightly, one can distinguish three quite different such traditions: economic statistics, economic calculus, and mathematical economics.

The economic statisticians already mentioned used statistics in a descriptive manner. They were highly familiar with their data and with the objects that they analyzed, but were reticent about generalizing or formalizing their observations in reference to a theory. This was the tradition of Moore, Mitchell, and also of Harvard's "econometric barometers" (Persons). Detecting statistical regularities rather than economic laws, this trend would be criticized by Frisch and Haavelmo, according to whom the relations it established were "confluent" and insufficiently "autonomous": these relations expressed nothing about profound economic mechanisms. These statisticians, who were skeptical in regard to advanced models, feared that they might lead to the disappearance of the intuitive knowledge of the situations analyzed—itself a result of the combination of different kinds of knowledge, incorporated in the analyst. In the America of the 1930s and 1940s, this point of view was generally held by members of the NBER or, in France, by the creators of short-term economic analysis such as Alfred Sauvy or, later on, Jacques Méraud. But the rejection of models based on theoretical hypotheses could also lead to increased sophistication in purely descriptive statistical techniques: calculus of index-numbers; breakdown of time series and correction of seasonal variation, in the 1930s; then, during the 1970s, the analysis of multidimensional data; and last, during the 1980s, a return to methods of analyzing series rejecting the distinction between explanatory and explained variables (vector autoregression). This tradition of descriptive economic statistics could therefore rely either on historical knowledge, intuition, and familiarity with the data, or on formalizations that were advanced, but careful to remain close to the data.

Economic calculus is a tradition of engineers, and has been represented in France since the nineteenth century (Dupuit, Cheysson, Colson, Gibrat, Massé, Boiteux). The intention was to optimize decisions involving, for example, investment or tariffs (Etner, 1987). Mathematical economics belonged more to an academic tradition (except in France: Allais was also an engineer) in the wake of Walras, Pareto, and Marshall. The French who, during the 1940s, organized the first seminars of econometrics were in general representatives of these two traditions, rather than statisticians (Bungener and Joël, 1989).

The problem of closely linking an economic theory with statistical data did not really arise for any of these three groups, since the first rejected theory, and the two others used few statistics. However, the question of identifying an economic law from among the heap of observations had already been raised, during the 1910s and 1920s, by the men who had attempted to estimate the laws of supply and demand—Lenoir, Lehfeldt, Moore, Elmer Working—and by American agricultural statisticians (Christ, 1985). This "early econometrics," which was not yet called by that name, did not rely on probabilistic hypotheses to interpret the variability intrinsic to observations and the residuals of linear adjustments effected by the method of least squares. A general theoretical solution to these problems was proposed, in the early 1940s, by the Cowles Commission, and especially by Haavelmo's programmatic texts (1943 and 1944), with the notions of simultaneity and restriction of the space of the laws to be estimated, on the basis of theoretical hypotheses. The problem and its solution were introduced to France by Malinvaud during the 1950s; after a stay with the Cowles Commission he reformulated them in a clear and synthetic manner, and his manuals, which were translated into English, became bestsellers in the teaching of econometrics. Forty years apart, Lenoir and Malinvaud set themselves, in France, the same problem of the link between theories and observations, of which neither descriptive statisticians nor mathematical economists had any idea, though for different reasons. Both of them were officials of the statistical administration (the SGF, then INSEE) and not academics. This typically French situation may explain why they were the only ones to see this problem.

The very general and mathematically elegant solution provided by Haavelmo constituted, after 1945, the reference point for all subsequent controversies, concerning, for example, the primacy of a particular economic theory (NBER); the question of simultaneity (Wold and recursive models); or the distinction between exogenous and endogenous variables (Sims). From 1947 to 1949 the first such controversy pitted the Cowles Commission against the NBER, through the densely reasoned arguments between Koopmans and Vining, and can be reread in this perspective of the connection between theories and observations. The controversy originated in a severe criticism by Koopmans of the final book coauthored in 1946 (with Arthur Burns) by Wesley Mitchell (the founder of the NBER): *Measuring Business Cycles.* Koopmans's article (1947) was entitled "Measurement without Theory." It attacked the school of descriptive economic statisticians, who did not refer their statistical investigations to any theory and therefore could not generalize their observations with a view to making a forecast. He compared Mitchell's position to that of Kepler who, according to Koopmans, accumulated measurements without being able to prove any law, whereas Newton identified the laws of universal gravitation from Kepler's observations: the Cowles Commission was thus the collective Newton of the economy. Mitchell was ill at the time this article appeared, and died in 1948. The following year a young member of the NBER, Rutledge Vining, presented a response to Koopmans. Two further replies were exchanged, and published together in the *Review of Economics and Statistics.* The whole affair presents a fine picture of the different ways then envisaged of using statistical records to support a theoretical construct or historical analysis of an economic situation.

This quarrel has been interpreted in terms of a competitive pursuit of grants from the Rockefeller Foundation; or again, of an opposition between two epistemological positions on the very subject of social sciences, methodological individualism or holism (Mirowski, 1989b). It is true that certain parts of the exchange lend themselves to this second interpretation. Thus, when Vining argued, in a manner reminiscent of Durkheimian sociology, that "the aggregate has an existence apart from its constituent particles and behavior characteristics of its own not deducible from the behavior characteristics of the particles," Koopmans responded:

> If a theory formulates precisely (although possibly in probability terms) the determination of the choices and actions of each individual in a group or population, in response to the choices and actions of other individuals or the consequences thereof (such as prices, quantities, states of expectation), then the set of these individual behavior characteristics is *logically* equivalent to the behavior characteristics of the group. Such a theory does not have an opening wedge for essentially new group characteristics. Any deus ex machina who should wish to influence the outcome can only do so by affecting the behavior of individuals. This does not deny the existence of essentially social phenomena, based on imitation, such as fads and fashions, waves of optimism and pessimism, panics and runs; or based on power struggles, such as price wars, trust formation, lobbying; or based on a social code or sense of responsibility, such as the acceptance of personal sacrifice for a common objective. It is maintained only that such social phenomena are necessarily acted out by individuals as members of a group. (Koopmans, 1949, pp. 86–87)

When Koopmans criticized the NBER for its "measurements without a theory," Vining replied that his group was no less exacting in regard to theoretical reference, but that its theory was not the same, and that other constructs existed besides the one the Cowles Commissions was referring to. Thus several accounts of the controversy are possible: an externalist narrative couched in terms of the battle to obtain grants; or an internalist narrative couched in the terms—not specific to the problem of the link between data and theories—of a recurrent epistemological quarrel between holism and individualism. While these two narratives are enlightening, neither of them accounts for what separates the two groups on the question of the status and the use of statistical material in a scientific construct.

Koopmans suggested that it was not possible to deduce generalizable conclusions on the basis of a set of statistical series, if the examination of these series was not supported by theoretical hypotheses on individual behaviors. In any case, the cycle could not in itself constitute a pertinent unity of analysis. In the absence of such hypotheses, "the movements of economic variables are described as eruptions of a mysterious volcano, whose burning cauldron can never be entered." The assembled data must be summarized in "statistics" (in the sense of inferential statistics) less numerous than the observations, with a view to estimating parameters or testing hypotheses. The set of relations must be estimated by a probabilistic model with simultaneous equations. Only this approach allows one to proceed beyond an accumulation of data in the manner of Kepler, and to construct a theory supported by observations, as did Newton.

Vining replied by opposing the phase of exploration and discovery to that of proof and demonstration. The "statistical efficiency" that Koopmans claimed was more an attribute of the processes of estimation and testing than a characteristic of inventive research:

> Discovery has never been a field of activity in which elegance of conception and equipment is of prime consideration; and foreign to the nature of exploration is the confinement involved in the requirements that the procedure followed shall be characterized by theoretical preconceptions having certain prescribed forms and shall yield findings that are directly subject to the rather restricted tests provided by ideas included in the Neyman-Pearson theory of estimation. (Vining, 1949, p. 79)

Vining feared that emphasis on the formal procedures of the theory of statistical influence might cause people to lose interest in the actual object and its multiple aspects: "Koopmans' emphasis on the hypothesis of distribution is based more on the concerns of estimation than on a real interest in the distribution itself." To support his defense of a finely descriptive statistics, Vining referred to a discussion on the "future of statistics" (published in 1942 by the *Journal of the Royal Statistical Society*) between Yule and Kendall, the two authors of the manual of statistics most often used at the time. Kendall had maintained that "estimating the properties of a population on the basis of a sample has long been the most important practical problem in statistics." Yule, who was seventy in 1942, replied by minimizing the importance of inferential statistics and probabilistic samplings, formalized during the 1930s, notably by Neyman and Egon Pearson. He gave a summary in his way of the history of statistics as he had known it for half a century:

> The initial problem of the statistician is simply the description of the data presented; to tell us what the data themselves show. To this initial problem the function of sampling theory is in general entirely secondary . . . The development of theory during my own lifetime followed at first the natural course suggested by this fact. Primarily, it was new methods that were developed, and investigations of the "probable errors" involved only followed in their train. More recently methods, with few exceptions (time-series in economics, factor-methods in psychology), have been almost neglected, while there has been a completely lopsided development—almost a malignant growth—of sampling theory. I hope there may be a swing back towards the study of method proper, and as methods only develop in connection with practical problems, that means a swing back to more practical work and less pure theory . . .
>
> If the investigator possesses caution, common sense and patience, those qualities are quite likely to keep him more free of error in his conclusions than the man of little caution who guides himself by a mechanical application of sampling rules. He will be more likely to remember that there are sources of error more important than fluctuations of samplings . . . No: I cannot assign the place of highest importance to sampling theory—a high place perhaps, but not the highest. (Yule, 1942, quoted by Vining, 1949, p. 84)

Beyond Yule's manifest incomprehension for techniques of sampling with their decided advantages (if only in terms of the cost of the surveys), his insistence on the "method of description" to which Vining referred in 1949 draws attention to a way of conceiving statistical work different from one that aims at testing theories. This concept was later developed by specialists in sampling surveys, from which they drew ever more sophisticated descriptions.

In 1909, at the Paris congress of the International Statistical Institute, Yule, Bowley, and March presented the methods of statistics resulting from biometrics: partial correlation, multiple regression. Some economists, Lenoir and Moore, took hold of them and tried to identify the laws of supply and demand on the basis of statistical observations; or even to find explanations of business cycles. Economic theory was beginning to rub up against empirical records. Forty years later, in 1949, the possible conditions of this interaction were explored in depth, mainly through the adoption, after the 1930s, of a new way of speaking of statistics, by writing them into probabilistic models. At that point, however, as is shown by the controversy between Koopmans and Vining, statistical tools could still be connected to different rhetorics, with the support of various intellectual, social, or political constructions. There is no single good way of making numbers speak and of using them as a basis for argument. That is why it is indispensable to have a sociology of statistics in which each of these discourses is taken equally seriously, and reintroduced into the network that it supports and that supports it.

Conclusion: Disputing the Indisputable

Among the traits characteristic of the historical line of research begun during the 1930s by the *Annales* school, reference to statistical objectifications has been significant. From this point of view quantitative history has inherited, via Simiand, Halbwachs, and Labrousse, elements of the Durkheimian school and, even closer to the source, of the mode of thinking centered on averages engendered by Quetelet, who opposed macrosocial regularities to the random, unpredictable, and always different accidents of particular events. It sought, by this technique, to overcome individual or factual contingencies in order to construct more general things, characterizing social groups or the long run, depending on the case. This attempt to give form to the chaos of countless singular observations involves having recourse to previous sources or to specific encodings, of which historians ask two questions: Are they available? Are they reliable? In this perspective, the question of the reality and consistency of objects is assimilated to the question of the reliability of their measurements. The cognitive tools of generalization are presumed to have been acquired and firmly constituted. All that matters is the controlled collection and the technical treatment—eventually automated—of the data.

I have tried, in this book, to reverse this traditional relationship between history and statistics by reconstructing what Jean-Claude Perrot (1992) terms a "concrete history of abstraction." The modes of thought and the material techniques operative during the different stages of the history of the social sciences (especially history) are themselves related to the arts of doing and saying studied by general history, and to the political and scientific debates concerning ways of adding. Thus the dominant statistical and macroeconomic constructs of the years between 1940 and 1970 are now opposed by microhistorical (Ginzburg, 1980) or microsociological research, as if some doubt existed in regard to global history or social sciences reasoned on the basis of large ensembles. But is this evolution purely epistemological? Does it involve only cognitive techniques that can be conceived independently of the history they are intended to account for?

The reintroduction of statistical reasoning as a mode of abstraction into a more general social or political history poses a particular problem, for this technique has become virtually synonymous with proof, with almost uncontestable standards of reference. The mental reversal involved in paying attention to the metamorphoses of statistical argument is almost as difficult for researchers as it is for ordinary citizens, henceforth accustomed to grasp the social world through a dense network of indices and percentages. Both the scientific and the social debate have been expressed in a language that is now established, the transformations of which we have followed in these pages. Within this perspective of the study of "debatability" I shall now compare a few of the results obtained, examining how statistical tools have helped to fashion a "public sphere," in the Habermassian meaning of a context of collective debate. I shall also attempt to extend my analysis of the role of these techniques beyond the 1940s—the cut-off point for the preceding chapters—by briefly describing the relative crisis in statistical calculation apparent since the 1970s.[1]

A Cognitive Space Constructed for Practical Purposes

The public sphere, a sphere within which social questions are open to public debate, goes hand in hand with the existence of statistical information accessible to everyone. Claude Gruson, one of the founding fathers of French official statistics, describes this as a condition necessary for democracy and for enlightened debate, and an indispensable reference in discerning society's "heavy trends" (Ladrière and Gruson, 1992). But the bonds between the public sphere and statistical reason are probably deeper than Gruson suggests. The construction of a statistical system cannot be separated from the construction of equivalence spaces that guarantee the consistency and permanence, both political and cognitive, of those objects intended to provide a reference for debates. The space of representativeness of statistical descriptions is only made possible by a space of common mental representations borne by a common language, marked mainly by the state and by law.

From this point of view the public sphere is not only a sometimes vague performative idea, but a sphere that is historically and technically structured and limited. Statistical information did not fall from the sky like some pure reflection of a preexisting "reality." Quite the contrary: it can be seen as the provisional and fragile crowning of a series of conventions of equivalence between entities that a host of disorderly forces is continually trying to differentiate and disconnect. Because it holds its persuasive power from a double reference to two generally distinguished principles of solidification—that of science and that of the state—the space of statistical information is especially significant if one wishes to study whatever makes a public sphere both possible and impossible. The tension between the fact that this information aspires to be a reference in the debate and that it can nonetheless always be called into question and thus itself become a subject of the debate contains one of the major difficulties in thinking through the conditions in which such a sphere would be possible.

One could compare this tension with the more general tension resulting from the fact that many debates bear simultaneously on substantial objects and on the very rules and modalities of the debate: the constitution, the functioning of the assemblies, the means for designating the representatives. Any constitution lays down the rules of its own modification. But that is the point: statistical information does not present itself in the same way. The "indisputable facts" that it is called upon to provide (but that it has also helped to certify) do not contain the modalities of their own debate. This is often perceived as intolerable, at least, far more so than the debate on the modalities of the debate. What is needed, therefore, is to work out a scale of the levels of "debatability" of the objects. The gap between technical objects and social objects—dating back to the seventeenth century—is now a deep one. From this standpoint statistical objects, whose good repute is linked to that of these two categories of objects, are admirably suited to a reexamination of this essential manner of dividing up the modern world; also, to a reflection on the political consequences of the fact that it is both difficult and indispensable to conceive these objects as simultaneously constructed and real, conventional, and solid. For want of such an endeavor, statistical information runs the risk of oscillating endlessly between opposing and complementary states: an undebatable reference situated above these debates, and a target of polemical denunciations destroying the complex pyramid of equivalences.

In its current architecture, statistics presents itself as a combination of two distinct types of tool, the historical paths of which did not converge and lead to a robust construction until about the middle of the twentieth century. The first is politico-administrative: since the eighteenth century, systems for recording, encoding, tabulating, and publishing statistics have gradually been instituted, meaning "statistics" in the sense of the numerical description of various aspects of the social world. The second is cognitive and involves the formalizing of scientific schemes (averages, dispersion, correlation, and probabilistic sampling) designed to summarize, mainly through mathematical tools, a diversity presumed unmanageable. A trace of this double historical origin survives in the fact that, depending on the context, the word "statistics" has different connotations: sometimes quantified results and descriptions, sometimes methods, mathematical formalism, and mode of reasoning. This double path must be traced if we are to understand how a cognitive space of equivalence and comparability came to be constructed for practical purposes. These purposes, these ad hoc means, have been the subject of public judgments and debates that could draw on comparisons. But comparisons are far from constituting the only available resources: statistical reasoning can always enter into conflict with other forms of reasoning that are incommensurate with it, linked for example to the irreducible singularity of the individual. The explanation of this tension is one of two major challenges when considering the place of the statistical argument in the social debate.

The ideas of comparison and juxtaposition for purposes of judgment enable us to follow shifts in the meaning of the word "statistics." In seventeenth-century Germany, statistics was a formal framework used in describing the separate states. A complex form of classification was designed to make facts "easier to remember, teach, and use for men of government." This statistics was not an activity of quantification (numbers were absent from it), but of taxonomy. Memorizing, teaching, governing: things had to be externalized, written down in books, in order to be used again later or transmitted to others. This activity of organizing and ordering created a common language. It was as essential to modern statistics as its arithmetical branch, which we are more likely to think of today.

But the canvas of German statistics contained a further potential. It suggested *comparing* descriptions by means of tables, cross-referencing states and the various items thus standardized so that an observer could take in at a single glance the multiplicity of situations and points of view, thanks to the two dimensions of the book's page. This distinguished written from oral evidence, graphic reasoning from that of speech. But this use of cross-tabulation gave rise to some resistance. It forced people to imagine spheres of comparison, and common criteria. It fell open to the reproach that it reduced and impoverished situations by sacrificing their singularity. Cross-tabulation prompted all the more criticism in that, by its very logic, it tended henceforth to include numbers that could be compared, manipulated, and translated into diagrams. The tabular form, which encouraged the production and comparison of numbers, created a space of calculation, thus ushering in a new form of statistics that dealt with quantities.

Resistance to the construction of such a space was both political and cognitive. It was subsequently overtaken by the establishment of increasingly unified state areas, and by mathematical formulations allowing singularities to be abstracted into normalized relationships that could be transposed from one case to another. The nomenclatures and mathematical formulations enabled the initial task of reducing singular cases to be encapsulated, and new things to be created. These latter were both the black boxes and the elements, singular in their turn, of larger mechanisms. The "reality" of these objects differs depending on whether it is viewed from the standpoint of their genesis or of their use.

Averages, Regularities; Scales, Distributions

Debate over an action involves making clear relationships between objects or events that are a priori incommensurate, in a framework of reference that allows us to conceive them simultaneously. This debate can take shape between several people or, in the case of one person, between moments or between alternative actions. Inner consistency raises problems of the same type as the production of a framework of objectification common to several subjects. The history of probability calculus from the seventeenth to the eighteenth centuries illustrates this duality. How could a relationship be established between future and uncertain events, such as the safe return or the wreck of a ship, the results of a game of chance, the consequences of inoculation? The idea of expectation, which preceded the idea of probability, allowed such frames of reference to be constructed; it also allowed the choices and decisions of one person, or the judgments of several people, to be made coherent and commensurate.

This mode of reasoning had the characteristic of being situated at the junction of two radically distinct interpretations. One, termed subjective or epistemic, was linked to states of mind, and treated probability as a measurement of lack of knowledge. It described "reasons for believing" and was predominant in the eighteenth century. The other, in contrast, termed objective or frequentist, was linked to states of the world and to the observed regularities of their occurrences. The "law of large numbers" and the ambiguity of its philosophical interpretation symbolize this frequentist branch of the probabilistic scheme, dominant during the nineteenth century.

The central instrument of this transformation was the calculus of averages and the examination of their stabilities, formulated by Quetelet around 1830. Questions of public health, epidemiology, and delinquency involved measurements of the goals to be reached and the means devoted to that end. The alchemy that transforms random free individual actions into determined and stable aggregates provided reference points for the debate: objects that were transmittable, because they were external to people. It formed the kernel of the statistical instrumentation of a public sphere. This has been demonstrated regarding the transformation of the social treatment of work-related accidents during the nineteenth century, which saw a shift from individual responsibility as defined by the Civil Code to the firm's responsibility to insure its employees, a responsibility grounded on probability calculus and averages. Systems of insurance and social protection were based on this transformation of individual chance into stable collective objects, capable of being publicly assessed and debated. But instead of paying attention to unpredictable individuals, Quetelet focused on an average on which a controllable action hinged. His reasoning thus did not provide debates with the necessary apparatus for discussing distribution and ordering among individuals. Bent on reducing heterogeneities, Quetelet was not interested in objectifying them—something which was necessary once a debate hinged precisely on them. This objectification occurred when a Darwinian, hereditarian problem concerning the inequalities between individuals was imported from the animal world into the human world, by Galton.

Quetelet's construct did, however, contain a form adequate for this objectification: the "normal law" of a Gaussian distribution of the physical or mental attributes of the human race, such as size or intelligence. But, concerned with the description (and idealization) of his *average man*, Quetelet had not used this form to group and classify different individuals. As it happened, the normal distribution lent itself well to the constitution of ordered one-dimensional scales, and thus to the construction of spaces or reference allowing a simple comparison of individuals—something Galton and his eugenicist successors did, beginning in the 1870s. This reversal in the interpretation of the normal law—from the average toward the dispersion of individual attributes—led Galton to forge tools for creating equivalence between different populations, linked partly by a statistical (nondeterministic) correlation: the height of the son was *partly* "explained" by that of the father. Thus, in opening up a new continent for the objectification of causality—that of a partial, statistical causality—Galton and Pearson offered argumentative rhetorics, with their highly checkered proceedings (one can no longer tell if they were "social" or "technical"), an entirely new tool box, of which the decision makers and their experts in the twentieth century made considerable use. More generally, the tools of mathematical statistics born of biometrics (regression, correlation, tests of adjustments of distribution, variance analysis, probabilistic econometric models) helped instrument the socio-technical space of the debates on the forecasts and decisions directing the social world: they provided solid objects that were reference points for the actors, endowing them with a common language that possessed an elaborate and complex grammar.

The English eugenicist biometricians of the late nineteenth century, inventors of mathematical techniques destined to enjoy a brilliant future, did not conceive of themselves as "statisticians." First and foremost they were activists for a political cause that was hereditarian and meritocratic; according to them, physical and mental attributes were hereditary, and it was therefore desirable to encourage the birth of the "fittest" and to limit birth rates in the poorest classes. This scientifico-political war machine was directed first, against the landed gentry and the clergy, who were hostile to modern science and to Darwinism; and second, against the reformers who held that the causes of poverty were economic and social rather than biological, and who campaigned in favor of welfare systems. For the eugenicists these social measures were noxious because they favored the reproduction of the "unfit." This polemics marked the context of political debates on poverty and its remedies in early twentieth-century England.

What was striking about this episode in the history of statistics was not so much the fact that these new techniques had been invented by partisans of an absurd political philosophy—criminal, in the uses it was subsequently put to—but that these techniques quickly became (from the 1920s and 1930s on) almost obligatory checkpoints for proponents of other views. They thus structured the very terms of debates, and the language spoken in the space of the politico-scientific debate, although the problematics of economic inequalities and familial socio-cultural handicaps have largely replaced the problem of biological inequalities since the 1940s. The statistical causality equipped on the one hand by the surveys of statistical bureaus in existence since the 1830s and on the other by the mathematical formulations imported from biometrics during the 1920s, became the reference of debates on economic and social policies, especially in the United States, after the crisis of 1929.

Roosevelt, elected president in 1932, established a policy of global intervention. To this end he transformed the organization of federal statistics, which previously had been a rather unimportant administration. Not only were population and business censuses developed, but the methods of mathematical statistics were definitively adopted by economists and statisticians. Above all, the three essential tools of social statistics and modern economics were created: surveys based on representative samplings; national accounts; and, slightly later, computers. This ensemble of techniques and their uses in the socio-political debate were devised and assumed their current faces in the United States between 1935 and 1950. They were then adopted and transposed, by joining with other specific traditions, after 1945, first in Western Europe, then in the rest of the world.

The method of *random sampling* involved the application of mathematical tools that had been available since the start of the nineteenth century. But it was not used until around 1900, experimentally at first in Norway and England, to describe the living conditions of the various social classes. It was subsequently used on a large scale in the United States during the 1930s for unemployment measurement, market studies, and election forecasting. This century-long gap between the mathematical formalization and its application by the social sciences can be interpreted in various ways. Whereas probabilities have long been associated with the idea of an incomplete knowledge, statistics, which was linked to the activity of the state, involved exhaustive censuses, covering the entire country without gaps or approximations. The use of probabilistic formulas presupposed a territorial homogeneity (the urn with balls drawn by chance) that nothing could ensure: the potential equivalence of the country's various regions was by no means certain. Finally, the social surveys of the nineteenth century were intended to account for the social relationships and causes of poverty at the *local* level of parishes or urban districts, because the political treatment of these questions resulted from the initiative of local communities.

In the early years of the twentieth century, social questions gradually ceased to be a matter of charity and local paternalism and became a concern of the law, debated in Parliament, to be applied uniformly throughout the entire country. A national space was thus created for debating the causes of poverty and the judicial and legal remedies to apply. The period witnessed the simultaneous creation of institutions for dealing with these problems (unemployment bureaus), administrative records connected to this management (written lists), and methods of measuring these new objects: the poor were replaced by the unemployed. These complex mechanisms of recording, measurement, and treatment were conceived, discussed, and administered according to uniform standards throughout the country. References to a nationally measured unemployment rate appeared around 1930 in public debate in America, and around 1950 in France. Thus the public sphere for dealing with social relationships became increasingly national (but with subtle variations, depending on the degrees and types of centralization in the various countries). In the United States, the spaces of political representation and expression extended and became "nationalized" (thanks mainly to radio), as did markets for consumer goods (thanks to railways and the growth of large firms). This created the right conditions not only for the federal territory to become more uniform, but also for it to be *thought* of as a pertinent totality, a space of equivalence, in both a political and a logical sense.

A Space for Negotiation and Calculation

Analogous transformations in the space of economic information and its use in the public debate took place in France during the 1950s, although here the old administrative centralization had long been preparing this evolution in a typically French fashion. These transformations were linked to the establishment of national procedures for negotiating salaried relationships and making them relatively uniform (description of workers as defined by the "Parodi decrees" in 1946; statute of civil service in 1947), and to the creation of a welfare state: health insurance, family allowances, and retirement pensions. Moreover, the state's economic policy was changing in nature. On the one hand the short-term macroeconomic balance, described through Keynesian categories of national accounts, was henceforth considered as deriving from corrective public actions. On the other hand, indicative planning of major public investments and more generally of major tendencies in the long- and medium-term development of the relationships between economic agents was supposed to provide these latter with a framework of reference for the microeconomic decisions that strongly affected the future (Gruson, 1968). Both from the standpoint of its basic records and taxonomies and of its uses in a public debate directed mainly by this planning, the system of statistical information erected between 1950 and 1970 was strongly connected to these egalitarian structures of a welfare system; to the mode of negotiating salaried relationships (role of the national price index); and to a macroeconomic policy that was Keynesian in inspiration.

Making a space for the contradictory debate on policy options presupposed the existence of a minimum of referential elements common to the diverse actors: a language in which to express things, to state the aims and means of the action, and to discuss its results. This language did not exist prior to the debate. It was negotiated, stabilized, then distorted and gradually destroyed in the course of interactions proper to a given space and historical period. Nor was it a pure system of signs reflecting things existing outside it: the history of unemployment, of how it was defined and measured, and of the institutions designed to reduce it and to help the unemployed offers an example of the interactions between statistical measurements and the institutional procedures for identifying and encoding the objects. This close connection between the nature of a measurement and the functioning of the network of connections and records that led to it can collide with a realist epistemology, spread across the field of economic information by virtue of its use in the social debate. Polemics on the assessment of unemployment have been triggered at regular intervals, couched in more or less the same terms, every two or three years, from the mid-1970s on. They show that the idea of a clearly definable and measurable unemployment—and also the idea of a rate of inflation or a gross domestic product, valid for the whole French economy—were henceforth firmly integrated into the network of common representations. From that point of view, these certainly were realities.

Inscribing a measurement in a system of negotiations and stabilized institutions (for example, through rules of indexing) can provide arguments for denying the objectivity and consistency of certain statistical indicators. That is often the case with unemployment. It was formerly the case with the retail price index that served as a reference for wage negotiations. These polemics on the realism of the equivalences created at a given moment by an institutional and cognitive network show that such networks are never definitively fixed. They can be attacked and unraveled. But this debate on the indicators is ambiguous. It only has an impact to the extent that the realism of the object deemed poorly assessed is not itself called into question. It is in referring to a "real," unknown (if not knowingly concealed) unemployment figure that the polemics assumes its meaning. On the other hand, the statement that measurement always results, one way or another, from a conventional procedure modifies the space of the debate too much—that is, the language expressing it—to be used in the course of it. Thus reflection on encoding procedures can be evoked in different rhetorics, depending on whether or not doubt is cast on the realism of the object and the political and cognitive language in which it is expressed.

In fact, an object's reality depends on the extent and robustness of the broadest network of the objects in which it is inscribed. This network is made up of stabilized connections, of routinized equivalences, and words to describe them. It forms a language: that is, a discernable set of bonds that make things hold. These things are designated by words, themselves linked by a specific grammar. My preferred hypothesis for analyzing the place of statistical information in the space of public debate is that this language assumes, in certain countries and for certain periods, an original consistency, itself linked to the consistency of a form of regulation of social relationships. It is precisely this language that provides the reference points and the common meaning in relation to which the actors can qualify and express their reactions. From this point of view, during the period of 1950–1975 an effort was made—at least provisionally—to unify the economic and social debate around a common language. This was the language of planning and Keynesian macroeconomics, of growth and national accounts, of the sociology of social inequalities and its statistical indicators. It was also the language of state-supported collective bargaining between bosses and trade unions over salaries inscribed in conventional grids, and over an egalitarian and redistributive welfare system (Desrosières, 1989).

This ensemble of actors, procedures, and the words used to express them was relatively coherent, due mainly to the terminology and tools of a statistical system erected precisely during this period. During the 1950s and 1960s national accounts were explicitly presented by its actors as a language that allowed the social partners meeting in committees of the Plan or around tables of egalitarian bargaining to rely on stable, regularly measured objects. These latter were inscribed, on the one hand, in the coherent and exhaustive network of accountable relationships (Fourquet, 1980) and, on the other, in the econometric relations used, beginning in the 1970s, in large macroeconomic models (Malinvaud, 1991). This language was disseminated through the teaching of the École Nationale d'Administration (ENA), the École de Sciences Politiques, by universities, then by secondary education, especially in the new programs of economic and social sciences, the manuals of which were strongly inspired by the works and analytical categories of official statistics. It would even seem that this dissemination and widespread acceptance were more marked in France than in other countries, being situated within an older tradition that placed great importance on the state engineers, trustees of a science applied in managing a strong, long-centralized state. Thus mathematical economics was introduced into and developed in France by engineers, whereas in the Anglo-Saxon countries this research was conducted by universities. Keynesian macroeconomics and national accounts had a certain particular coloration, because they were implanted and promoted by high civil servants and engineers rather than professors. The legitimacy and authority of the state were subtly combined with those of science.

Thus for thirty or so years a cognitive space of negotiation and calculation existed, endowed with the legitimacy of science and the state. Within this space a large number of debates and technical studies were conducted, preceding or accompanying decisions in economic policy. However, this relatively coherent space, comprised of institutions, social objects, and words for debating them, itself entered into a crisis in the late 1970s. The networks of equivalences leading to political and statistical additions have been partly unraveled. The Plan has less importance as a place for concerting and making medium-term forecasts for major public decisions. The econometric models that simulated the development of relationships between the most central macroeconomic and macrosocial objects of this system of addition are often deemed incapable of predicting tensions and crises. Debates over the actual measurement of certain of these objects and their significance have become more bitter: active population, unemployment, monetary mass, poverty, so-called informal economics (in other words, that eluded administrative coding).

There is no one, general "explanation" for this evolution, precisely because the previous language is inadequate to account for its own crisis. No explanation is therefore more general than another. One can mention a few. The halt of growth makes it harder to assemble the social partners for debates, the topic of debate no longer being how to share benefits, but how to divide the effects of the crisis. The greater integration of the French economy into global trading henceforth inhibits the use of Keynesian models valid for an autonomous economy. A decrease in the representativeness of trade unions and political organizations, which were partly responsible for adding claims and projects in a unified and stable language, makes their spokesmen more vulnerable, whereas previously those spokesmen were intermediaries allowing a relatively well delineated public space to function. As a place for accumulating information and producing representations adequate for political action, the national state finds itself increasingly torn between local communities—whose importance increased with laws of decentralization—and European institutions and regulations. The action of the state is less voluntarist and macroeconomic, and more oriented toward the production of rules facilitating a free market and competition. Firms are now less often managed in a centralized manner according to Taylorian and Fordian principles which, in favoring the standardizing of tasks and products of large consumption, were well suited to the construction of integrated systems of industrial statistics. In contrast, "Japanese" modes of decentralized management are based on the local circulation of information through direct horizontal contacts between people rather than by a hierarchic path, thus diminishing the pertinence of previous statistical syntheses.

All the preceding hypotheses (one could add others) do not "explain" the crisis in the model of public space that occurred between 1950 and 1970; but the very fact that they are widely circulated and reinforce one another helps make this model less credible, threatening its status as a system of reference accepted without discussion. This is simply the trend, and large pieces of the model survive—or at least a large number of the debates concern it, since it still constitutes a widespread framework of thought, if not the only one. For this reason the observed debates display great heterogeneity. They range from those positioned completely within the realist epistemology suggested by the wide diffusion of the model, to those denounced by the networks for creating equivalence—denounced either as deceptive from the point of view of knowledge, or even as oppressive by virtue of their intrusions into individual freedoms.

Statistical Reason and Social Debate

The controversies about statistics offer an original combination of two forms of opposition that are generally conceived separately, in distinct contexts of discussion. The first separates two linguistic registers: that of description and science *(there is)*, and that of prescription and action *(we must)*. This distinction, affirmed in the seventeenth century by the autonomization of a scientific discourse, has left a profound mark on statistical practice. Statisticians constantly demand nothing but a description of the facts, as opposed to values and opinions. The second opposition distinguishes two attitudes in relation to the question of reality, one realist (or objectivist), the other relativist (or historicist): in one case, the equivalence between the cases is presumed to exist prior to the chain of events; in the other, however, it is seen as conventional and constructed. This opposition runs through many controversies in the epistemology of sciences, but for statistics it offers some original aspects once it is combined with the former—which distinguishes the languages of science and of action—in such a way as to make visible four different attitudes in relation to statistical argument.[2]

Within the scientific language of description, the realist position postulates that there are objective things, existing independently of observers and exceeding singular contingencies (case 1). This is, typically, the language of Quetelet: *there are* regularities and stable relationships. Statistics aims at "approaching reality." It sets itself problems of "reliability of measurement." It speaks the language of description and causality formalized by the tools of mathematical statistics. This position is one that statistical discourse necessarily tends toward, once it seeks to purge itself of its birth and uses. In any case, it is the reference point for the three other positions resulting from this hybrid mixture. But, while remaining in the language of science, it is possible to reconstruct a genesis, and the social practices that have led to a solid statistical object (case 2). *There are* historical and social processes of constructing and solidifying equivalences and mental schemes. It is up to science to reconstitute them, by describing how social facts become things, through customs, law, or social struggles. The language of this position is that of social history, or of a constructivist sociology of knowledge.[3]

For its part, the political and administrative language of action and social debate either uses or denounces statistics. It derives support from one or another of the scientific rhetorics mentioned above, but is distinguished from them by its normativeness. In its objective version (case 3), it takes up the real objects described and analyzed in scientific language and makes the action bear upon them. *We must* have things that hold up well, independently of particular interests, in order to be able to act on them. These things are categories of action: poverty, unemployment, inflation, the trade deficit, monetary mass, fertility, causes of death. The language used is pragmatic: means toward an end. In the relativist version (case 4), the political language can have several modalities. It can be polemical and accusatory. *We must* open up the black boxes to show what they conceal. Statistical production results from power relationships. It is ideological. This modality is unstable because, being based on the language of denunciation, it implicitly refers to a potential positivity, either scientific or political. Nonetheless it has often been maintained in controversies, in particular during the 1970s.

A further modality of the use of statistics in the language of action can be considered. This is based on the idea that conventions defining objects really do give rise to realities, in as much as these objects resist tests and other efforts to undo them. This principle of reality affords an exit from the dead-ended epistemological opposition between these two complementary and complicitous enemies, the realist and the relativist. It does not deny the reality of things once numerous persons refer to them to guide and coordinate their actions. In this respect statistics is above all, by virtue of its objects, nomenclatures, graphs, and models, a conventional language of reference. Its existence allows a certain type of public sphere to develop, but its vocabulary and syntax can themselves be debated: debate on the referent of the debate, and the words used to conduct it, is an essential aspect of any controversy. But, just like other major investments that are almost irreversible, on account of their cost, the conventions of equivalence and permanence of the objects on which statistical practice is based are themselves the product of extremely expensive political, social, and technical investments.

A public debate that employs statistical reason, either in support of its investments or to discuss them, is thus circumscribed by contradictory constraints. On the one hand, controversy can lead to a questioning of the equivalence and permanence of the qualities of the objects. But on the other hand, the institution of other conventions is very costly. I would like the reflection offered in this book on the relationships between statistics and the public sphere to help clarify and analyze these spaces of durably solidified forms, which must simultaneously remain undebated so that life may follow its course, and debatable, so that life can change its course.

Notes

Unless otherwise specified, translations of quoted passages are mine.—Trans.

Introduction

1. But not all scientists make this choice; for example, Stephen Stigler's very useful work (1986) on the history of mathematical statistics in the nineteenth century is mainly internalist.
2. In modified form, this chapter reproduces a text published in a collective work edited by J. Mairesse (1988).
3. Although the two perspectives, internal and external, are connected as closely as possible considering the available sources, Chapters 2, 3, 4, and 9 deal more with changes in cognitive systems, whereas Chapters 1, 5, 6, and 7 are closer to social and institutional history. Chapter 8, which deals with classification, is in a certain sense at the intersection of these two points of view.

1. Prefects and Geometers

1. The comparison of Bottin and Barème—tools of standardization destined to disappear into anonymity—is significant.
2. Only after the 1830s did these two groups begin to be thought of as a single class, united by the struggle of the "workers' movement" (Sewell, 1980).

2. Judges and Astronomers

1. This chapter deals with the birth of probability theory and its uses in the natural sciences. It is slightly difficult for readers with a purely literary background. But it is not necessary to understand it completely in order to read the chapters that follow.
2. This quarrel offers more than mere anecdotal interest, for it has to do with

what causes a scientific tool to solidify and be able to be transmitted and transposed into other contexts.

3. The network of these successive comparisons is suggested later in this book, in the form of a genealogical tree of econometrics (Chapter 9).

3. Averages and the Realism of Aggregates

1. One can estimate an indicator p of this variation in the composition of the urn by observing that $R^2 = r^2 + p^2$. The decomposition of the total variance between intraclass and interclass variance thus becomes explicit.
2. In 1896 Livi, an Italian (quoted by Stigler), declared that this result was an artifact resulting from an error in the conversion of inches into centimeters. Bertillon's text (1876, pp. 287–291) does not seem to justify this criticism. He suggests, in contrast, another point: the height corresponding to the second maximum was what gave conscripts access to the cavalry and to genius, thus escaping from the infantry.

4. Correlation and the Realism of Causes

1. We find an echo of this model in the Popperian idea of falsifiability. Popper belonged, at least from this point of view, to the same intellectual tradition as Mach and Pearson.
2. In part of his work, MacKenzie (1981) does not completely avoid the danger of this macrosocial reading—as for example when he tries to relate the rival expressions of correlation coefficient almost directly to the social origins of the people upholding them. (The more technical part of his work, however, is very rich, and provides essential data on the links between different kinds of arguments.)

5. Statistics and the State: France and Great Britain

1. Yet Moreau de Jonnès was close to the medical world: see on this point Chapter 3 and the debates on cholera.
2. Yet the Fondation pour l'Étude des Problèmes Humains, founded in 1942 by Alexis Carrel, provided a vehicle for this eugenicist concept before being transformed and integrated into the nascent Institut National d'Études Démographiques (INED) after the Liberation (Drouard, 1983; Thévenot, 1990).
3. It was also consistent with the position taken by the doctors of the GRO in the debate described in Chapter 3, in which the "contagionists" were pitted against the partisans of the "miasma theory." William Farr favored the miasma theory, and was anticontagionist.
4. I shall further discuss and interpret the surveys of Booth, Rowntree, and Bowley in Chapter 7.

5. Another pertinent indicator was the more or less anonymous nature of the publication. A signature was associated with scientific competition: anonymity, with administrative practice.

6. Statistics and the State: Germany and the United States

1. Later on, they themselves became a source of inspiration for the American institutionalist economists, major users of statistics who helped found econometrics between 1910 and 1930.
2. An extract has been published in French and analyzed, in the context of the Verein, by Michael Pollak (1986).
3. This remark is not intended to invalidate the philosophical debate on account of its eclectic opportunism; rather, it is meant to suggest that such transfers require precautions. Otherwise, the original model may find itself subject to opposing reinterpretations, as in the case of the *average man*.
4. In a study entitled "L'homme probable: Robert Musil, le hasard, et la moyenne" Jacques Bouveresse (1993) shows that the theme of the "man without qualities" was influenced in part by the German debates about the average man and individual freedom.
5. The irony of this story is that economics has retained the name of "Engel's law" to denote the relationship he established between revenue and expenditure on food.
6. My analysis of American statistics owes much to the works of Duncan and Shelton (1978) and Anderson (1988).

8. Classifying and Encoding

1. For France, see Salais, Baverez, and Reynaud (1986), and for Great Britain, see Mansfield (1988).
2. In this section, I rely mainly on the reference work by Anne Fagot-Largeault (1989).

9. Modeling and Adjusting

1. This chapter refers mainly to the remarkable synthesis by Mary Morgan (1990).
2. This judgment could itself be made according to two different perspectives—subjects of lively controversy between Fisher on one side and Neyman and Pearson on the other. Fisher's test was designed to promote truth and science: a theoretical hypothesis is judged plausible, or it is rejected, after consideration of the observed data. Neyman and Pearson's test, on the other hand, was designed to promote decision making and action: one evaluated the respective costs of accepting a false hypothesis and of rejecting a true one, described as errors of types I and II (Gigerenzer and Murray, 1987).

3. An account of the quarrel—rather unfavorable, in regard to Keynes—is provided by Mary Morgan (1990). Patinkin (1976) and Bateman (1990) emphasize and comment on the epistemological problems it raised.
4. This theme was typical of Karl Pearson's main interest: he wanted to show that these aptitudes were hereditary, and not linked to the parents' way of life.
5. At least, Keynes is so demanding in regard to the conditions of legitimate induction that he seems to discredit the first sketches of macroeconomic constructs.
6. The word "multicollinearity," as used by Frisch and Haavelmo, covered the two ideas of simultaneity of a system of equations, and of multicollinearity in the current sense, expressing the fact that certain variables could themselves be combinations of other variables.
7. These three perspectives correspond in part to successive historical forms of the goals of state action, as described particularly by Musgrave (1959), under the name of allocation, distribution, and stabilization (or regulation).

Conclusion

1. Here I reproduce material presented in greater detail in two previous articles. One appeared in the *Courrier des statistiques,* and deals with the recent history of French official statistics (Desrosières, 1989). The other appeared in *Raisons pratiques* in an issue devoted to the public sphere (Desrosières, 1992).
2. This formulation owes much to discussions with Luc Boltanski and Nicolas Dodier. Dodier has conducted an accurate study of the practices of recording and statistical encoding in the case of occupational medicine (Dodier, 1993).
3. Each of these two positions was adopted at one time or another by Durkheim in his attempt to make sociology into a science. In *Suicide* he bases his observations, as did Quetelet, on macrosocial statistical regularities. Subsequently, with Mauss, in *Some Primitive Forms of Classification,* he relates taxonomy and society in primitive societies. But he does not do so for Western societies and their statistical mechanisms. Admittedly, these mechanisms were less developed than they are today.

References

Abrams, P. 1968. *The Origins of British Society, 1834–1914.* Chicago: University of Chicago Press.

Affichard, J. 1987. Statistiques et mise en forme du monde social: Introduction à l'histoire des statistiques écrite par les statisticiens. In *INSEE 1987: Pour une histoire de la statistique,* vol. 2: *Matériaux,* pp. 9–17. Paris: INSEE, Economica.

Aldrich, J. 1989. Autonomy. *Oxford Economic Papers,* 41, 1 (January): 15–34.

Anderson, M. J. 1988. *The American Census: A Social History.* New Haven: Yale University Press.

Armatte, M. 1991. Une discipline dans tous ses états: la statistique à travers ses traités (1800–1914). *Revue de synthèse,* 2: 161–205.

——— 1992. Conjonctions, conjoncture, et conjecture: les baromètres économiques (1885–1930). *Histoire et mesure,* 7, 1/2: 99–149.

Balinski, M., and H. P. Young. 1982. *Fair Representation: Meeting the Ideal of One Man, One Vote.* New Haven: Yale University Press.

Bateman, B. W. 1990. Keynes, Induction, and Econometrics. *History of Political Economy,* 22, 2: 350–379.

Bayart, D. 1992. La quantifiction du contrôle qualité dans l'industrie: un point de vue sociologique et historique, pp. 26–27. Paper given at the conference La qualité dans l'agro-alimentaire, October, Société française économie rurale. Paris.

Bayes, T. 1764 (1970). An Essay towards Solving a Problem in the Doctrine of Chances. *Philosophical Treatise of the Royal Society of London,* 53, pp. 370–418. Reprinted in Pearson and Kendall (1970), pp. 131–153.

Benzécri, J. P. 1973. *L'analyse des données.* Paris: Dunod.

——— 1982. *Histoire et préhistoire de l'analyse des données.* Paris: Dunod.

Bertillon, A. 1876. La théorie des moyennes en statistiques. *Journal de la Société statistique de Paris:* 265–308.

Bertillon, J. 1895. *Cours élémentaire de statistique administrative.* Paris: Société d'éditions statistiques.

Besnard, P. 1976. Anti- ou anté-durkheimisme? Contribution au débat sur les

statistiques officielles du suicide. *Revue française de sociologie,* 17, 2 (April-June): 313–341.
Bienaymé, I. J. 1855. Sur un principe que M. Poisson avait cru découvrir et qu'il avait appelé loi des grands nombres. *Séances et travaux de l'Académie des sciences morales et politiques,* 31: 379–389.
Bloor, D. 1982. *Socio-logie de la logique, ou les limites de l'épistémologie.* Paris: Editions Pandore.
Board of Trade. 1909. *Cost of Living in French Towns: Report on an Inquiry by the Board of Trade.* London: Darling.
Boltanski, L. 1982. *Les cadres: la formation d'un groupe social.* Paris: Minuit.
Boltanski, L., and L. Thévenot. 1983. Finding one's way in social space: a study based on games. *Social Science Information,* 22, 4–5: 631–679.
——— 1991. *De la justification: les économies de la grandeur.* Paris: Gallimard.
Booth, C. 1889. *Labour and Life of the People.* London.
Borel, E. 1920. La statistique et l'organisation de la présidence du Conseil. *Journal de la Société statistique de Paris:* 9–13.
Bourdelais, P., and J. Y. Raulot. 1987. *Une peur bleue: Histoire du choléra en France, 1832–1854.* Paris: Payot.
Bourdieu, P. 1979. *La distinction.* Paris: Minuit.
——— 1980. *Le sens pratique.* Paris: Minuit.
Bourguet, M. N. 1988. *Déchiffrer la France: la statistique départementale à l'époque napoléonienne.* Paris: Éditions des archives contemporaines.
Bouveresse, J. 1993. *L'homme probable: Robert Musil, le hasard, la moyenne, et l'escargot de l'histoire.* Combas 30250: L'éclat.
Bowley, A. 1906. Presidential address to the Economic Section of the British Association. *Journal of the Royal Statistical Society:* 540–558.
——— 1908. The improvement of official statistics. *Journal of the Royal Statistical Society:* 460–493.
——— 1919. *The Division of the Product of Industry: An Analysis of National Income before the War.* Oxford: Clarendon.
Bowley, A., and J. Stamp. 1927. *The National Income, 1924.* Oxford: Clarendon.
Box, G. 1980. Sampling and Bayes' Inference in Scientific Modelling and Robustness. *Journal of the Royal Statistical Society,* ser. A, 143: 383–430.
Bravais, A. 1846. Analyse mathématique sur les probabilités des erreurs de situation d'un point. *Mémoires présentés par divers savants à l'Académie Royale des Sciences de l'Institut de France,* 9: 255–332.
Brian, E. 1989. Statistique administrative et internationalisme statistique pendant la seconde moitié du XIXe siècle. *Histoire et mesure,* 4, 3/4: 201–224.
——— 1991. Le prix Montyon de statistique à l'Académie royale des sciences pendant la Restauration. *Revue de synthèse,* 2: 207–236.
Bru, B. 1988. Estimations Laplaciennes. Un exemple: la recherche de la population d'un grand Empire, 1785–1812. In J. Mairesse, ed., *Estimation et sondages—cinq contributions à l'histoire de la statistique,* pp. 7–8. Paris: Economica.

Buckle, H. 1857. *History of Civilisation in England.* London.

Bulmer, M., Bales, K., and K. Sklar. 1991. *The Social Survey in Historical Perspective, 1880–1914.* Cambridge: Cambridge University Press.

Bungener, M., and Joël, M. E. 1989. L'essor de l'économétrie au CNRS. *Cahiers pour l'histoire de la CNRS,* no. 4: 45–78.

Bunle, H. 1911. Relation entre les variations des indices économiques et le mouvement des mariages. *Journal de la Société statistique de Paris* (March): 80–93.

Callon, M., ed. 1989. *La science et ses réseaux: genèse et circulation des faits scientifiques.* Paris: La Découverte.

Chang, W. C. 1976. Statistical theories and sampling practice. In D. Owen, ed., *On the History of Statistics and Probability.* New York: Marcel Dekker.

Chateauraynaud, F. 1991. *La faute professionnelle: une sociologie des conflits de responsabilité.* Paris: Métaillié.

Cheysson, E. 1890. *Les budgets comparés de cent monographies de famille.* With A. Toque. Rome: Botta.

Christ, C. F. 1985. Early progress in estimating quantitative economic relationship in America. *American Economic Review,* 75, 6: 39–52.

Cicourel, A. 1964. *Method and Measurement in Society.* New York: Free Press of Glencoe.

Clero, J. P. 1986. Remarques philosophiques sur l'essai de Bayes en vue de résoudre un problème de la doctrine des chances. *Publications de l'Institut de recherches mathématiques de Rennes,* IRMAR, Université de Rennes I.

Converse, J. 1987. *Survey Research in the United States: Roots and Emergence.* Berkeley: University of California Press.

Coumet, E. 1970. La théorie du hasard est-elle née par hasard? *Annales ESC,* no. 3 (May-June): 574–598.

Cournot, A. 1843 (1984). *Exposition de la théorie des chances et des probabilités.* Paris: Hachette. Reprinted in 1984 in Cournot, *Oeuvres Complètes,* vol. 1, ed. Bernard Bru. Paris: Vrin.

Crombie, A. C. 1981. Philosophical Presuppositions and Shifting Interpretations of Galileo. In J. Hintikka, D. Gruender, and E. Agazzi, eds., *Theory Change, Ancient Axiomatics, and Galileo's Methodology: Proceedings of the 1978 Pisa Conference and the History and Philosophy of Science,* p. 284. Dordrecht: Reidel.

Daston, L. J. 1987. The Domestication of Risk: Mathematical Probability and Insurance, 1650–1830. In L. Kruger, L. Daston, and L. Heidelberger, eds., *The Probabilistic Revolution,* vol. 1, *Ideas in History,* pp. 237–260. Cambridge, Mass.: MIT Press.

——— 1988. *Classical Probability in the Enlightenment.* Princeton: Princeton University Press.

——— 1989. L'interprétation classique du calcul des probabilités. *Annales ESC,* 3 (May-June): 715–731.

Delaporte, F. 1990. *Le savoir de la maladie: essai sur le choléra de 1832 à Paris.* Paris: PUF.

De Marchi, N. 1991. League of Nations economists and the ideal of peaceful change in the decade of the thirties. In C. D. Goodwin, ed., *Economic and National Security: A History of Their Interaction,* pp. 143–178. Durham: Duke University Press.

De Marchi, N., and C. Gilbert, eds. 1989. History and Methodology of Econometrics. *Oxford Economic Papers,* 1 (January).

Desrosières, A. 1977 (1987). Eléments pour l'histoire des nomenclatures socioprofessionnelles. In *Pour une histoire de la statistique,* vol. 1. Reprinted in 1987; Paris: INSEE, Economica, pp. 155–231.

——— 1985. Histoire de formes: statistiques et sciences sociales avant 1940. *Revue française de sociologie,* 26, 2: 277–310.

——— 1986. L'ingénieur d'état et le père de famille: Émile Cheysson et la statistique. *Annales des mines,* series *Gérer et comprendre,* no. 2: 66–80.

——— 1988a. Masses, individus, moyennes: la statistique sociale au XIXe siècle. *Hermès,* 2, CNRS, pp. 41–66.

——— 1988b. La partie pour le tout: comment généraliser? La préhistoire de la contrainte de représentativité. In J. Mairesse, ed., *Estimation et sondages: Cinq contributions à l'histoire de la statistique,* pp. 97–116. Paris: Economica.

——— 1989. Les spécificités de la statistique publique en France: une mise en perspective historique. *Courrier des statistiques,* 49: 37–54.

——— 1992. Discuter l'indiscutable. Raison statistique et espace publique. *Raisons pratiques,* 3:131–154.

Desrosières, A., and M. Gollac. 1982. Trajectoires ouvrières, systèmes d'emploi, et comportements sociaux. *Economie et statistique,* 147 (September): 43–66.

Desrosières, A., J. Mairesse, and M. Volle. 1977 (1987). Les temps forts de l'histoire de la statistique française. *Pour une histoire de la statistique,* vol. 1: *Contributions.* Reprinted in 1987; Paris: INSEE, Economica, pp. 509–518.

Desrosières, A., and L. Thévenot. 1988. *Les Catégories socioprofessionnelles.* Paris: La Découverte.

Dodier, N. 1993. *L'expertise médicale: essai de sociologie sur l'exercice du jugement.* Paris: Métaillié.

Droesbeke, J. J., and P. Tassi. 1993. *Histoire de la statistique.* Paris: PUF, "Que sais-je?" no. 2527.

Drouard, A. 1983. Les trois âges de la Fondation française pour l'étude des problèmes humains. *Population,* 6: 1017–1037.

Dugé de Bernonville, L. 1916. Enquête sur les conditions de la vie ouvrière et rurale en France en 1913–1914. *Bulletin de la Statistique générale de la France,* vol. 4.

Dumont, L. 1983. *Essai sur l'individualisme.* Paris: Seuil.

Duncan, J. W. and W. C. Shelton. 1978. *Revolution in United States Government Statistics, 1926–1976.* Washington, D.C.: U.S. Department of Commerce.

Dupaquier, J. 1985. *Histoire de la démographie.* Paris: Perrin.

Durkheim, E. 1893. *De la division du travail social.* Translated as *The Division of Labor in Society.* Trans. George Simpson. New York: Macmillan, 1933.

——— 1894. *Les règles de la méthode sociologique.* Translated as *The Rules of*

Sociological Method. Trans. Sarah A. Solovay and John H. Mueller. New York: Collier-Macmillan, 1938.

——— 1897. *Le suicide: etude de sociologie.* Translated as *Suicide: A Study in Sociology.* Trans. John A. Spaulding and George Simpson. New York: Collier-Macmillan, 1951.

Durkheim, E., and M. Mauss. 1903 (1968). De quelques formes primitives de classification. *Année sociologique,* reprinted in M. Mauss, *Essais de sociologie;* Paris: Minuit, 1968.

Duvillard, E. 1806 (1989). Mémoire sur le travail du Bureau de statistique. Reprinted in *Études et documents,* no. 1, 1989; Paris: Comité pour l'histoire économique et financière de la France, Imprimerie Nationale.

Edgeworth, F. 1881. *Mathematical Psychics: An Essay on the Application of Mathematics to the Moral Sciences.* London: Kegan Paul.

——— 1885. Calculus of Probabilities Applied to Psychical Research, I. *Proceedings of the Society for Psychical Research* 3: 190–208.

——— 1892. The Law of Error and Correlated Averages. *Philosophical Magazine,* 5th ser., 34: 429–438, 518–526.

Elias, N. 1973. *La civilisation des moeurs.* Paris: Calmann-Lévy.

Epstein, R. J. 1987. *A History of Econometrics.* Amsterdam: North-Holland.

Etner, F. 1987. *Histoire du calcul économique en France.* Paris: Economica.

Ewald, F. 1986. *L'état providence.* Paris: Grasset.

Eymard-Duvernay, F. 1986. La qualification des produits. In R. Salais and L. Thévenot, eds., *Le travail: marchés, règles, conventions.* Paris: INSEE, Economica.

——— 1989. Conventions de qualité et formes de coordination. *Revue économique,* 40, 2: 329–359.

Fagot-Largeault, A. 1989. *Les causes de la mort: histoire naturelle et facteur de risque.* Paris: Vrin.

——— 1991. Réflexions sur la notion de qualité de la vie. *Archives de philosophie du droit,* vol. 36, *Droit de science,* pp. 135–153.

Feldman, J., G. Lagneau, and B. Matalon, eds. 1991. *Moyenne milieu, centre: histoires et usages.* Paris: Éditions de l'école des hautes études en sciences sociales.

Fisher, R. 1925. *Statistical Methods for Research Workers.* Edinburgh: Oliver and Boyd.

Foucault, M. (1966) 1970. *Words and Things.* Pantheon, 1970. English translation of *Les mots et les choses.* Paris: Gallimard, 1966.

Fourquet, F. 1980. *Les comptes de la puissance: histoire de la comptabilité nationale et du Plan.* Paris: Encres.

Fox, K. 1989. Some Contributions of U.S. Agricultural Economists to Statistics and Econometrics, 1917–1933. *Oxford Economic Papers,* 41 (January): 53–70.

Fréchet, M., and M. Halbwachs. 1924. *Le calcul des probabiliés à la portée de tous.* Paris: Dunod.

Frisch, R. 1931. A method of decomposing an empirical series into its cyclical and progressive components. *Journal of the American Statistical Association,* 26: 73–78.

——— 1933. Propagation problems and impulse problems in dynamic economics. In *Economic Essays in Honor of Gustav Cassel.* London: Allen and Unwin.

——— 1938. Statistical versus theoretical relations in economic macrodynamics. In *Autonomy of Economic Relations,* University of Oslo, 1948.

——— 1948. Repercussion studies at Oslo. *American Economic Review,* 39: 367–372.

Galton, F. 1869. *Hereditary Genius: An Inquiry into Its Laws and Consequences.* London: Macmillan.

——— 1886. Regression towards Mediocrity in Hereditary Stature. *Journal of the Anthropological Institute,* 15: 246–263.

——— 1889. *Natural Inheritance.* London: Macmillan.

——— 1909. *Essays in Eugenics.* London: Eugenics Education Society.

Gigerenzer, G., and D. J. Murray. 1987. *Cognition as Intuitive Statistics.* Hillsdale, N.J.: Lawrence Erlbaum Associates.

Gigerenzer, G., et al. 1989. *The Empire of Chance: How Probability Changed Science and Everyday Life.* Cambridge: Cambridge University Press.

Gille, B. 1964. *Les sources statistiques de l'histoire de France: des enquêtes du XVIIe siècle à 1870.* Geneva: Droz.

Gini, C. 1928. Une application de la méthode représentative aux matériaux du dernier recensement de la population italienne (1er décembre 1921). *Bulletin de l'Institut international de statistique,* 23, 2: 198–215.

Ginzburg, C. 1980. Signes, traces, pistes: racines d'un paradigme de l'indice. *Le débat,* no. 6 (November): 3–44.

Goody, J. 1979. *La raison graphique.* Paris: Minuit.

Gould, S. J. 1981. *The Mismeasure of Man.* New York: Norton.

Gruson, C. 1968. *Origines et espoirs de la planification française.* Paris: Dunod.

Guibert, B., J. Laganier, and M. Volle. 1971. Essai sur les nomenclatures industrielles. *Économie et Statistique,* 20: 23–36.

Haavelmo, T. 1943. The statistical implications of a system of simultaneous equations. *Econometrica,* 11: 1–12.

——— 1944. The probability approach in econometrics. Supplement to *Econometrica,* 12: 1–118.

Hacking, I. 1975. *The Emergence of Probability.* Cambridge: Cambridge University Press.

——— 1987. Prussian Numbers 1860–1882. In L. Kruger, L. Daston and L. Heidelberger, eds., *The Probabilistic Revolution,* vol. I, pp. 377–394. Cambridge, Mass.: MIT Press.

——— 1990. *The Taming of Chance.* Cambridge: Cambridge University Press.

——— 1991. Statistical language, statistical truth, and statistical reason: the self-authentication of a style of scientific reasoning. In McMullen, ed., *Social Dimensions of Sciences.* Notre Dame, Ind.: University of Notre Dame Press.

Halbwachs, H. 1912. *La classe ouvrière et les niveaux de vie.* Paris: Alcan.

——— 1913. *La théorie de l'homme moyen: essai sur Quetelet et la statistique morale.* Paris: Alcan.

——— 1914. Budgets de familles ouvrières et paysannes en France, en 1907. *Bulletins de la Statistique générale de la France,* 4, fasc. 1: 47–83.

Hansen, M. H., and W. G. Madow. 1976. Some important events in the historical development of sample surveys. In D. Owen, ed., *On the History of Statistics and Probability.* New York: Marcel Dekker.

Hecht, J. 1977 (1987). L'idée de dénombrement jusqu'à la révolution. In *INSEE 1977: Pour une histoire de la statistique,* vol. 1, pp. 21–81. Reprinted in 1987; Paris: INSEE, Economica.

Hecht, J., ed. 1979. *"L'Ordre divin," aux origines de la démographie,* by J. P. Sussmilch (1707). Paris: INED.

Hennock, E. P. 1976. Poverty and social theory in England: the experience of the eighteen-eighties. *Social History (I),* pp. 67–69.

——— 1987. The measurement of poverty: from the metropolis to the nation, 1880–1920. *Economic History Review,* 2nd ser., 40, 2: 208–227.

Henry, L. 1963. Réflexions sur l'observation en démographie. *Population,* 2: 233–262.

Héran, F. 1984. L'assise statistique de la sociologie. *Economie et statistique,* 168 (July-August): 23–35.

——— 1987. La seconde nature de l'habitus: tradition philosophique et sens commun dans le langage sociologique. *Revue française de sociologie,* 28: 385–416.

Hoock, J. 1977 (1987). D'Aristote à Adam Smith: quelques étapes de la statistique allemande entre le XVIIe et le XIXe siècle. In *INSEE 1977: Pour une histoire de la statistique,* vol. 1, pp. 477–492. Reprinted in 1987; Paris: INSEE, Economica.

Hooker, R. H. 1901. Correlation of the marriage rate with trade. *Journal of the Royal Statistical Society,* 64: 485–492.

Huber, M. 1937. Quarante années de la Statistique générale de la France, 1896–1936. *Journal de la Société statistique de Paris,* pp. 179–214.

INSEE. 1977 (1987). *Pour une histoire de la statistique,* vol. 1: *Contributions.* Reprinted in 1987; Paris: INSEE, Economica.

INSEE. 1987. *Pour une histoire de la statistique,* vol. 2: *Matériaux,* ed. J. Affichard. Paris: INSEE, Economica.

Jensen, A. 1925a. Report on the representative method in statistics. *Bulletin de l'Institut international de statistique,* 22, bk. 1: 359–380.

——— 1925b. The representative method in practice, annexe B. *Bulletin de l'Institut national de statistique,* vol. 22, bk. I: 381–489.

John, V. 1884. *Geschichte der Statistik.* Stuttgart: Enke.

Jorland, G. 1987. The Saint-Petersburg Paradox, 1713–1937. In L. Kruger, L. Daston, and L. Heidelberger, eds., *The Probabilistic Revolution,* vol. 1, *Ideas in History,* pp. 157–190. Cambridge, Mass.: MIT Press.

Juglar, C. 1862. *Des crises commerciales et de leur retour périodique en France, en Angeleterre, et aux Etats-Unis.* Paris.

Kalaora, B., and Savoye, A. 1985. La mutation du mouvement le playsien. *Revue française de sociologie,* 26, 2: 257–276.

——— 1987. *Les inventeurs oubliés: Frédéric Le Play et ses continuateurs.* CERFISE.

Kang, Z. 1989. *Lieu du savoir social: La Société de Statistique de Paris au XIXe siècle (1860–1910).* Paris: EHESS thesis.

Kendall, M. 1942. The future of statistics. *Journal of the Royal Statistical Society,* 105, pt. 2: 69–72.

Kendall, M., and R. L. Plackett, eds. 1977. *Studies in the History of Statistics and Probability,* vol. 2. London.

Kendrick, J. 1970. The Historical Development of National-Income Accounts. *History of Political Economy,* 2: 284–315.

Keverberg, Baron de. 1827. Notes sur Quetelet. *Nouveaux Mémoires de l'Académie royale des sciences et belles-lettres de Bruxelles,* 4: 175–192.

Keynes, J. M. 1910. Review of K. Pearson: a first study of the influence of parental alcoholism on the physique and ability of the offspring. *Journal of the Royal Statistical Society* (July): 769–773; (December): 114–121.

——— 1921. *A Treatise on Probability.* London.

——— 1939. Professor Tinbergen's method. *Economic Journal,* 49: 558–568; 50: 141–156.

Kiaer, A. N. 1895. Observations et expériences concernant les dénombrements représentatifs. *Bulletin de l'Institut international de statistique,* 9: 176–178.

——— 1897. Sur les méthodes représentatives ou typologiques appliquées à la statistique. *Bulletin de l'Institut international de statistique,* 11: 180–185.

——— 1901. Sur les méthodes représentatives ou typologiques. *Bulletin de l'Institut international de statistique,* 13, bk. 1: 66–70.

Kocka, J. 1989. *Les employés en Allemagne, 1850–1980: histoire d'un groupe social.* Paris: Éditions de l'EHESS.

Koopmans, T. 1947. Measurement without theory. *Review of Economic Statistics,* 29: 161–172.

Kramarz, F. 1991. Déclarer sa profession. *Revue française de sociologie,* 32, 1:3–27.

Kruger, L., L. J. Daston, and M. Heidelberger, eds. 1987. *The Probabilistic Revolution,* vol. 1: *Ideas in History.* Cambridge, Mass.: MIT Press.

Kruger, L., G. Gigerenzer, and M. S. Morgan, eds. 1987. *The Probabilistic Revolution,* vol. 2: *Ideas in Sciences.* Cambridge, Mass.: MIT Press.

Kruskal, W., and Mosteller, F. 1980. Representative sampling IV: the history of the concept in statistics, 1895–1939. *International Statistical Review,* 48: 169–195.

Kuisel, R. 1984. *Le capitalisme et l'état en France: modernisation et dirigisme au XXe siècle.* Paris: Gallimard.

Kuznets, S. 1934. *National Income, 1929–1932.* New York: NBER, Bulletin 49.

Ladrière, P., and C. Gruson. 1992. *Ethique et gouvernabilité: un projet européen.* Paris: PUF.

Laplace, P. S. 1810. Mémoire sur les approximations des formules qui sont fonctions de très grands nombres et sur leur application aux probabilités. *Mémoire de l'Académie des Sciences de Paris,* pp. 353–415, 559–565.

Largeault, J. 1971. *Enquête sur le nominalisme.* Paris.

Latour, B. 1984. *Les microbes: guerre et paix,* followed by *Irréductions.* Paris: A. M. Métaillié.

——— 1989. *La science en action.* Paris: La Découverte.

——— 1991. *Nous n'avons jamais été modernes.* Paris: La Découverte.

Lawson, T. 1989. Realism and Instrumentalism in the Development of Econometrics. *Oxford Economic Papers,* 41, 1 (January): 236–258.

Lazarsfeld, P. 1970. Notes sur l'histoire de la quantification en sociologie: les sources, les tendances, les grands problèmes. In *Philosophie des sciences sociales,* pp. 317–353. Paris: Gallimard.

Le Bras, H. 1987. La statistique générale de la France. In P. Nora, ed., *Les lieux de mémoire,* vol. 2: *La nation,* pp. 317–353. Paris: Gallimard.

Lécuyer, B. 1982. Statistiques administratives et statistiques morales au XIXe siècle. In *Actes de la journée d'étude "Sociologie et statistique,"* pp. 155–165. Paris: INSEE.

——— 1987. Médecins et observateurs sociaux: les *Annales d'hygiène publique et de médecine légale* (1820–1850). In *INSEE 1977: Pour une histoire de la statistique,* vol. 1: *Contributions.* Reprinted in 1987; Paris: INSEE, Economica, pp. 445–476.

Legendre, A. M. 1805. *Nouvelles méthodes pour la détermination des orbites des comètes.* Paris: Courcier.

Le Goff, J. 1962. *Marchands et banquiers au Moyen-Age.* Paris: PUF.

Le Mée, R. 1975. *Statistique de la France: la Statistique générale de la France de 1833 à 1870.* Paris: Service de Microfilm.

Lenoir, M. 1913. *Études sur la formation et le mouvement des prix.* Paris: Giard and Brière.

Lévy, M. 1975. L'information statistique. Paris: Seuil.

Livi, R. 1896. Sulla interpretazione delle curve seriali in antropometria. *Atti della Societa Romana di Antroplogia,* 3: 21–52.

Lottin, J. 1912. *Quetelet, statisticien et sociologue.* Paris.

Luciani, J., and R. Salais. 1990. Matériaux pour la naissance d'une institution: l'Office du travail (1890–1900). *Genèses, Sciences sociales et histoire,* 3: 83–108.

Mach, E. 1904. *La Mécanique: exposé historique et critique de son développement.* Paris: E. Bertrand, Hermann.

MacKenzie, D. 1981. *Statistics in Britain, 1865–1930: The Social Construction of Scientific Knowledge.* Edinburgh: Edinburgh University Press.

Mairesse, J., ed. 1988. *Estimations et sondages: Cinq contributions à l'histoire de la statistique.* Paris: Economica.

Malinvaud, E. 1988. Econometric methodology at the Cowles Convention: rise and maturity. *Econometric Theory,* 4: 187–209.

——— 1991. *Voies de recherche macroéconomique.* Paris: Odile Jacob.

Mansfield, M. 1988. *La construction sociale du chômage: l'émergence d'une catégorie en Grande-Bretagne.* Paris: IEPE, document de travail no. 8803.

March, L. 1905. Comparaison numérique de courbes statistiques. *Journal de la Société statistique de Paris:* 255–277.

——— 1910. Essai sur un mode d'exposer les principaux éléments de la théorie statistique. *Journal de la Société statistique de Paris:* 447–486.

Marietti, P. G. 1947. *La statistique générale en France.* Imprimerie de Gouvernement, Rufisque.

Maroussem, P. du. 1900. *Les Enquêtes: pratique et théorie.* Paris.

Ménard, C. 1977 (1987). Trois formes de résistance aux statistiques: Say, Cournot, Walras. In *Pour une histoire de la statistique,* vol. 1. Reprinted in 1987; Paris: Economica, pp. 417–429.

Merllié, D. 1987. Le suicide et ses statistiques: Durkheim et sa postérité. *Revue philosophique,* no. 3: 303–325.

Metz, K. 1987. Paupers and Numbers: The Statistical Argument for Social Reform in Britain during the Period of Industrialization. In L. Kruger, L. Daston, and L. Heidelberger, eds., *The Probabilistic Revolution,* vol. 1, *Ideas in History,* pp. 337–350. Cambridge, Mass.: MIT Press.

Meusnier, N. 1987. *Jacques Bernoulli et l'Ars conjectandi.* Rouen: Université de Haute Normandie, IREM.

Mirowski, P. 1989a. The Probabilistic counter-revolution, or how stochastic concepts came to neoclassical economic theory. *Oxford Economic Papers,* 41, 1: 217–235.

——— 1989b. The Measurement without Theory Controversy. *Economie et Sociétés,* "Oeconomia," 11: 65–87.

Mitchell, W. 1913. *Business Cycles and Their Causes.* Berkeley: University of California Memoirs, vol. 3.

——— 1923. Business Cycles. In *NBER,* ed., *Business Cycles and Unemployment.* New York.

Mols, R. 1954. *Introduction à la démographie historique des villes d'Europe du XIVe au XVIIIe siècle.* Paris: Duculot.

Moore, H. 1914. *Economic Cycles: Their Law and Cause.* New York: Macmillan.

Moreau de Jonnès, A. 1847. *Eléments de statistique, comprenant les principes généraux de cette science et un aperçu historique de ses progrès.* Paris: Guillaumin.

Morgan, M. S. 1990. *The History of Econometric Ideas.* Cambridge: Cambridge University Press.

Morrisson, C. 1987. L'enseignement des statistiques en France du milieu du XIXe siècle à 1960. In *INSEE 1987: Pour une histoire de la statistique,* vol. 2, pp. 811–823. Paris: INSEE, Economica.

Murphy, T. 1981. Medical knowledge and statistical methods in early nineteenth-century France. *Medical History,* 25: 301–319.

Musgrave, R. 1959. *The Theory of Public Finance: A Study in Public Economy.* New York: McGraw-Hill.
Musil, R. 1985. *Théâtre.* Translated from the German into French by Philippe Jacottet. Paris: Seuil.
NBER. 1921. *Income in the United States: Its Amount and Distribution, 1909–1919,* vols. 1 and 2. New York.
Neyman, J. 1934. On the different aspects of the representative method: the method of stratified sampling and the method of purposive selection. *Journal of the Royal Statistical Society,* 97: 558–606; debate, pp. 607–625.
Neyman, J., and E. Pearson. 1928. On the use and interpretation of certain test criteria for purposes of statistical inference. *Biometrika,* 20A: 175–240 and 263–294.
Nisbet, R. A. 1984. *La tradition sociologique.* Paris: PUF.
Nissel, M. 1987. *People Count: A History of the General Registry Office.* London: Office of Population Censuses and Surveys.
Norton, B. 1978. Karl Pearson and statistics: the social origins of scientific innovation. *Social Studies of Sciences,* 8, 1 (February): 3–34.
O'Muircheartaigh, C., and T. Wong. 1981. The impact of sampling theory on survey sampling practice: a review. *Bulletin de l'Institut international de statistique,* 49: 465–493. Buenos Aires.
Owen, D. ed., 1976. *On the History of Statistics and Probability.* New York: Marcel Dekker.
Ozouf-Marignier, M. V. 1986. De l'universalisme constituant aux intérêts locaux: le débat sur la formation des départments en France (1789–1790). *Annales ESC,* 6: 1193–1213.
Patinkin, D. 1976. Keynes and econometrics: on the interaction between the macroeconomic revolutions of the interwar period. *Econometrica,* 44: 1091–1123.
Paty, M. 1991. Mach (Ernst), 1838–1916. *Encyclopaedia universalis.*
Pearson, E. 1938. *Karl Pearson: An Appreciation of Some Aspects of His Life and Work.* Cambridge: Cambridge University Press.
Pearson, E., and M. Kendall, eds. 1970. *Studies in the History of Statistics and Probability,* vol. 1. London: Griffin.
Pearson, K. 1896. Mathematical contributions to the theory of evolution, III: regression, heredity, and panmixia. *Philosophical Transactions of the Royal Society of London* A, 187: 253–318.
——— 1912. *La grammaire de la science.* Trans. from the English by Lucien March. Paris: Alcan. U.S. edition: *The Grammar of Science.* New York: Meridian, 1956.
——— 1920. Notes on the history of correlation. *Biometrika,* 12: 25–45.
Pearson, K., and D. Heron. 1913. On theories of association. *Biometrika,* 9: 159–315.
Perrot, J. C. 1977. La statistique régionale à l'époque de Napoléon. In *INSEE 1977: Pour une histoire de la statistique,* vol. 1: 233–253. Paris: INSEE.

——— 1992. *Une histoire intellectuelle de l'économie politique.* Paris: EHESS.

Perrot, M. 1974 (1987). Premières mesures des faits sociaux: les débuts de la statistique criminelle en France (1780–1830). In *INSEE 1977: Pour une histoire de la statistique,* vol. 1, *Contributions,* pp. 125–137. Reprinted in 1987; Paris: INSEE Economica.

Peuchet, J. 1805. *Statistique élémentaire de la France.* Paris.

Piquemal, J. 1974. Succès et décadence de la méthode numérique en France, à l'époque de Pierre Louis. *Médecine de France,* 250 (March): 11–22.

Plackett, R. L. 1972. The Discovery of the Method of Least Squares. *Biometrika,* 59: 239–251.

Poisson, S. D. 1837. *Recherches sur la probabilité des jugements en matière criminelle et en matière civile, précédées des règles générales de calcul des probabilités.* Paris: Bachelier.

Polanyi, K. 1983. *La grande transformation: aux origines politiques et économiques de notre temps.* Paris: Gallimard.

Pollak, M. 1985. Sociologie de la science. In *Encyclopaedia universalis: les enjeux,* pp. 625–629.

——— 1986. Un texte dans son contexte: l'enquête de Max Weber sur les salariés agricoles. *Actes de la recherche en sciences sociales,* 65: 69–75.

——— 1993. *Une identité blessée.* Paris: Métaillié.

Popkin, R. 1964. *The History of Scepticism from Erasmus to Descartes.* Assen, Netherlands: Van Gorkum.

Porte, J. 1961. Les catégories socio-professionnelles. In G. Friedmann and P. Naville, *Traité de sociologie de travail,* vol. 1. Paris: Armand Colin.

Porter, T. M. 1986. *The Rise of Statistical Thinking.* Princeton: Princeton University Press.

Quetelet, A. 1832. Sur la possibilité de mesurer l'influence des causes qui modifient les éléments sociaux: lettre à M. de Villermé. *Correspondance mathématique et publique,* 7: 321–346.

——— 1835. *Sur l'homme et le développement de ses facultés, ou Essai de physique sociale.* Paris: Bachelier.

——— 1846. *Lettres à S.A.R. le Duc Régnant de Saxe Cobourg et Gotha, dur la théorie des probabilités, appliquée aux sciences morales et politiques.* Brussels: Hayez.

Reinhart, M. 1965. La population de la France et sa mesure, de l'Ancien Régime au Consulat. In *Contributions à l'histoire démographique de la Révolution française.* 2nd ser. Paris: Commission d'histoire démographique et sociale de la Révolution.

Rosch, E., and B. B. Lloyd, eds. 1978. *Cognition and Categorization.* New York: Erlbaum.

Saenger, K. 1935. Das Preussische Statistische Landesamt, 1805–1934. *Allgemeines Statisches Archiv,* 24: 445–460.

Salais, R., N. Baverez, and B. Reynaud. 1986. *L'invention du chômage.* Paris: PUF.

Samuelson, P. 1947. *Foundation of Economic Analysis.* Cambridge, Mass.: Harvard University Press.
Sauvy, A. 1975. Statistique générale et service national de statistique de 1919 à 1944. *Journal de la Société statistique de Paris:* 34–43.
Savage, L. J. 1954. *The Foundations of Statistics.* New York.
Savoye, A. 1981. Les continuateurs de Le Play au tournant du siècle. *Revue française de sociologie,* 22, 3: 315–344.
Schiller, J. 1963. Claude Bernard et la statistique. *Archives internationales de l'histoire des sciences,* 65: 405–418.
Schultz, H. 1928. *Statistical Laws of Demand and Supply with Special Application to Sugar.* Chicago: University of Chicago Press.
Schumpeter, J. 1983. *Histoire de l'analyse économique,* vol. 1. Paris: Gallimard.
Seng You Poh. 1951. Historical Survey of the Development of Sampling Theories and Practice. *Journal of the Royal Statistical Society,* ser. A, 114: 214–231.
Serverin, E. 1985. *De la jurisprudence en droit privé: théorie d'une pratique.* Lyon: Presses universitaires de Lyon.
Sewell, W. H. 1980 (1983). *Work and Revolution in France.* Cambridge: Cambridge University Press. French trans., *Gens de métier et révolutions,* Paris, Aubier-Montaigne, 1983.
SGF. 1847. *Industrie 1847.* Paris.
SGF. 1873. *Industrie, enquête de 1861–1865.* Paris.
SGF. 1896. *Recensement de 1896: Industrie.* Paris.
Shafer, G. 1990. The unity and diversity of probability. *Statistical Science,* 5, 4: 435–462.
Simiand, F. 1908. Le salaire des ouvriers des mines de charbon. *Journal de la Société statistique de Paris:* 13–29.
Slutsky, E. 1927. The summation of random causes as the source of cyclic processes. *The Problems of Economic Conditions,* 3, 1: 34–64, 156–161. Moscow.
Spearman, C. 1904. General intelligence objectively determined and measured. *American Journal of Psychology,* 15: 201–293.
Stephan, C. 1948. History of the uses of modern sampling procedures. *Journal of the American Statistical Association,* 43: 12–39.
Stigler, S. M. 1978. Francis Ysidro Edgeworth, statistician. *Journal of the Royal Statistical Society (A),* 141: 287–322.
——— 1986. *The History of Statistics: The Measurement of Uncertainty before 1900.* Cambridge, Mass.: Belknap Press of Harvard University Press.
Stockmann, R., and A. Willms-Herget. 1985. *Erwerbsstatistik in Deutschland.* Frankfurt: Campus Verlag.
Sutter, J. 1950. *"L'Eugénique: Problèmes, méthode, résultats,"* Paris: Cahiers de l'INED, 11, PUF.
Szreter, S. 1984. The Genesis of the Registrar-General's Social Classification of Occupations. *British Journal of Sociology,* 35, 4 (December): 529–546.
——— 1991. The GRO and the Public Health Movement in Britain, 1837–1914. *Social History of Medecine,* 4, 3 (December): 435–463.

Thévenot, L. 1984. L'enregistrement statistique: une mesure décisive. Conference of the Conseil national de la statistique.

——— 1986. Les investissements de forme. In *Les Conventions économiques, Cahiers du Centre d'études de l'emploi,* 29, pp. 21–71. Paris: PUF.

——— 1990. La politique des statistiques. Les origines sociales des enquêtes de mobilité sociale. *Annales ESC,* 6: 1275–1300.

Tinbergen, J. 1937. *An Economic Approach to Business Cycle Problems.* Paris: Hermann.

——— 1939. *Statistical Testing of Business Cycle Theories,* vol. 1: *A Method and Its Application to Investment Activity;* vol. 2: *Business Cycles in the United States of America, 1919–1932.* Geneva: League of Nations.

Tocqueville, A. de. 1856 (1967). *L'Ancien Régime et la Révolution.* Reprinted in 1967; Paris: Gallimard, Folio.

Topalov, C. 1991. La ville, terre inconnue: l'enquête de Charles Booth et le peuple de Londres, 1889–1891. *Genèses, sciences sociales et histoire,* 5: 5–34.

Venn, J. 1866. *The Logic of Chance.* London: Macmillan.

Vignaux, P. 1991. Nominalisme. *Encyclopaedia universalis.*

Villermé, L. R. 1845. Sur l'institution par le gouvernement belge d'une commission centrale de statistique, et observations sur les statistiques officielles publiées en France par les divers ministères. *Journal des économistes,* 15: 130–148.

Villey, M. 1975. *La formation de la pensée juridique moderne: cours d'histoire de la philosophie du droit.* Paris: Montchrestien.

Vining, R., and T. C. Koopmans. 1949. Methodological issues in quantitative economics. *Review of Economics and Statistics,* 31: 77–94.

Volle, M. 1982. *Histoire de la statistique industrielle.* Paris: Economica.

Ward, R., and T. Doggett. 1991. *Keeping Score: The First Fifty Years of the Central Statistical Office.* London: CSO.

Westergaard, H. L. 1932. *Contributions to the History of Statistics.* London: King.

Woolf, S. 1981. Contributions à l'histoire des origines de la statistique en France, 1789–1815. In *Statistique en France à l'époque napoléonienne.* Brussels: Centre Guillaume Jacquemyns.

X-Crise, 1982. *De la récurrence des cycles économiques.* Paris: Economica.

Yule, G. U. 1897. On the theory of correlation. *Journal of the Royal Statistical Society,* 60: 812–854.

——— 1899. An investigation into the causes of changes in pauperism in England, chiefly during the last two intercensal decades, I. *Journal of the Royal Statistical Society,* 62: 249–295.

——— 1909. Les applications de la méthode de corrélation aux statistiques sociales et économiques. *Bulletin de l'Institut international de statistique,* first delivery: 265–277.

——— 1911. *An Introduction to the Theory of Statistics.* London: Griffin.

——— 1927. On a model of investigating periodicities in disturbed series, with special reference to Wolfer's sunspot numbers. *Philosophical Transactions of the Royal Society of London,* ser. A, 226: 267–298.

——— 1942. The future of statistics. *Journal of the Royal Statistical Society,* 105, pt. 2: 83ff.

Zucker-Rouvillois, E. 1986. La politique des anges: démographie, eugénisme, et immigrations. *Milieux,* 23–24: 100–107.

Index